World Sustainability Series

Series Editor

Walter Leal Filho, European School of Sustainability Science and Research, Research and Transfer Centre "Sustainable Development and Climate Change Management", Hamburg University of Applied Sciences, Hamburg, Germany

Due to its scope and nature, sustainable development is a matter which is very interdisciplinary, and draws from knowledge and inputs from the social sciences and environmental sciences on the one hand, but also from physical sciences and arts on the other. As such, there is a perceived need to foster integrative approaches, whereby the combination of inputs from various fields may contribute to a better understanding of what sustainability is, and means to people. But despite the need for and the relevance of integrative approaches towards sustainable development, there is a paucity of literature which address matters related to sustainability in an integrated way.

Notes on the quality assurance and peer review of this publication

Prior to publication, the works published in this book are initially assessed and reviewed by an in-house editor. If suitable for publication, manuscripts are sent for further review, which includes a combined effort by the editorial board and appointed subject experts, who provide independent peer-review. The feedback obtained in this way was communicated to authors, and with manuscripts checked upon return before finally accepted. The peer-reviewed nature of the books in the "World Sustainability Series" means that contributions to them have, over many years, been officially accepted for tenure and promotion purposes.

Vikas Sharma · Tejinder Singh ·
Andreia de Bem Machado ·
Ramona Rupeika-Apoga
Editors

FinTech for Sustainable Finance and a Well-Blue Economy

Springer

Editors
Vikas Sharma
University School of Business
Chandigarh University
Mohali, Punjab, India

Tejinder Singh
University School of Business
Chandigarh University
Mohali, Punjab, India

Andreia de Bem Machado
Federal University of Santa Catarina
Florianópolis, Brazil

Ramona Rupeika-Apoga
Faculty of Economics and Social Sciences
University of Latvia
Riga, Latvia

ISSN 2199-7373 ISSN 2199-7381 (electronic)
World Sustainability Series
ISBN 978-3-031-92389-0 ISBN 978-3-031-92390-6 (eBook)
https://doi.org/10.1007/978-3-031-92390-6

This Springer imprint is published by the registered company Springer Nature Switzerland AG
The registered company address is: Gewerbestrasse 11, 6330 Cham, Switzerland

This book is dedicated to the visionary innovators and thought leaders in FinTech, whose relentless pursuit of innovation and transformative ideas are redefining the financial landscape, paving the way for a more sustainable, inclusive, and resilient future.

To our families and loved ones, especially Seshan and Danika Sharma, whose unwavering support, patience, and boundless encouragement have been the foundation of our strength throughout this journey. Your love and belief in us have made this endeavor possible and meaningful.

And to the future generations of changemakers, researchers, and leaders, may you continue to challenge boundaries, harness technology with responsibility, and uphold the principles of sustainability, equity, and stewardship, ensuring the prosperity of humanity and the flourishing of our planet's blue economy.

Preface I

FinTech stands at the forefront of transforming financial systems to support sustainability and inclusivity. *FinTech for Sustainable Finance and a Well-Blue Economy* examines how innovative financial technologies are advancing sustainable finance and empowering the growth of the blue economy.

This book brings together expert perspectives on how digital solutions enhance transparency, promote financial inclusion, and foster environmental responsibility. Through practical insights and case studies, it offers a roadmap for leveraging FinTech to create a more equitable, resilient, and sustainable future.

We hope this work inspires innovation and action toward sustainable finance and ocean prosperity.

Mohali, India — Prof. Vikas Sharma
Mohali, India — Dr. Tejinder Singh
Florianópolis, Brazil — Andreia de Bem Machado
Riga, Latvia — Prof. Dr. Ramona Rupeika-Apoga

Preface II

FinTech for Sustainable Finance and a Well-Blue Economy explores the transformative role of financial technology in advancing global sustainability and supporting the development of a thriving blue economy. This book examines how FinTech innovations can drive responsible financial practices, promote environmental sustainability, and foster economic growth, particularly in ocean-based industries and coastal communities.

The volume highlights the potential of FinTech to streamline sustainable finance by improving transparency, efficiency, and accessibility. It addresses key areas such as sustainable investment, green bonds, carbon credits, and financial mechanisms that directly support marine conservation, renewable ocean energy, sustainable fisheries, and other blue economy sectors.

Furthermore, the book delves into the challenges and opportunities faced in aligning FinTech solutions with sustainability goals. It critically examines regulatory landscapes, ethical considerations, and the risk of greenwashing, emphasizing the importance of measurable, genuine impact in sustainable financial initiatives.

Through theoretical insights, case studies, and practical applications, the book presents a clear picture of how financial technology can be harnessed to support sustainable development goals (SDGs) and promote economic resilience, especially in relation to ocean resources and coastal communities.

We hope this work provides valuable guidance to researchers, policymakers, and financial practitioners dedicated to leveraging FinTech for a sustainable future and a flourishing blue economy.

Mohali, India — Vikas Sharma
Mohali, India — Tejinder Singh
Florianópolis, Brazil — Andreia de Bem Machado
Riga, Latvia — Ramona Rupeika-Apoga

Acknowledgments

We would like to express our heartfelt gratitude to all who contributed to creating this book, *FinTech for Sustainable Finance and a Well-Blue Economy.*

First, we sincerely thank the esteemed contributors who have shared their valuable knowledge, expertise, and research. Your thoughtful insights and dedication have been instrumental in shaping this volume, enriching it with diverse perspectives on how FinTech can drive sustainable finance and adopt a thriving blue economy. We are deeply grateful for your commitment to advancing the discourse in this crucial area.

Our deepest appreciation also goes to my family (Seshan, Danika) and loved ones, whose unwavering support, patience, and encouragement have constantly motivated me. Your understanding throughout this journey has allowed us to dedicate the time and effort needed to complete this work.

We are immensely thankful to our publisher for believing in the significance of this project and providing the essential guidance and resources that have helped bring our vision to life. Your professionalism and support have been invaluable at every stage of the process.

A special note of thanks goes to the peer reviewers whose insightful feedback and constructive critiques have greatly enhanced the quality and coherence of the manuscript. Your careful evaluation has been pivotal in refining the outcome.

Finally, we acknowledge the pioneering efforts of FinTech innovators, sustainability advocates, and researchers worldwide whose groundbreaking work continues to inspire. Your relentless pursuit of financial innovation, environmental stewardship, and social equity has laid the foundation for the ideas explored in this book. We hope this volume serves as a meaningful contribution to the ongoing efforts toward sustainable finance and a flourishing blue economy.

To everyone who has played a role in the development of this book, we offer our deepest gratitude. Your collective contributions have made this endeavor possible, and together, we aspire to inspire further innovation, collaboration, and positive change.

About This Book

FinTech for Sustainable Finance and a Well-Blue Economy offers a deep exploration of how financial technology (FinTech) is driving sustainable finance and fostering the growth of a thriving blue economy. This forward-thinking volume assembles leading scholars, industry experts, and practitioners who collectively examine how innovative FinTech solutions can support environmental, social, and economic sustainability.

The book emphasizes the transformative potential of FinTech in advancing sustainable finance practices enhancing transparency, promoting green investments, and enabling efficient financial services aligned with sustainable development goals (SDGs). It also highlights FinTech's crucial role in empowering communities through inclusive financial services, bridging socioeconomic divides, and catalysing equitable growth.

Special focus is given to the blue economy, exploring how FinTech can support marine conservation, sustainable fisheries, ocean-based renewable energy, and responsible maritime trade. By integrating digital technologies such as blockchain, AI, and digital payments, the book demonstrates how FinTech can create synergies between economic prosperity and ocean health.

Through real-world case studies, practical frameworks, and critical analysis, readers will uncover actionable insights into applying FinTech innovations to foster a resilient and sustainable financial ecosystem. The book also navigates the regulatory, ethical, and technological challenges in achieving sustainability goals within the financial sector.

FinTech for Sustainable Finance and a Well-Blue Economy is an indispensable resource for policymakers, financial professionals, academics, and sustainability advocates aiming to leverage FinTech for building a sustainable future and thriving blue economy. It encourages further research, collaboration, and dialogue at the intersection of technology, finance, and environmental stewardship.

Vikas Sharma
Tejinder Singh
Andreia de Bem Machado
Ramona Rupeika-Apoga

Contents

Contributors

Shalini Aggarwal Management Department, Chandigarh University, Mohali, Punjab, India

Shakil Ahmad School of Business, University of Creative Technology Chittagong, Chittagong, Bangladesh

Shamsul Alam University School of Business, Chandigarh University, Punjab, India

Silvio Andrae Independent Researcher, Berlin, Germany

M. S. Asmaa Begum Department of Information Technology, Hindustan Institute of Technology and Science, Bay Range Campus, Padur, Chennai, India

Jyotirmoy Banerjee Amity Law School, Amity University, Bengaluru, India

Rajkumar Bhola Management Department, Chandigarh University, Mohali, Punjab, India

Lucius J. S. Botes Development Studies and Director Research Development, North-West University, Potchefstroom, South Africa

Andreia de Bem Machado Federal University of Santa Catarina (UFSC), Florianópolis, SC, Brasil

Amruta Deshpande Director Research, Indira School of Business Studies, Indira Group of Institutes, Pune, India

Prerna Dhawan Management Department, Chandigarh University, Mohali, Punjab, India

João Rodrigues dos Santos Instituto Superior de Educação E Ciências (ISEC Lisboa), Lisboa, Portugal

Alwin Estareja Faculty of Economics and Social Sciences, University of Latvia, Riga, Latvia

Andris Fomins BA School of Business and Finance, Riga, Latvia

Nitin Goyal Lovely Professional University, Phagwara, Punjab, India

Wynand C. J. Grobler School of Economic Sciences, North-West University, Potchefstroom, South Africa

Ewelina Idziak Department of Law and Economics, Uniwersytet Kazimierza Wielkiego, Bydgoszcz, Poland;
Azerbaijan State University of Economics, Baku, Azerbaijan;
Vilniaus Kolegija, Higher Education Institution, Vilnius, Lithuania;
Women Researchers Council, Azerbaijan State University of Economics, Baku, Azerbaijan

Harshad.L. Jadhav Department of Economics and Research Centre, Karmaveer Bhaurao Patil College, Navi Mumbai, Maharashtra, India

Katarzyna Kachel Department of Law and Economics, Uniwersytet Kazimierza Wielkiego, Bydgoszcz, Poland

Bharti Kapur University School of Business, Chandigarh University, Mohali, Punjab, India

Simeon Karafolas Department of Accounting and Finance, University of Western Macedonia, Koila Kozanis, Greece

Rupinder Katoch Mittal School of Business, Lovely Professional University, Phagwara, Punjab, India

Komalpreet Kaur University School of Business, Chandigarh University, Mohali, Punjab, India

Samoon Khan Mittal School of Business, Lovely Professional University, Phagwara, Punjab, India

Attila Kurucz Kautz Faculty of Business and Economics, Széchenyi István University, Győr, Hungary

Jekaterina Kuzmina BA School of Business and Finance, Riga, Latvia

Reshma Pervin Lima Department of Accounting, Mawlana Bhashani Science and Technology University, Tangail, Bangladesh

Diya Mahendru Department of Political Science, School of Social Sciences and Languages, Lovely Professional University, Phagwara, Punjab, India

Suresh N. Maind Mumbai School of Economics and Public Policy, University of Mumbai, Mumbai, India

Kavita Mathad Dean Presidency University, Bangalore, India

Inese Mavlutova BA School of Business and Finance, Riga, Latvia

Gitana Mečėjienė Vilniaus Kolegija/Higher Education Institution, Faculty of Economics, Banking Department, Vilniaus, Lithuania

Kaushal Kishore Mishra University School of Business, Chandigarh University, Mohali, Punjab, India

Vivek Mishra Department of Pure and Applied Mathematics, Alliance University, Bengaluru, India

Amitava Pal ICFAI Management School, ICFAI University Jharkhand and Associate Dean International Institute of Sports and Management, Mumbai, India

Priyanka P. Pandey Department of Economics and Research Centre, Karmaveer Bhaurao Patil College, Navi Mumbai, Maharashtra, India

Samriddhi Pandey Student, BCOM (Applied Finance and Analytics), School of Commerce, Finance and Accountancy, CHRIST (Deemed to Be University), Bengaluru, India

Pawan Pant University School of Business, Chandigarh University, Mohali, Punjab, India

Gaurav G. Pawar Department of Economics, Abasaheb Marathe Arts and New Commerce, Science College, Ratnagiri, Maharashtra, India

Antònio Miguel Pesqueira Instituto Universitário de Lisboa (Iscte), Lisboa, Portugal

Fazla Rabby Stanford Institute of Management and Technology, Sydney, Australia

Saloni Raheja Lovely Professional University, Phagwara, Punjab, India

Kazi Saifur Rahman Department of Finance and Banking, Jatiya Kabi Kazi Nazrul Islam University, Mymensingh, Bangladesh

Marc François Richter State University of Rio Grande Do Sul (Uergs), Porto Alegre, RS, Brasil

Ramona Rupeika-Apoga Faculty of Economics and Social Sciences, University of Latvia, Riga, Latvia

Ashutosh Rath Sharma Birla Global University, Bhubaneswar, Odisha, India

Vikas Sharma University School of Business, Chandigarh University, Mohali, Punjab, India

Pooja Singh Freelancer, Researcher, Gujarat, India

Ram Singh Symbiosis University of Applied Sciences, Indore, India

Shalini Singh Assistant Professor, School of Commerce, Finance and Accountancy, CHRIST (Deemed to Be University), Bengaluru, India

Tejinder Singh Department of Law and Economics, Uniwersytet Kazimierza Wielkiego, Bydgoszcz, Poland

Nargis Sultana Department of Finance and Banking, Comilla University, Comilla, Bangladesh

C. V. Suresh Babu Department of Information Technology, Hindustan Institute of Technology and Science, Bay Range Campus, Padur, Chennai, India

Amanjot Singh Syan Canadore Stanford International College, Mississauga, ON, Canada

Atis Verdenhofs BA School of Business and Finance, Riga, Latvia

Anjali Vyas University School of Business, Chandigarh University, Punjab, India

Shubhanker Yadav Assistant Professor, Jaipuria Institute of Management, Lucknow, India

Azmee Zaheer University School of Business, Chandigarh University, Mohali, Punjab, India

Abbreviations

ACS ES&T	American Chemical Society
AI	Artificial intelligence
AML	Anti-Money Laundering
API	Application Programming Interface
AR	Anticipated Regret
BI	Business intelligence
BOC	Beautiful Ocean Coin
CAGR	Compound Annual Growth Rate
CBD	Convention on Biological Diversity
CCR	Climate Change Resilience
CERES	Clouds and the Earth's Radiant Energy System
CRRC	Climate-Related Risk Coverage
CSR	Corporate Social Responsibility
CT	Counterfactual Thinking
DeFi	Decentralised Finance
EPLI	Environmental Pollution Liability Insurance
ERB	Earth's Radiation Budget
ES&T	Environmental Science and Technology
ESG	Environmental, Social and Governance
EU	European Union
EXIM	Export-Import Bank of India
FAO-	Food and Agriculture Organisation
FAO	Food and Agriculture Organization of the United Nations
FinTech	Financial Technology
FOMO	Fear of missing out
FRPS	Facial Recognition Payment Systems
FTW	Floating Treatment Wetlands
GDP	Gross Domestic Product
GDPR	General Data Protection Regulation
GE	General Electric
GFBR	Global Forum on Bioethics in Research

GRRSBB	Global Recognition and Replicability Seychelles' Blue Bond
GW	Gigawatt
HAB	Harmful Algal Bloom
IFC	International Finance Corporation
IMF	International Monetary Fund
IMO	International Maritime Organization
Insurtech	Insurance Technology
IoT	Internet of Things
IPCC	Intergovernmental Panel on Climate Change
ITU	Intention to Use
IUU	Illegal, Unreported, and Unregulated
IUUF	Illegal, unreported, and unregulated fishing
MARPOL	International Convention for the Prevention of Pollution from Ships
MDGs	Millennium Development Goals
ML	Machine Learning
MPAs	Marine Protected Areas
MTF	Markov transition field
NBBBI	Nordic-Baltic Blue Bond Initiative
NBI	Nordic and Baltic Industries
NGOs	Non-Governmental Organizations
NIB	Nordic Investment Bank
NITK	National Institute of Technology Karnataka
NLP	Natural Language Processing
NOAA	National Oceanic and Atmospheric Administration
OECD	Organisation for Economic Co-operation and Development
P2P	Peer-to-peer
PEU	Perceived Ease of Use
PLI	Professional Liability Insurance
PMMSY	Pradhan Mantri Matsya Sampada Yojana
PoA	Proof of Authority (Blockchain consensus mechanism)
PoS	Proof of Stake (Blockchain consensus mechanism)
PoS	Proof of Stake
PoW	Proof of Work
PPP	Public-Private Partnership
PU	Perceived Usefulness
RAS	Recirculating Aquaculture Systems
RE	Renewable Energy
RETF	Recipient Executed Trust Funds
RFMOs	Regional Fisheries Management Organizations
RL	Reinforcement Learning
RRR	Reuse, Recycle and Rebuild
SC	Social Comparison
SCADA	Supervisory Control and Data Acquisition
SDGs	Sustainable Development Goals
SEG	Sustainable Economic Growth

SIDS	Small Island Developing States
SMEs	Small and Medium Enterprises
SSBB	Seychelles' Sovereign Blue Bond
SSCM	Sustainable Supply Chain Management
TAM	Technology Acceptance Model
UN	United Nations
UNCED	United Nations Conference on Environment and Development
UNCLOS	United Nations Convention on the Law of the Sea
UNEP FI	United Nations Environment—Finance Initiative
UNEP	United Nations Environment Programme
UNESCO	United Nations Educational, Scientific and Cultural Organization
UPI	Unified Payments Interface
USD	United States Dollar
WB	World Bank
WTO	World Trade Organization
WWF	World Wildlife Fund
XAI–e	Explainable Artificial Intelligence

The Role of Fintech in Promoting Sustainable Blue Economy

Komalpreet Kaur, Bharti Kapur, and Attila Kurucz

Abstract The sustainable blue economy lays a strong emphasis on utilising ocean resources responsibly in order to enhance livelihoods, preserve ecosystem health, and grow the economy. Financial technology, or fintech, has emerged as a major enabler, offering creative ways to eliminate financial gaps, enhance monitoring, and more efficiently manage ocean resources. In this chapter discusses how fintech may stimulate sustainable growth in the blue economy by concentrating on key issues including transparent resource management, scalable investment mechanisms, and inclusive financial systems. This chapter presents a conceptual framework for addressing economic, environmental, and social issues by fusing sustainable marine practices with technological advancements like blockchain, artificial intelligence, and the internet of things. Fintech's practical implications in advancing sustainability are demonstrated through in-depth case studies, such as blockchain for Indonesian fisheries and crowdfunding for the restoration of coral reefs in the Caribbean. It also focuses on practical policy recommendations for incorporating fintech into nationally and globally blue economy strategies, like developing regulatory framework and strengthening cross-sector collaborations. The chapter concludes by highlighting fintech's critical role in promoting sustainability, inclusion, and resilience in ocean-based economies, widening up opportunities for the accomplishment of more general sustainable development goals.

Keywords Fintech · Blue economy · Sustainability · Marine conservation · Financial inclusion · Sustainable development goals · Ocean resources

K. Kaur (✉) · B. Kapur
University School of Business, Chandigarh University, Mohali, Punjab, India
e-mail: komalpreet.r217@cumail.in

B. Kapur
e-mail: bharti.e8454@cumail.in

A. Kurucz
Kautz Faculty of Business and Economics, Széchenyi István University, Győr, Hungary

V. Sharma et al. (eds.), *FinTech for Sustainable Finance and a Well-Blue Economy*,
World Sustainability Series, https://doi.org/10.1007/978-3-031-92390-6_1

1 Introduction

The ocean is a key part of earth's ecosystem and offers necessary resources that aid human existence (Costello et al. 2020). It covers more than 70% of the earth's surface and holds about 97% of its water. This large body is an important source of economic activity and environmental balance (Laffoley et al. 2021). It is often called the 'blue planet'. The ocean offers necessary resources, affects climate next to sustains different aquatic life forms. The maritime economy includes a broad set of financial actions. These actions rely on oceanic materials. These actions include fisheries, aquaculture, sea tourism, transport businesses and renewable energy creation at sea (Education at a Glance 2016 2016; World Development Report 2017 2017). The specified fields not only produce financial benefit but also have a key task in offering income, food supply along with cultural history for shore towns across the world.

Despite its substantial capacity for economic expansion and progress, the blue economy has many problems. They place its lasting viability at risk. The resource faces a rising threat. This comes from practices which cannot be sustained, climate variations, contaminants next to overuse (Halpern et al. 2019; Worm et al. 2006). Fishing methods that cannot be sustained cause too much fishing. This reduces fish numbers and disrupts marine food chains. Marine contamination mainly from synthetic materials and compounds, is a notable hazard. It affects marine ecosystems and people's well-being (Jambeck et al. 2015). Climate variations create impacts. Ocean acidification sea rise next to severe weather make these problems worse. This threatens the future of the blue economy. It also puts communities at risk that rely on it (Intergovernmental Panel on Climate Change (Ipcc) 2023).These problems require fresh strategies. This is to secure the lasting health and output of marine ecosystems. At the same time, it is important to support economic expansion and societal well-being.

Financial technology has come forward as a factor that changes the traditional financial world. It gives new answers to make access, how well things work, being open as well as lasting power better in diverse areas (Philippon, NBER working paper series; Schueffel 2016). In recent times fintech found acceptance as something that helps lasting growth plans in many fields (World Development Report 2017). Using tools such as blockchain for being open in supply networks or artificial intelligence for data study and choices in how sources are handled as well as the internet of things and big data study, fintech can deal with key problems in the ocean-based financial system. (Bughin et al. 2013). These comprise money needs, resource handling that does not achieve desired results and not being able to track things (Gomber et al. 2018).

Financial technology gives new answers which aid durable funding for ocean preservation projects, while growing financial access among shore towns. As an example: Public funding websites aid the move of funds toward ocean preservation work. They achieve this goal by linking project builders to single funders who feel sustainability is important (Burg et al. 2017). Distributed ledger tech can improve tracking within seafood distribution routes, giving customers sure data concerning where items started (Patro et al. 2022). Artificial intelligence-based analysis helps to refine how tools get used, achieved by looking at big data sets taken from many origins.

This paper explores the transformative potential of fintech in driving sustainable development within the blue economy, focusing on key areas such as sustainable financing, resources monitoring, financial inclusion, and policy framework. The paper argues that fintech can act as a catalyst for change, fostering a more resilient, inclusive, and sustainable ocean- based economy that benefits both current and future generations. Furthermore, this paper aims to comprehensively analyse the benefits, challenges, and future directions of integrating fintech solutions within the blue economy context, supported by empirical evidence and case studies.

2 Methodology

This study adopts a qualitative analysis to provide a comprehensive understanding of the role of the fintech in the blue economy. A thorough Systematic review of academic journals, industry reports, policy documents, and literature to identify key trends, challenges, and opportunities related to fintech and the sustainable blue economy. Databases such as Scopus, web of science, and Google Scholar are utilized to identify relevant publications. Search Strings include "fintech", "blue economy", "marine conservation", "sustainable finance"," sustainable development".

In-depth case study analysis exanimating of real-word examples where fintech solutions have been implemented to promote marine sustainability. Cases are identifying through literature reviews, and online databases of sustainable projects. Developing a conceptual framework that integrates fintech innovations with sustainable blue economy practices, emphasizing the synergies between financial technology and marine ecosystem management. This framework serves as a guide for policymakers, and researchers seeking to promote sustainable development in the blue economy. The framework refined based on the findings from the literature review and case study analysis.

3 Review of Literature

3.1 Fintech or Financial Technology

Fintech or financial technology highlights its various transformative potential across various sectors: Blockchain provides a decentralized ledge system that enhances transparency in transactions while reducing fraud risks (Kshetri and Voas 2018). In particular within the seafood supply chain, blockchain technology can follow items from capture to consumer. The technology also helps to ensure adherence to standards for sustainability (Galvez et al. 2018). Algorithms using Artificial Intelligence or AI, examine large data sets to improve processes of decision-making. These processes relate to the control of resources. AI can predict future directions based on past data (Drogkoula et al. 2023). Devices using the Internet of Things or IoT, gather current data on environmental states. These states can include water quality or temperature variations. This allows for better observation of marine ecosystems. It also informs flexible approaches to control (Gubbi et al. 2013). Banking on mobile devices and platforms for payments done digitally can aid with financial access. They allow availability of banking options to populations who are poorly served. Smart contracts systems for payment digitally and platforms for finance that are decentralized, known as DeFi, have all made financial resources more within reach, mostly for groups that are marginalized. Platforms for crowdfunding allow people or groups to gather money for projects. This is achieved by asking for payments from numerous backers. This provides a different means of finance for efforts in marine preservation that might not be able to obtain typical sources of funds (Sharma et al. 2024).

3.2 Fintech and Sustainable Development Goals

Financial technology shows its capacity to cause large changes in moving sustainable progress in many fields. (Agrawal et al. 2013). The link between finance tech and sustainability has obtained more focus as the globe struggles with environment issues and inequalities in society and economics. Fintech, characterized by its innovative application of technology to financial services, offers unprecedented opportunities to drive sustainable development across various sectors, including the blue economy (Dorfleitner et al. 2022). AI-driven financial models further optimize resource allocation, helping business align with sustainability objectives by minimizing environmental risks (Arner et al. 2020).

Fintech innovations are playing an increasingly important role in promoting sustainability and advancing the United Nations Sustainable Goals (SDGs), Including those related to marine conservation and the blue economy, Sustainable Development Goals (SDGs) has been extensively explored, particularly in promoting economic growth (SDG 8), reducing inequalities (SDG10), and fostering innovations (SDG 9) (Raihan 2024). Finance tech can add greatly to sustainable progress via better inclusion in finance, easier digital payments along with backing efforts related to green funding. (Mhlanga 2022). The joining of finance tech with sustainable progress goals has been observed to grow investment chances in projects for the environment and to raise the range and involvement from savers and lenders. This better market state can result in higher production and lasting growth because of spread of investments, tech improvements as well as new ideas (Krishnan et al. 2024).

3.3 Fintech and Blue Economy

The blue economy means the use of ocean resources in a sustainable way. This supports economic expansion, better ways of living next to the preservation of ecosystems. In the blue economy along with marine protection, financial technology can be very important. It can direct money to the sustainable administration of ocean resources, also to protection work (Hasan et al. 2024). Although it has potential for economic growth, this area has notable money-related problems. These include not enough access to capital plus problems in how money is given out (Thiele and Gerber 2017). Financial technology innovations, like blockchain and artificial intelligence, came about as answers to these issues. These can make finances easier to get and assure resources from the sea are handled sustainably (Choudhary et al. 2021). Blockchain programs make supply chains in fishing clearer, which lowers illegal and unregulated fishing. Data study using artificial intelligence helps watch the variety of life in the ocean as it happens. (Elias et al. 2024). This also helps with work to protect ecosystems. (Tolentino-Zondervan et al. 2023).

Decentralized financial platforms enable crowdfunding for ocean conservation initiatives, allowing small-scale fisher and local businesspeople to raise finding through tokenized asset systems (Bakker 2022). The ocean protocol, demonstrate how blockchain-based data-sharing ecosystem can maximize resource management while providing economic inclusivity (Merrill et al. 2019). Moreover, Mobile fintech solutions enhance financial inclusion for coastal communities by offering digital financial services specifically designed for their needs (Fanea-Ivanovici and Baber 2024).

4 Integrating Sustainable Marine Practices with Emerging Technologies

This chapter addresses the biggest environmental, economic and social challenges of the blue economy through the integration of fintech innovations such as blockchain, artificial intelligence (AI), and internet of thing (IoT). These technologies advance innovative solutions for enhancing financial access, environmental protection, and social inclusion in ocean resource management.it emphasizes the supportive role of these technological innovations in promoting sustainability, economic resilience, and social inclusion in ocean-based economies.

4.1 Technological Innovations and Financial Accessibility

Financial access under the blue economy is crucial for vulnerable populations, coastal entrepreneurs, and small-scale fishers because it ensures sustainable development and economic resilience. Blockchain technology facilitates secure, transparent, and decentralized financial transactions with low rates of fraud and increased trust in ocean investments (Danladi et al. 2023). This is particularly beneficial to small-scale fisheries and coastal entrepreneurs that are normally prone to challenges of accessing formal finance due to empowering peer-to-peer lending, tokenized assets, and secure microfinance transactions, thereby promoting economic stability among coastal communities (Sarker et al. 2021).

Artificial intelligence (AI) has a key role to play in the upgrading of investment decision-making for sustainable fisheries and blue economy marine conservation (Kleih et al. 2013). All of this enables smooth sustainable investments by aligning financial flows and environmentally sustainable actions. Additionally, Internet of Things (IoT) help to govern financial risk by providing real-time monitoring of sea operations, which can reduce financial losses due to environmental hazards, illegal fishing, or unexpected operational breakdown (Ranganathan et al. 2023).

4.2 *Sustainable Resource Management*

Blue economy's responsible utilization of resources is vital for long-term preservation of marine biodiversity, preventing overfishing, and minimizing pollution. Fintech innovations, especially AI, blockchain, and IoT, are key drivers of sustainable marine practice. AI-assisted data analytics enable biodiversity protection by tracking fish populations, detecting illicit fishing practices, and isolating environmental risks in marine environments (Cavanaugh et al. 2025). AI also makes predictive modelling for fish population behaviour possible, which enables policymakers to set sustainable fish quotas and conservation policies (Isabelle and Westerlund 2022).

Blockchain records immutable transactions that trace the journey of seafood from catch to consume, ensuring transparency and compliance with sustainability stands (Mandal and Ghosh 2024a). This reduces illegal, unreported, and unregulated (IUU) fishing and promotes sustainable harvesting practices. In addition, IoT devices support environmentally friendly marine resource management through automating environmental observation. IoT sensors used in ocean environments gather up-to-date data on water conditions, pollution concentration, and marine habitats, providing a basis for active conservation as well as aquaculture conservation practices (Glaviano et al. 2022).

4.3 *Social and Economic Inclusion*

Social and economic inclusion in the blue economy should be promoted in order to secure equal access to financial services, economic opportunities, and decision making structures by marginalized groups, fintech innovation like blockchain, AI, IoT have proven good potential to accelerate social and economic inclusion in coastal communities blockchain-based decentralized finance (DeFi) platforms provide microfinance options for small-scale fishers, coastal business people, and minority communities by providing safe, affordable, and transparent financial transactions. (Glaviano et al. 2022). These platforms cut out intermediaries, lower transaction costs, and offer financial services to traditionally excluded populations in formal financial systems (Cisneros-Montemayor et al. 2019).

Traditional methods of credit scoring usually have intrinsic biases, preventing small-scale fishers and minorities from accessing services from financial institutions. However, AI-driven models utilize alternative data sources such as truncation histories, mobile usage patterns, and marine operational performance to assess creditworthiness, reducing biases and improving financial accessibility (Tirumala and Tiwari 2022). Additionally, IoT-based community-led monitoring systems empower local communities to participate in marine conservation and decision-making processed. By equipping coastal communities with IoT-enabled devices, stakeholders can monitor local marine environments, report illegal activities, and contribute to biodiversity conservation, ensuring community participation in sustainable marine practices (Omokhoa et al. 2024).

4.4 Policy and Governance Integration

Robust policy and governance frameworks are required to bring the maximum potential of fintech solutions to promoting sustainable marine practices. Governments, regulators, and global institutions have a fundamental responsibility to establish enabling regulatory conditions that promote the uptake of fintech solutions in the blue economy. (Sarjana et al. 2024). Governments would also have to formulate easy or simple and steady policies favouring sustainable marine finance, such as a tax allowance for sustainable fisheries gear, preferential loans for small-scale fishermen, and framework rules or principles for blockchain-traceable supply chain traceability (Shih 2024).

Moreover, AI and blockchain technologies must be synchronized with Environmental, Social, and Governance (ESG) principles to promote sustainable investment in blue economy management (Sumaila et al. 2021). Lastly, collaborative policymaking via public–private partnership is necessary in upscaling successful fintech applications. Public–private partnership has the ability to leverage the capacities and resources of both sides in order to enhance innovation, boost knowledge exchange, and develop innovative fintech products that address marine resource management's systemic problems (Mandal and Ghosh 2024b).

4.5 Stakeholder Engagement and Collaborative Action

Successful implementation of fintech innovations in the blue economy requires active involvement and collaboration of stakeholders. Stakeholder like fintech companies, government agencies, non-governmental organizations (NGOs), research centres, and coastal communities must work together to design and implement sustainable fintech solution. Collaboration facilitates knowledge sharing, resource mobilization, and the development of tailored solution that address the particular financial, environmental, and social issues in the blue economy (Salas et al. 2022). In addition, public–private partnership is essential in facilitating mass fintech adoption by bridging public sector regulatory ability and private sector innovation ability as well as investment ability.

Community participation is also critical in ensuring that fintech interventions address the distinctive needs and priorities of coastal communities. By involving local stakeholders in decision-making, fintech innovations can be framed to build social equity, promote financial inclusion, and support sustainable marine resource management (Zak et al. 2022).

The conceptual framework stresses how synergies among blockchain, AI, and IoT can lead to synergistic effects in advocating for environmental sustainability, economic resilience, and social inclusion in the blue economy. Blockchain technology, for example, enables supply chain traceability, which promotes responsible seafood sourcing and reduces illegal fishing (Kraemer 2017). Concurrently, predictive analytics by AI maximize management of marine resources through maximizing fishery operations while reducing operational risk. (Isabelle and Westerlund 2022).

5 Fintech Application in the Blue Economy

5.1 Sustainable Financing for Marine Conservation

Fintech platforms have revolutionized sustainable financing for marine conservation by offering innovative financial instruments, enabling transparency, and attracting private sector investments. Traditional funding sources such as government grants and philanthropic donations have long been insufficient to meet the rising demands of marine conservation efforts (Howson 2020). Fintech platforms, however, have introduced mechanisms like green and blue bonds, blockchain transparency, and impact investing, enhancing financial sustainability.

5.2 *Green and Blue Bonds*

Green bonds are debt instruments aimed at funding environmentally friendly projects, while blue bonds focus specially on marine and coastal conservation project (Macchiavello and Siri 2022). Fintech platforms facilitate the issuance, trading, and monitoring of these bonds, allowing institutional investors to support marine biodiversity conservation. Blockchain technology, in particular, enhance transparency by tracking fund allocation and ensuring investor confidence (Thompson 2022) Blockchain technology in fintech has reduced administrative costs and mitigated fraud risks, significantly improving green fiancé mobilization (Bennett et al. 2024).

5.3 *Crowdfunding for Marine Conservation*

Crowdfunding platforms have emerged as essential tools for raising funds for marine conservation projects such as coral reef restoration and marine debris removal. Social media marketing, coupled with fintech platforms, has enhanced fundraising efficiency, reaching wider audiences and boosting small donations. Digital crowdfunding also allows individuals to contribute to localized conservation projects, reducing financial dependency on traditional model (Singh and Kaunert 2024a).

5.4 *Peer-to-Peer Lending*

Peer-to-peer (P2P) lending platforms facilitate direct loans from individual investors to small-scale marine business and conservation initiatives. This enhances financial inclusion and supports sustainable aquaculture, eco-tourism, and community-led marine conservation. Fintech platforms employing AI-powered credit scoring have increased accessibility to funds, reducing loan processing costs and time (Trivedi et al. 2024).

5.5 *Impact Investing*

Impact investing, where investors seek financial returns alongside positive environmental and social impacts, has gained momentum in the blue economy. Fintech platforms play a critical role by connecting investors with marine conservation projects and providing impact assessment tools. AI-based predictive analytics, have enabled more accurate assessments of social and environmental outcomes, fostering responsible investments (Azmeh 2025).

5.6 Digital Monitoring and Compliance

Digital monitoring and compliance facilitated by fintech technologies are transforming marine resources management. Traditional resource monitoring approaches often lacked efficiency and were prone to manipulation. However, emerging fintech solution have increased traceability, real-time monitoring, and automated compliance, particularly in fisheries management.

5.7 Blockchain-based Tracking

Blockchain can secure the traceability of seafood products from catch to consumer, combating illegal, unreported, and unregulated (IUU) fishing. Blockchain unsure that all data related to origin, processing, and distribution is immutable and verifiable, fostering sustainable fisheries practices (Oyewole et al. 2024).

5.8 AI-driven Analytics

AI-powered monitoring tools analyse vast datasets from satellite imagery, acoustic sensors, and vessel tracking systems to detect illegal activities and optimize resource management. For instance, AI can identify vessels engaged in IUU fishing, monitor marine mammal populations, and predict harmful algal blooms, enabling pre-emptive conservation strategies (Sarjana et al. 2024).

5.9 IoT Sensors

Internet of Thing (IoT) sensors collect real-time data om water quality, pollution levels, and biodiversity health. This data is transmitted to cloud-based fintech platforms, enabling policymakers and conservation agencies to make data-driven decision (Azmeh 2025).

5.10 Financial Inclusion for Coastal Communities

Financial Inclusion for coastal communities, often marginalized in traditional financial systems, is being addressed through fintech platforms. By leveraging mobile banking, microfinance, and digital insurance, fintech is transforming the economic landscape for coastal populations.

5.11 Mobile Banking and Digital Payments

Mobile banking has brought financial services to remote coastal settlements, enabling citizens to conduct transactions, access credit, and receive payments. Mobile payment systems have empowered citizens by making micro-lending and saving easier, minimizing financial exclusion (Singh and Kaunert 2024b).

5.12 Microfinance and Digital Credit

Fintech platforms have transformed microfinance through lending application automation, lowered processing costs, and improved credit scoring. This has facilitated access to finance for small-scale aquaculture enterprises, ocean-based tourism operations, and people-led conservation initiatives (Tirumala and Tiwari 2022).

5.13 Digital Insurance

Digital insurance products like parametric insurance, in which the trigger for paying claims is environmental thresholds, insures coastal communities against climate catastrophes. Blockchain-backed insurance insures the speed and openness of claim settlements, promoting resilience in coastal communities (Pomeroy et al. 2020).

5.14 Enhancing Investment in Blue Economy Initiatives

Fintech has contributed substantially towards investment in blue economy projects through decreased transaction fees, increased transparency, and opened up more areas of investment opportunities.

5.15 Blue Bonds

Fintech platforms create an opportunity to issue blue bonds to finance blue carbon ecosystem marine conservation and blue carbon ecosystem coastal restoration projects. Distributed ledger technology makes fund usage transparent and kept the risk of fund misuse minimal (Singh and Kaunert 2024a).

5.16 Carbon Credit Trading Platforms

Fintech platforms offer carbon credit markets generated by the blue carbon ecosystem (mangroves, seagrass, and salt marshes). Blockchain technology guarantees the integrity and traceability of carbon credits, stimulating corporate investment in blue carbon projects (Thompson 2022).

6 Case Studies on Fintech Applications in the Blue Economy

The convergence of fintech like blockchain, AI, and IoT has proven to have revolutionary capacity to enhance sustainability, financial inclusion, and effective management of marine resources. This chapter provide in-depth case studies that show how fintech tools are being effectively applied in the blue economy to overcome some of the major challenges faced, including illegal fishing, paucity of funds for ocean conservation, and financial exclusion of coastal communities.

6.1 Blockchain for Fisheries Management in Indonesia

Indonesia, a global top seafood producer, has long struggled to combat illegal, unreported and, unregulated (IUU) fishing and achieve traceability of seafood products. To address this, the Fishcoin Initiative was established, utilizing blockchain technology to improve traceability along the seafood supply chain. By leveraging a decentralized system, fishers can document their catch on a blockchain ledger, making it transparent from catch to consumer (Rumson and Hallett 2019).

The Fishcoin Initiative also provide tokenized incentives to small-scale fishermen, promoting sustainable fishing practices while making them more financially resilient. By eliminating middlemen and allowing buyers to track the origin of their seafood themselves, Fishcoin provides market value for sustainably caught seafood products (Cai 2018). Blockchain technology also reduce paperwork and transaction fees, ensuring that local fishers get a fair price for sustainable fishing. This case study illustrates how Blockchain-enabled fintech can be used to stimulate sustainable fisheries management while facilitating financial inclusion for coastal fishers. It underscores the importance of fintech in countering IUU fishing, enhancing transparency, and aiding the economic resilience of coastal communities (Baliker et al. 2024).

6.2 Crowdfunding for Coral Reef Restoration in the Caribbean

Coral reefs are critical ecosystems that support marine biodiversity, protect coastlines, and sustain coastal economies. However, climate change, overfishing, and pollution have significantly degraded coral reefs worldwide. The ReefChain Initiative, a blockchain-powered crowdfunding platforms, was introduced to address this challenge by enabling direct financial contributions to coral reef restoration efforts.

ReefChain utilizes Decentralized Finance (DeFi) principles global donors with marine conservation projects, ensuring complete transparency in fund allocation. Also, ReefChain incorporates AI-based monitoring of biodiversity so that ocean scientists can track the health of coral reefs in real-time (Senyo and Osabutey 2020).

Through blockchain-based crowdfunding, ReefChain has financed over 30 Caribbean coral reef restoration projects and collected millions of dollars from global donors. This initiative demonstrates how fintech can direct capital into marine conservation efforts and enhance transparency, accountability, and civic participation as well (Syafira et al. 2024).

6.3 AI-driven Smart Aquaculture in Norway

Norway's aquaculture industry, a top global producer of salmon, employs AI-driven predictive analytics to enhance fisheries management, reduce its environmental impact, and maximize efficiency. AquaByte Project is a collaborative venture between the Norwegian government and fintech firms utilizing AI-enhanced IoT sensors to track fish health, water quality, feeding activity, and general marine environment conditions (Rizwan and Mustafa 2022).

AI systems analyze sensor data in real-time, enabling fish farmers to optimize feeding times, reduce fish mortality rates, and minimize feed wastage. This has led to 30% decrease in the waste of fish feed owing to the reduction of operational costs, 20% rise in survival rates of fish owing to real-time disease forecasting, increased compliance with environmental sustainability laws. Through the combination of fintech with AI-based IoT sensors, the AquaByte project has also greatly increased efficiency, sustainability, and profitability of Norway's aquaculture sector. The integration of fintech and Ai has proven an effective model for encouraging sustainable aquaculture and minimizing economic losses (Maiti and Ghosh 2023).

6.4 Financial Inclusion for Coastal Communities in Bangladesh

Fishing and aquaculture-based coastal communities in Bangladesh tend to be devoid of formal financial services like credit, insurance, and saving. To offset this issue, WavePay, mobile fintech platforms, was launched to extend financial inclusion through mobile banking and micro-lending.

WavePay facilitates micro-lending for small fishers, offering low-interest loans to buy fishing gear or scale up small enterprises. Parametric insurance policies with instant payouts during climate calamities, facilitating rapid relief from financial damage. Mobile banking products enabling fishers to deposit and withdraw funds using mobile phones without having to access physical bank branches. The program has significantly enhanced financial inclusion of coastal communities, lowered dependency on unorganized lending, and improved economic resilience in exposed coastal locations (Sanyaolu et al. 2024).

6.5 Carbon Credit Trading for Blue Carbon Ecosystems in Kenya

Blue carbon ecosystems like seagrass beds, mangroves, and salt marshes are of significant importance for carbon sequestration in Kenya. Yet, these ecosystems are under immense risk due to deforestation, coastal development, and pollution. To counter this, Blue Carbon Exchange was launched, a blockchain-based carbon credit exchange.

The exchanges employ blockchain technology to authenticate blue carbon sequestration credits earned from blue carbon ecosystems and links local conservation projects with international carbon markets. It allows companies to buy authenticated carbon credits, creating revenue streams for coastal communities in terms of maintaining mangroves forests and seagrass meadows (Cai 2018). Blue Carbon Exchange success shows how fintech, intensive blockchain-based carbon credit trading, can promote marine conservation and build economic rewards for local communities.

As these case studies show, fintech will remain a transformational force to support sustainable finance, conservation footprint, and financial inclusion in the blue economy. Governments, conservation groups, and private investors need to capitalize on the potential of fintech in order to fulfil long-term ocean sustainability targets.

7 Implications of Fintech in the Blue Economy

The fusion of fintech in the blue economy has far-reaching implications in economic, social, environmental, and governance aspects. Financially, fintech facilitates financial inclusion, lowers the cost of operations, and mobilizes private capital for sustainable marine conservation initiatives. Blockchain-based financing systems have significantly enhanced investment in aquaculture, fisheries, and marine resource management, building economic resilience in coastal communities (Cai 2018). Moreover, predictive analytics with the help of AI streamline the management of resources, reducing financial losses for aquaculture businesses (Monirul Islam et al. 2014).

Socially, fintech enables financial inclusion by providing micro-lending, mobile banking, and digital insurance to poor coastal communities. Mobile payment technologies have enabled small-scale fishers and marine entrepreneurs to access saving, credit, and market services independent of conventional banks (Li and Chen 2024). In addition, such innovations enhance gender equality through greater access to financial services among women entrepreneurs in coastal economies.

Environmental wise, fintech aid blue carbon ecosystems conservation by financing for them, restoring coral reefs, and sustainably managing fisheries. Carbon credit marketplaces that leverage blockchain, for example, incentivize coastal populations to preserve and restore important marine ecosystems (Chitalu and Odoyo 2024). IoT sensors with AI help track marine biodiversity, thwart blacklisted boats from fishing in prohibited areas, and optimize usage in a aquaculture facility, which saves millions of overfishing incidents and habitat destruction.

From a governance perspective, fintech promotes accountability and transparency in the management of marine resources. Blockchain technology minimizes the risk of corruption through open financial flows in fisheries supply chains. These implications all show how fintech can boost sustainable development in the blue economy.

8 Discussion and Conclusion

Fintech technologies, especially blockchain, artificial intelligence (AI), and the Internet of Thing (IoT), have become game-changing instruments in solving financial, environmental, and social issues in the blue economy. These technologies enhance economic resilience, sustainable management of ocean resources, and financial access by increasing inclusiveness, efficiency, and transparency. Blockchain facilitates decentralized finance (DeFi), enabling small-scale fishers and coastal entrepreneurs to access money using non-conventional financial mechanisms (Shih et al. 2023). AI-powered analytics enhance real-time decision-making in ocean conservation, optimizing fisheries management and avoiding overfishing risks (Tolentino-Zondervan et al. 2023). In addition, IoT-driven monitoring systems boost the preservation of marine biodiversity with real-time data supporting policymakers and conservation bodies (Glaviano et al. 2022). Collectively, these technologies cumulatively address budget shortfalls, achieve peak efficiency with resources, and support vulnerable populations, underpinning fintech's function to foster ocean economies' sustainability (Agbemabiese et al. 2024).

Fintech lies at the center of promoting ocean-based economy resilience, inclusiveness, and sustainability and steering towards broader Sustainable Development Goals (SDGs). By lowering barriers to entry for financial inclusion, fintech guarantees that coastal communities make an active contribution to sustainable marine industries while benefiting from economic growth and environmental protection (Thompson 2022). Furthermore, cross-industry partnerships and set policy guidelines are required to guarantee that the benefits of fintech are optimized while reducing risks associated with cybersecurity attacks and regulatory uncertainty. The seamless convergence of blockchain, AI, and IoT in maritime sectors will be crucial in maintaining sustained economic activities while protecting marine ecosystems (Oyewole et al. 2024).

Several fields require further research in an effort to expand the application of fintech in blue economy. To enable global decentralized finance (DeFi) platforms with the ability to support large-scale marine conservation projects and fishery management, blockchain-based financial systems need to scale (Shih et al. 2023). Research needs to focus on developing blockchain-based financial products that enhance coastal sustainability projects' funding frameworks while enabling transparency and accountability in transactions. Second, AI-based marine resources management research should be prioritized to facilitate greater real-time monitoring of marine ecosystems and biodiversity. AI-based solutions can be employed to enhance predictive analytics for assessing ocean health, forecasting fish stock availability, and maximizing sustainable fishing (Thompson 2022). Future research should explore how AI models can be utilized to integrate climate data for environmental risk mitigation and support log-term marine conservation. Third, the regulatory and policy environment will be critical in facilitating the innovation of fintech solutions in the blue economy. Governments must design robust policy regulations to address issues of data security, AI ethics, and blockchain regulation (Cai 2018). This framework should encourage private–public partnerships that incentivize green investment and inclusive financial participation for small-scale fishers and coastal entrepreneurs. Additionally, the expansion of IoT-based monitoring systems can promote smart ocean governance through enabling real-time tracking of marine pollution, illegal fishing activity, and climate change impacts.

References

Agbemabiese WT, McKibbin WJ, Muchnick M et al (2024) Factors influencing the intent of Ghanaian fishery enterprises to adopt blockchain-based IOT-enabled traceability technologies

Agrawal A, Catalini C, Goldfarb A (2013) Some simple economics of crowdfunding

Arner DW, Buckley RP, Zetzsche DA, Veidt R (2020) Sustainability, FinTech and financial inclusion. Eur Bus Org Law Rev 21:7–35. https://doi.org/10.1007/s40804-020-00183-y

Azmeh C (2025) Financial globalization, poverty, and inequality in developing countries: the moderating role of fintech and financial inclusion. Res Glob 10:100277. https://doi.org/10.1016/j.resglo.2025.100277

Bakker K (2022) Smart oceans: artificial intelligence and marine protected area governance. Earth Syst Gov 13:100141. https://doi.org/10.1016/j.esg.2022.100141

Baliker C, Baza M, Alourani A et al (2024) On the applications of blockchain in FinTech: advancements and opportunities. IEEE Trans Eng Manage 71:6338–6355. https://doi.org/10.1109/TEM.2022.3231057

Bennett M, March A, Failler P (2024) Blue economy financing solutions for the fisheries and aquaculture sectors of Caribbean island states. Fishes 9:305. https://doi.org/10.3390/fishes9080305

Bughin J, Chui M, Manyika J (2013) Ten IT-enabled business trends for the decade ahead

Cai CW (2018) Disruption of financial intermediation by FinTech: a review on crowdfunding and blockchain. Account Financ 58:965–992. https://doi.org/10.1111/acfi.12405

Cavanaugh KC, Bell TW, Aerni KE et al (2025) New technologies for monitoring coastal ecosystem dynamics. Ann Rev Mar Sci 17:409–433. https://doi.org/10.1146/annurev-marine-040523-020221

Chitalu MC, Odoyo CO (2024) Impact of implementation models on financial inclusion and resilience: a study of rural women in Zambia. Int J Entrepreneurship Bus Innov 7:87–103. https://doi.org/10.52589/IJEBI-LUKIRE8O

Choudhary P, Venkata Subhash G, Khade M et al (2021) Empowering blue economy: from underrated ecosystem to sustainable industry. J Environ Manage 291:112697. https://doi.org/10.1016/j.jenvman.2021.112697

Cisneros-Montemayor AM, Moreno-Báez M, Voyer M et al (2019) Social equity and benefits as the nexus of a transformative blue economy: a sectoral review of implications. Mar Policy 109:103702. https://doi.org/10.1016/j.marpol.2019.103702

Costello C, Cao L, Gelcich S et al (2020) The future of food from the sea. Nature 588:95–100. https://doi.org/10.1038/s41586-020-2616-y

Danladi S, Prasad MSV, Modibbo UM et al (2023) Attaining sustainable development goals through financial inclusion: exploring collaborative approaches to fintech adoption in developing economies. Sustainability 15:13039. https://doi.org/10.3390/su151713039

de Salas K, Scott JL, Schüz B, Norris K (2022) The super wicked problem of ocean health: a socio-ecological and behavioural perspective. Philos Trans R Soc B Biol Sci 377:20210271. https://doi.org/10.1098/rstb.2021.0271

Dorfleitner G, Forcella D, Nguyen QA (2022) The digital transformation of microfinance institutions: an empirical analysis. JAAR 23:454–479. https://doi.org/10.1108/JAAR-02-2021-0041

Drogkoula M, Kokkinos K, Samaras N (2023) A comprehensive survey of machine learning methodologies with emphasis in water resources management. Appl Sci 13:12147. https://doi.org/10.3390/app132212147

Education at a Glance 2016 (2016). In: OECD. https://www.oecd.org/en/publications/education-at-a-glance-2016_eag-2016-en.html. Accessed 13 Mar 2025

Elias O, Awotunde OJ, Oladepo OI et al (2024) The evolution of green fintech: leveraging AI and IoT for sustainable financial services and smart contract implementation. World J Adv Res Rev 23:2710–2723. https://doi.org/10.30574/wjarr.2024.23.1.2272

Fanea-Ivanovici M, Baber H (2024) Alternative finance: a framework for innovative and sustainable business models, 1st edn. Routledge, London

Galvez JF, Mejuto JC, Simal-Gandara J (2018) Future challenges on the use of blockchain for food traceability analysis. TrAC Trends Anal Chem 107:222–232. https://doi.org/10.1016/j.trac.2018.08.011

Glaviano F, Esposito R, Cosmo AD et al (2022) Management and sustainable exploitation of marine environments through smart monitoring and automation. J Mar Sci Eng 10:297. https://doi.org/10.3390/jmse10020297

Gomber P, Kauffman RJ, Parker C, Weber BW (2018) On the fintech revolution: interpreting the forces of innovation, disruption, and transformation in financial services. J Manag Inf Syst 35:220–265. https://doi.org/10.1080/07421222.2018.1440766

Gubbi J, Buyya R, Marusic S, Palaniswami M (2013) Internet of things (IoT): a vision, architectural elements, and future directions

Halpern BS, Frazier M, Afflerbach J et al (2019) Recent pace of change in human impact on the world's ocean. Sci Rep 9:11609. https://doi.org/10.1038/s41598-019-47201-9

Hasan M, Hoque A, Abedin MZ, Gasbarro D (2024) FinTech and sustainable development: a systematic thematic analysis using human- and machine-generated processing. Int Rev Financ Anal 95:103473. https://doi.org/10.1016/j.irfa.2024.103473

Howson P (2020) Building trust and equity in marine conservation and fisheries supply chain management with blockchain. Mar Policy 115:103873. https://doi.org/10.1016/j.marpol.2020.103873

Intergovernmental Panel on Climate Change (Ipcc) (2023) Climate change 2021—the physical science basis: working group I contribution to the sixth assessment report of the intergovernmental panel on climate change, 1st edn. Cambridge University Press

Isabelle DA, Westerlund M (2022) A review and categorization of artificial intelligence-based opportunities in wildlife, ocean and land conservation. Sustainability 14:1979. https://doi.org/10.3390/su14041979

Islam MM, Sallu S, Hubacek K, Paavola J (2014) Limits and barriers to adaptation to climate variability and change in Bangladeshi coastal fishing communities. Mar Policy 43:208–216. https://doi.org/10.1016/j.marpol.2013.06.007

Jambeck JR, Geyer R, Wilcox C et al (2015) Plastic waste inputs from land into the ocean. Science 347:768–771. https://doi.org/10.1126/science.1260352

Kleih U, Linton J, Marr A et al (2013) Financial services for small and medium-scale aquaculture and fisheries producers. Mar Policy 37:106–114. https://doi.org/10.1016/j.marpol.2012.04.006

Kraemer RA (2017) A sustainable ocean economy, innovation and growth: a G20 initiative

Krishnan S, Anand AJ, Kumar R (2024) Sustainable development goals: technologies and opportunities, 1st edn. CRC Press, Boca Raton

Kshetri N, Voas J (2018) Blockchain-enabled e-voting. IEEE Softw 35:95–99. https://doi.org/10.1109/MS.2018.2801546

Laffoley D, Baxter JM, Amon DJ et al (2021) Evolving the narrative for protecting a rapidly changing ocean, post-COVID-19. Aquat Conserv 31:1512–1534. https://doi.org/10.1002/aqc.3512

Li Z, Chen P (2024) Sustainable finance meets FinTech: amplifying green credit's benefits for banks. Sustainability 16:7901. https://doi.org/10.3390/su16187901

Macchiavello E, Siri M (2022) Sustainable finance and fintech: can technology contribute to achieving environmental goals? A preliminary assessment of 'green fintech' and 'sustainable digital finance.' Eur Company Financ Law Rev 19:128–174. https://doi.org/10.1515/ecfr-2022-0005

Maiti M, Ghosh U (2023) Next-generation internet of things in fintech ecosystem. IEEE Internet Things J 10:2104–2111. https://doi.org/10.1109/JIOT.2021.3063494

Mandal A, Ghosh AR (2024a) Role of artificial intelligence (AI) in fish growth and health status monitoring: a review on sustainable aquaculture. Aquacult Int 32:2791–2820. https://doi.org/10.1007/s10499-023-01297-z

Mandal A, Ghosh AR (2024b) AI-driven surveillance of the health and disease status of ocean organisms: a review. Aquacult Int 32:887–898. https://doi.org/10.1007/s10499-023-01192-7

Merrill RK, Schillebeeckx SJD, Blakstad S (2019) Sustainable digital finance in Asia: creating environmental impact through bank transformation

Mhlanga D (2022) The role of financial inclusion and FinTech in addressing climate-related challenges in the industry 4.0: lessons for sustainable development goals. Front Clim 4:949178. https://doi.org/10.3389/fclim.2022.949178

Omokhoa HE, Odionu CS, Azubuike C, Sule AK (2024) AI-powered fintech innovations for credit scoring, debt recovery, and financial access in microfinance and SMEs. Gulf J Adv Bus Res 2:411–422. https://doi.org/10.51594/gjabr.v2i6.55

Oyewole AT, Adeoye OB, Addy WA et al (2024) Promoting sustainability in finance with AI: a review of current practices and future potential. World J Adv Res Rev 21:590–607. https://doi.org/10.30574/wjarr.2024.21.3.0691

Patro PK, Jayaraman R, Salah K, Yaqoob I (2022) Blockchain-based traceability for the fishery supply chain. IEEE Access 10:81134–81154. https://doi.org/10.1109/ACCESS.2022.3196162

Philippon T. NBER working paper series

Pomeroy R, Arango C, Lomboy CG, Box S (2020) Financial inclusion to build economic resilience in small-scale fisheries. Mar Policy 118:103982. https://doi.org/10.1016/j.marpol.2020.103982

Raihan A (2024) Financial technology optimization using artificial intelligence (AI) to accomplish sustainable development goals (SDGs)

Ranganathan CS, Raman R, Parikh S et al (2023) IoT applications in marine monitoring: protecting ocean health and biodiversity. 2023 international conference on sustainable communication networks and application (ICSCNA). IEEE, Theni, India, pp 305–310

Rizwan A, Mustafa F (2022) Fintech attaining sustainable development: an investor perspective of crowdfunding platforms in a developing country. Sustainability 14:7114. https://doi.org/10.3390/su14127114

Rumson AG, Hallett SH (2019) Innovations in the use of data facilitating insurance as a resilience mechanism for coastal flood risk. Sci Total Environ 661:598–612. https://doi.org/10.1016/j.scitotenv.2019.01.114

Sanyaolu TO, Adeleke AG, Azubuko CF, Osundare OS (2024) Exploring fintech innovations and their potential to transform the future of financial services and banking. Int J Scholarly Res Sci Technol 5:054–072. https://doi.org/10.56781/ijsrst.2024.5.1.0033

Sarjana S, Claudia SA, Ramadhina AT, Suyanti L (2024) A sustainable blue economy: blockchain and internet of things integration in tourism villages. IOP Conf Ser Earth Environ Sci 1324:012020. https://doi.org/10.1088/1755-1315/1324/1/012020

Sarker S, Henningsson S, Jensen T, Hedman J (2021) The use of blockchain as a resource for combating corruption in global shipping: an interpretive case study. J Manag Inf Syst 38:338–373. https://doi.org/10.1080/07421222.2021.1912919

Schueffel P (2016) Taming the beast: a scientific definition of fintech

Senyo P, Osabutey ELC (2020) Unearthing antecedents to financial inclusion through FinTech innovations. Technovation 98:102155. https://doi.org/10.1016/j.technovation.2020.102155

Sharma V, Jangir K, Gupta M, Rupeika-Apoga R (2024) Does service quality matter in FinTech payment services? An integrated SERVQUAL and TAM approach. Int J Inf Manage Data Insights 4:100252. https://doi.org/10.1016/j.jjimei.2024.100252

Shih Y-C (2024) Ocean governance in practice: a study of the application of marine science and technology research techniques to maritime law enforcement in Taiwan. Mar Policy 163:106081. https://doi.org/10.1016/j.marpol.2024.106081

Shih CM, Gwizdalski A, Deng X (2023) Building a sustainable future: exploring green finance, regenerative finance, and green financial technology. World Sci Ann Rev Fintech 01:2350002. https://doi.org/10.1142/S2811004823500021

Singh B, Kaunert C (2024a) Revealing green finance mobilization: harnessing FinTech and blockchain innovations to surmount barriers and foster new investment avenues. In: Harnessing blockchain-digital twin fusion for sustainable investments. IGI Global Scientific Publishing, pp 265–286

Singh B, Kaunert C (2024b) Blue ocean and machine learning trajectories SDG 14: life below water for handling ocean pollution: metaverse conserve ocean health sustainability through the lens of transboundary legal-policy regulations as articulating space for futuristic changes. In: De D, Sengupta D, Tran TA (eds) Artificial intelligence and edge computing for sustainable ocean health. Springer Nature Switzerland, Cham, pp 77–99

Sumaila UR, Walsh M, Hoareau K et al (2021) Financing a sustainable ocean economy. Nat Commun 12:3259. https://doi.org/10.1038/s41467-021-23168-y

Syafira T, Jackson S, Tambunan A (2024) Fintech integration with crowdfunding and blockchain in industry 4.0 era. Startupreneur Bus Digit (SABDA J) 3:10–18. https://doi.org/10.33050/sabda.v3i1.433

Thiele T, Gerber LR (2017) Innovative financing for the high seas. Aquat Conserv 27:89–99. https://doi.org/10.1002/aqc.2794

Thompson BS (2022) Blue bonds for marine conservation and a sustainable ocean economy: status, trends, and insights from green bonds. Mar Policy 144:105219. https://doi.org/10.1016/j.marpol.2022.105219

Tirumala RD, Tiwari P (2022) Innovative financing mechanism for blue economy projects. Mar Policy 139:104194. https://doi.org/10.1016/j.marpol.2020.104194

Tolentino-Zondervan F, Ngoc PTA, Roskam JL (2023) Use cases and future prospects of blockchain applications in global fishery and aquaculture value chains. Aquaculture 565:739158. https://doi.org/10.1016/j.aquaculture.2022.739158

Trivedi P, Ravisankar A, Kumar GP et al (2024) Finance and technology-integrated digital economy: circular age of a sustainable digital revolution. In: Navigating the circular age of a sustainable digital revolution. IGI Global Scientific Publishing, pp 91–120

Van Den Burg SWK, Stuiver M, Bolman BC et al (2017) Mobilizing investors for blue growth. Front Mar Sci 3. https://doi.org/10.3389/fmars.2016.00291

World Development Report 2017 (2017) Governance and the law. In: World Bank. https://www.worldbank.org/en/publication/wdr2017. Accessed 13 Mar 2025

Worm B, Barbier EB, Beaumont N et al (2006) Impacts of biodiversity loss on ocean ecosystem services. Science 314:787–790. https://doi.org/10.1126/science.1132294

Zak S, Chappel B, Pilger C et al (2022) Chapter 27: Finance approaches supporting social enterprise, non-profits and for-purpose organisations: a marine permaculture case study. In: Cadman T, Sarker T (eds) De Gruyter handbook of sustainable development and finance. De Gruyter, pp 621–650

Green Economy: Integration of Environmental, Social and Economic Component for Sustainable Development and Employee Retention

Shamsul Alam, Anjali Vyas, Nargis Sultana, and Shakil Ahmad

Abstract The phrase "green economy" refers to a number of concepts and has a connection to sustainability. The green economy is an intriguing framework for building more socially inclusive, resource-efficient, carbon-free, and environmentally friendly societies. It reduces environmental dangers and resource consumption while influencing the growth of social fairness and human well-being. New conditions are created to facilitate this transformation by nations who support changing their economies to more environmentally friendly models. The main topics of discussion in this chapter are the definitions of the green economy and other related terms and an assessment of these terms in light of both strong and weak sustainability criteria. The theories, methods, and tools relating to a "green economy" will be identified and described in this chapter. This chapter creates a framework for economic sustainability, demonstrating the ability of green economy concepts, techniques, and instruments. Such a framework can be used as a guide to incorporate many ideas and strategies into a structure that supports the green economy. This chapter identifies current deficiencies to enhance measurement for green economic transformations. It presents four proposals: More affordable, faster, and publicly accessible data for evaluating interactions among the economy, environment, and society.

Keywords Green economy · Social fairness · Economic stability · Economic transformation · Environmental protection · Sustainable development

S. Alam · A. Vyas
University School of Business, Chandigarh University, Punjab, India

N. Sultana
Department of Finance and Banking, Comilla University, Comila, Bangladesh

S. Ahmad (✉)
School of Business, University of Creative Technology Chittagong, Chittagong, Bangladesh
e-mail: s.a.srabon@gmail.com

V. Sharma et al. (eds.), *FinTech for Sustainable Finance and a Well-Blue Economy*, World Sustainability Series, https://doi.org/10.1007/978-3-031-92390-6_2

1 Introduction

Traditional thinking about economic growth has long promoted excessive consumption and contributed to the depletion of natural resources; this is no longer acceptable and is having a negative impact on the planet. An innovative new paradigm, the green economy aims to strike a balance between monetary expansion, social development, and ecological protection. Reversing the negative correlation between economic growth and ecological degradation is central to the green economy movement. It covers a lot of ground, including alternative energy, sustainable farming, cutting-edge transportation, and waste management (Abdillah et al. 2022). To ensure that future generations can flourish without endangering natural processes, the green economy seeks to balance societal requirements with those of our earth. At every point in the production and consumption cycle, it stresses the need of utilizing resources wisely and effectively while cutting down on pollution. To being good for the earth, this increases competitiveness and promotes innovation. Renewable power plays a crucial role in the modern economy (Zhang et al. 2022). Greenhouse gas emissions are decreased and the economy is stimulated by investments in renewable energy infrastructure. In an effort to diversify their energy portfolios, many countries are setting ambitious targets to boost the share of renewables. The green economy also recognizes the importance of agriculture and the need to ensure food security for a growing global population. Organic farming and regenerative agriculture are ways to improve soil, safeguard biodiversity and reduce the use of harmful fertilizers. In addition to preserving ecosystems, sustainable agriculture boosts local economies, increases food quality and makes farmers more lively (Gasparatos et al. 2017).

One of the major principles of the green economy is a society that is equal. This environmentally conscious hiring drive aims to provide opportunities for fair wages, a safe workplace and reduce employee turnover. Energy efficiency initiatives and renewable energy sources have the potential to generate new jobs in both urban and rural areas. With the help of training programs workers in manufacture industries may make a fair transition into the green economy. To ensure green economy businesses, individuals and the government must collaborate closely (Reganold et al. 1990). It demands legislative incentives and legislative frameworks that encourage sustainable behaviors and discourage harmful ones. It is crucial for governments to enforce regulations, financial incentives & innovation through research and development. Green technology and sustainable practices should be the focus of every company's investment strategy. In a green economy, people's standard of living is not compromised in the chase of economic expansion, but rather, the limited resources are considered. The willingness to engage in social welfare, green agriculture, renewable energy and resource efficiency determines our capacity to provide for future generations. Transitioning to a green economy is a once-in-a-generation chance to build a better, more equitable and more flourishing future for our next generation (Aminudin 2013).

An innovative illustration for long-term improvement has evolved: the green economy. This movement is in response to growing concerns about global warming and economic inequality. According to UN Environment Programme (UNEP), if we move towards a green economy, we can drastically cut down on environmental hazards and create 24 million new jobs by 2030. In order to achieve a balance between economic development and social equality, this shift is most important for nations. The green economy promotes social inclusion and carbon neutrality through an integrated approach. Businesses, Individuals and governments all over the world are looking into new green and modern economic models to fuel sustainable economic development as the number of pledges to achieve net-zero emissions grows. This chapter examines the theoretical foundations, major drivers, major motivators, limitations and actual implementations of the green economy. It analyses the impact of green and modern economic concepts on sustainable development and employee's retention, offering insights into their influence on future workforce and next corporate strategy for the manufacturing firms. This chapter seeks to enhance the dialogue on sustainable economic transformation by addressing current sketchy and suggesting enhancements in measurement and accomplishment.

2 Foundations of the Green Economy: Key Concepts and Illustrations

Green economy is a type of economic system that places a high priority on sustainable economic growth, environmental protection and country's welfare. It aims to separate ecological harm from economic advancement. By acknowledging the interconnectedness of monetary, social, and ecological systems, proponents of the "green economy" hope to push forward a dynamic and well-rounded strategy for development (Georgeson et al. 2017; Talebzadehhosseini et al. 2020).

2.1 Illustration—1: Renewable Energy Sector

A well-known example of the green economy is the rapid development of the renewable energy sector. All countries are investing in solar power plants hydropower projects and geothermal energy systems. These clean and sustainable energy sources can lower greenhouse gas emissions. Developing renewable energy infrastructure also boosts economic expansion and creates more jobs.

2.2 Illustration—2: Sustainable Agriculture

Sustainable agriculture methods can help farmers to maintain ecosystems, to improve soil quality and to produce more nutritious food. The methods such as organic farming, perm culture, soil conservation and a reduction in the use of synthetic pesticides and fertilizers. Sustainable agriculture utilizes less water, strengthens local economies and mitigates the effects of climate change on agriculture.

2.3 Illustration—3: Circular Economy

The green economy focuses on resource efficiency and waste reduction. To maintain assets in use for as long as possible, a sustainable economy seeks to reuse, recycle and rebuild objects. Companies can establish closed-loop systems to optimize resource utilization and reduce environmental impacts through strategic implementation.

2.4 Illustration—4: Green Building and Infrastructure

Designing environmentally friendly and energy-efficient buildings that use sustainable materials and integrate renewable energy sources is known as "green building." Features like rooftop solar panels, rainwater harvesting systems, natural lighting and effective insulation can be included into buildings. It helps to lower emissions, improve air quality and make cities more livable.

2.5 Illustration—5: Eco-Tourism

The eco-friendly tourism can help the economy through safeguarding the environment and history. It motivates local and foreign tourists to do things that help local communities, protect natural habitats and bring more attention to the need to protect the environment. Ecotourism projects in the green economy framework include eco-lodges, nature reserves and guided trips that focus on sustainability and giving locals more power.

3 The Augmentation of the Green Economy Concept

In light of rising calls for environmentally friendly growth and ethical consumption, the concept of a "green economy" has emerged. The risks of global warming, resource depletion, and social inequalities have catapulted it to prominence. The story of the green economy concept may be traced back to numerous major milestones (Poltarykhin et al. 2018; Spangenberg 2016; Zhang 2022).

3.1 Phase—1: The Brundtland Report (1987)

The United Nations organized the Brundtland Commission, which produced the report "Our Common Future." Sustainable development was emphasized, as well as the significance of meeting present demands without jeopardizing future generations' capacity to do so.

3.2 Phase—2: Rio Earth Summit (1992)

Sustainability got more attention at the UNCED in Rio de Janeiro, Brazil. Agenda 21, an in-depth plan for sustainable development was approved after the conference. It was a turning point that made the modern economy idea possible by recognizing how important it is to address both economic and environmental issues at the same time.

3.3 Phase—3: Millennium Development Goals (2000)

The UN established the MDGs to address issues of poverty, education, healthcare and environmental sustainability. Goal 7 of MDGs aimed to ensure environmental sustainability through aims for sustainable development, bio-diversity conservation and access to clean energy. The MDGs established a model for incorporating environmental aim within the overall development agenda.

3.4 Phase—4: UNEP and Green Economy Report (2011)

By launching the green economy initiative in 2008, the UNEP aimed to promote the transformation of the economy towards a framework that is low in carbon emissions, environmentally benign and socially inclusive. After its publication in 2011, the Green Economy Report conducts an in-depth research into the potential advantage and policy frameworks associated with a green and modern economy.

3.5 Phase—5: Sustainable Development Goals (2015)

The Sustainable Development Goals (SDGs) were ratified by the UN General Assembly in 2015. By 2030, the 17 objectives and associated targets provide a framework for sustainable growth and development. The importance of the green economy is expressly articulated in the sustainable development goals at 8 and 12.

3.6 Phase—6: Paris Agreement (2015)

Keeping global warming range below 2 degrees Celsius is the goal of the Paris Agreement. It highlights the significance of shifting to a low-carbon economy and encourages climate-zippy growth, opening new doors for green investments and modern innovations.

3.7 Phase—7: Global Green New Deal (2020s)

This framework encourages substantial public and private investments in renewable energy, efficiency and green jobs. This agreement is designed to ensure an equitable transition towards a safe and sustainable future, addressing the challenges of climate change while promoting economic recovery.

4 Contemporary Drivers of the Green Economy

There are different forces that push the green economy forward and make the shift towards spring and environmental care. These green forces push people to think differently to solve environmental issues and direct the world towards sustainable growth (Death 2015; Ivlev and Ivleva 2018; Kosoy et al. 2012; McAfee 2016).

4.1 Driver—1: Environmental Concerns

Awareness about pollution, resource depletion and environmental degradation is causing individuals and governments to take necessary action. The movement towards environment friendly behaviors and policies is being driven by worries about how these challenges will affect human well-being and our future generations.

4.2 Driver—2: Policy and Regulation

Rules and policies play a key role in advancing the green economy. These policies may include emissions requirements, renewable energy incentives, carbon pricing system and sustainable procurement practices. Governments foster a policy climate that encourages businesses and individuals to embosom sustainable practices.

4.3 Driver—3: Technological Innovation

Technological innovations and developments are one of the main drivers behind the green and sustainable economy. The availability of green solutions, cost and effectiveness of green solutions are being improved by developments in organic farming, power storage, energy conservation and waste management. Technological innovations upend established sectors and make it possible to adopt more ecologically friendly alternatives.

4.4 Driver—4: Corporate Sustainability Initiatives

Corporate sustainability initiatives include implementing eco-friendly practices, reducing greenhouse gas emissions and reducing waste. These initiatives are motivated by consumer demand for sustainable products and cost savings through resource efficiency.

4.5 Driver—5: Investor Preferences

Now-a-days investors are taking environmental, social and governance (ESG) aspects into account before making a financial commitment. Investor gives preference to businesses that exhibit long-term sustainability and ethical business methods. Businesses are encouraged by this trend to include sustainability in order to draw in investment and obtain funding.

4.6 Driver—6: Consumer Demand

The consumers demand for products and services that are socially and ecologically responsible is growing. Customers are looking for goods that support fair labor standards and have less of an impact on the environment.

4.7 Driver—7: International Agreements and Commitments

The Paris Agreement, MDGs and SDGs are global agreements and commitments that can serve as a skeleton for the attainment of sustainability for nations and int. organizations. These agreements establish collaboration among the nations and motivate governments and businesses to implement environmentally friendly practices.

4.8 Driver—8: Financial Incentives and Support

Financial institutions offer incentives and support to businesses for introducing green initiatives. This may include low-interest loans, grants and subsidies for sustainable committed firms and renewable energy projects.

5 Indicators of the Green Economy

Green economy indicators are the measurement tools that examine overall economy performance and sustainability of environment. These help to analyse progress of programs aimed at advancing development and reducing negative environmental impacts. These indicators also assist policymakers, researchers and organisations in measuring progress, setting goals and developing effective strategies to support a green and sustainable economy. Here are some regularly used indicators of the green economy (Adarina et al. 2019; Bozhkova et al. 2020; Rana et al. 2023; Vukovic et al. 2019).

5.1 Indicator—1: Renewable Energy Generation

This indicator measures the proportion of energy produced by renewable sources, including wind, solar, and geothermal energy. This facilitates the transition from petroleum and gas to sustainable and renewable energy sources.

5.2 Indicator—2: Energy Efficiency

This examines the efficiency across many sectors including transportation, construction and industrial activities. It elucidates the efficiency of energy utilisation by quantifying the energy used to complete an appointed task.

5.3 Indicator—3: Carbon Footprint

The carbon footprint determines all greenhouse gas emissions resulting from economic activities. A reduced carbon footprint introduces a lower environmental impact and a shift towards a carbon free or low-carbon economy.

5.4 Indicator—4: Greenhouse Gas Emission Intensity

This measures the volume of greenhouse gas emissions as a percentage of economic output. Decreasing emission intensity quantifies efficiency and lower emissions in correlation with countries' economic growth.

5.5 Indicator—5: Waste Management and Recycling Rates

Waste generation and recycling rates track the total volume of garbage produced by a countries economic activity and the proportion that is recycled by the society or diverted from landfills. Increasing rates of recycle and lowering trash output debunk progress towards a more circular economy.

5.6 Indicator—6: Biodiversity Conservation

Biodiversity conservation mainly focuses on the preservation and restoration of biodiversity, including the protection of ecosystems, habitats and species. It examines the efforts to save natural resources and defend ecological balance.

5.7 Indicator—7: Green Jobs and Economic Opportunities

Green jobs and economic opportunities follow the no. of jobs and quality in eco-friendly sectors, including sustainable agriculture and waste management. It reflects the transition to a sustainable labour market.

5.8 Indicator—8: Environmental Expenditure and Investment

This indicator assesses the level of public and private investment in environmentally friendly projects, technologies, and infrastructure. It reflects the commitment and financial resources allocated to green initiatives.

5.9 Indicator—9: Sustainable Consumption and Production

This indicator evaluates the adoption of sustainable practices in production processes and consumer behaviour. It measures resource efficiency, eco-design, sustainable supply chains, and responsible consumption patterns.

5.10 Indicator—10: Environmental Governance

This indicator assesses the strength and effectiveness of environmental policies, regulations, and institutions. It includes indicators related to environmental legislation, enforcement mechanisms, and stakeholder engagement in decision-making processes.

6 Innovativeness of Green Economy

The environmentally friendly economy is defined by its innovation and the pursuit of developing and implementing novel solutions that foster sustainability. The ingenuity of the sustainable economy is crucial for expediting the shift towards sustainability. It drives technological advancements, encourages new business models, spurs policy innovations, and facilitates collaborations that contribute to the overall transformation of economic systems toward greater environmental and social sustainability (Brodziński and Kurowska 2017; Sulich et al. 2019; Tarnawska 2013; Zeya et al. 2023).

6.1 Innovation—1: Technological Innovation

The green economy fosters technological innovation by encouraging the development of clean and sustainable technologies. It propels developments in green construction materials, sustainable transportation options, energy storage, energy-efficient appliances, and renewable energy systems. Innovations in these sectors aid in the shift to a resource- and carbon-efficient economy.

6.2 Innovation—2: Circular Economy Models

The green economy advocates for using sustainable economy models that seek to reduce waste and enhance resource efficiency. Innovative strategies, like recycling, reuse, remanufacturing, and sharing economies, are incorporated into business models to diminish the consumption of limited resources and mitigate environmental effects.

6.3 Innovation—3: Sustainable Agriculture and Food Systems

Innovations in sustainable agriculture are central to the green economy. Precision farming, agroecology, vertical farming, and aquaponics enable more efficient land, water, and resource use. Advancements in food manufacturing, packaging and distribution play a significant role in minimising food waste and enhancing the environmental responsibility of the food system.

6.4 Innovation—4: Green Finance and Investment

Green finance and investment encourages investors to new ways of investing and managing money. Modern financial tools like green bonds and sustainable investment funds have come into being that put social and environmental issues first. These modern financial tools move money from risky projects to green ones.

6.5 Innovation—5: Digital Technologies and Data Analytics

This tool is monumental to the green economy because this allow for greater efficiency, monitoring and decision-making. Smart grids and IoT devices are exert to optimise energy use and resource management. These modern technologies enable data-driven decision-making and real-time monitoring for increased sustainability.

6.6 Innovation—6: Social Innovation and Collaboration

Green economy flourishes social innovation and social collaboration to tackle challenges. It exhorts the growth of business models, social enterprises and community-driven initiatives that prioritise social and environmental requirements. Collaborative platforms annex the sharing of knowledge and the formation of partnerships aimed at achieving objectives.

6.7 Innovation—7: Policy and Governance Innovation

The development of new rules and standards by governments and int. organizations facilitates the shift to a sustainable economy. Carbon costs and environmentally friendly public procurement are just a few examples of innovative policy instruments.

6.8 Innovation—8: Behavioural and Cultural Change

The green economy fosters innovative approaches to driving behavioral and cultural change toward sustainability. Educational campaigns, awareness-raising initiatives, and social marketing strategies encourage individuals and communities to adopt sustainable behaviors and lifestyles. These innovative approaches aim to shift societal norms and promote a broader cultural acceptance of sustainability principles.

7 Advantages of a Green Economy

The green economy presents a variety of benefits that support sustainable development while tackling urgent environmental and social issues. Utilizing the advantages of a green economy enables societies to attain a future that is more environmentally friendly, prosperous, and resilient and social welfare (Ahmad et al. 2025; Barbiroli 2011; Sultana et al. 2023).

7.1 Advantage—1: Environmental Sustainability

One of the primary advantages of a green economy is its focus on environmental sustainability. Minimizing environmental damage and mitigating climate change are the goals of the green economy, which promotes energy innovation, energy efficiency, reducing waste, and resource conservation. Ecosystems, biodiversity, and the responsible use of resources from the environment for generations to come are all better off as a result.

7.2 Advantage—2: Job Creation and Economic Growth

The transition to a sustainable economy has the potential to boost economic growth and generate a large number of job opportunities. Jobs in a variety of industries are created by investments in energy efficiency, energy-saving technologies, organic farming, and green infrastructure. Green occupations reduce poverty and promote social inclusion since they are labour-intensive and frequently pay well.

7.3 Advantage—3: Improved Public Health

Public health benefits from the green economy's focus on sustainable transportation, renewable energy, and lower pollution. The green economy contributes to disease prevention, better air quality, and increased well-being by lowering air and water pollution, encouraging active transportation, and limiting exposure to dangerous substances.

7.4 Advantage—4: Enhanced Energy Security

A green economy encourages a variety of decentralized energy sources and lessens dependency on fossil fuels. By promoting renewable energy technology, nations can increase their ability to generate electricity and reduce their susceptibility to fluctuations in the value of fossil fuels and supply disruptions. Furthermore, localised renewable energy systems promote energy independence and resilience.

7.5 Advantage—5: Cost Savings and Efficiency

Eco-friendly methods and technologies often lead to enhanced resource efficiency and financial savings. Energy-efficient products and structures consume reduced energy and incur lower operational costs. Eco farming approaches enhance total yields, minimise water consumption and chemical consumption and lower the expenses.

7.6 Advantage—6: Innovation and Technological Advancement

Innovation promotes research and development in the areas of waste management, clean transportation, sustainable agriculture and renewable energy. This invention solves environmental issues, opens doors for companies and boosts economic competitiveness.

7.7 Advantage—7: Social Equity and Poverty Alleviation

Green economy advances social fairness by ensuring equitable sharing of environmental advantages and opportunities. It tries to establish fair access to sustainable resources and green jobs. The green economy can help reduce the country's poverty.

7.8 Advantage—8: Climate Change Mitigation and Adaptation

The green economy is essential for mitigating greenhouse gas emissions to address climate change and facilitate adaptation. Carbon emissions are made possible by renewable energy. Green economy aids in attempts to adapt to climate change by encouraging robust infrastructure, sustainable agriculture and ecosystem restoration.

8 Limitations of Green Economy

There are a number of restrictions and obstacles associated with the green economy, which must be taken into mind, despite the fact that it has a lot of advantages. In the event that these challenges are managed in a proactive manner, stakeholders have the potential to work towards a future that is more sustainable and equitable (Ahmad et al. 2024; Alam et al. 2025; Kenis and Lievens 2015).

8.1 Limitation—1: Transition Costs

Transitioning to a green economy needed substantial investments. Reorganizing existing infrastructure, implementing renewable energy initiatives and adopting sustainable practices can be costly especially for developing countries. Initial investment may hinder the implementation of eco-friendly technologies and activities.

8.2 Limitation—2: Employment Disruptions

The green economy can make new jobs possible, but it can also make it harder to find work in some areas. Traditional industries that depend on oil, coal, or methods that aren't sustainable may need to teach their workers or lose their jobs in order to adapt to the changing economic climate. For the green economy to work, workers and towns that are affected must have a fair transition.

8.3 Limitation—3: Technological Limitations

Despite progress in green technologies, certain sectors continue to encounter technological constraints. Certain renewable energy sources exhibit dependence on weather conditions. Energy storage technologies are in a developmental phase, facing challenges related to grid integration. Technological barriers may impede the speed of transition and restrict the scalability of certain green solutions.

8.4 Limitation—4: Limited Market Demand

While consumer demand for environmentally friendly products and services is expanding, market demand may still be limited. Green products can carry a price premium, making them less accessible to certain groups of society. Furthermore, consumer knowledge and desire to prioritise sustainability can differ between areas and demographics, influencing market penetration of green options.

8.5 Limitation—5: Policy and Regulatory Challenges

Supportive laws and regulations are essential to the eco-friendly economy's success. The development of the green economy may be hampered by inconsistent or insufficient policies, a lack of long-term policy commitments, and policy gaps. Progress towards sustainability can also be aided by stronger enforcement methods, more regulatory frameworks, and uniform standards.

8.6 Limitation—6: Global Cooperation and Inequality

Coordination and international collaboration are necessary to create a truly green economy. International cooperation on environmental concerns, however, may be hampered by variations in national economies, political disputes, and priorities. Global inequality may worsen if developing nations have trouble obtaining green technology and funding.

8.7 Limitation—7: Trade-Offs and Unintended Consequences

Pursuing particular green projects can occasionally have unforeseen negative effects in other areas. Rare minerals may need to be gathered in order to produce renewable energy systems, which could have negative social and environmental effects. Assessing and minimising any trade-offs is essential when implementing a holistic approach to sustainability.

8.8 Limitation—8: Behaviour and Cultural Change

The transition to a green economy requires a change in both individual and community conduct. Changes in lifestyle choices, societal values, and purchasing patterns are frequently involved. It can be difficult to overcome resistance to change and encourage sustainable practices; this calls for awareness-raising, educational initiatives, and incentives to encourage behavioural change.

9 Sustainable Development and Green Economy

Sustainable development and the green economy complement one another. Achieving sustainable development goals can be achieved through the green economy, which prioritizes economic growth alongside environmental conservation and social well-being. Using it, we may discover a middle ground that satisfies the needs of both the economy and the environment while also uniting people. This will lead to a better and more stable future. This is how the green economy and sustainable growth are linked (Kasztelan 2017; Lavrinenko et al. 2019; Uddin et al. 2023; Zhironkin and Cehlár 2022).

9.1 Relationship—1: Environmental Sustainability

Conservation of natural resources and postponement of climate change's impacts are two ways the green economy works to improve environmental sustainability.

It seeks to separate economic development from the exploitation of resources and degradation of the environment in order to protect landscapes and the natural world for the long term.

9.2 Relationship—2: Economic Growth and Prosperity

Sustainable development emphasizes the importance of economic advancement to improve living standards and eradicate poverty. The green economy fosters economic growth by promoting environmentally sustainable sectors, such as renewable energy, eco-friendly technologies, and organic agriculture. It creates new job opportunities, promotes innovation, enhances productivity, ensures resource efficiency, and reduces negative externalities.

9.3 Relationship—3: Social Inclusion and Equity

Sustainable development, by definition, ensures equal sharing of growth gains among all societal members. The green economy strengthens this notion by emphasising green jobs, social entrepreneurship, and community involvement. It focusses on generating economic opportunities, eliminating poverty, and resolving social disparities, which contribute to social well-being and cohesiveness.

9.4 Relationship—4: Resilience and Climate Change Adaptation

The green economy supports climate change adaptation efforts by promoting sustainable land use, resilient infrastructure, and ecosystem-based approaches. It encourages investment in climate-resilient technologies, infrastructure, and practices to enhance adaptive capacity and reduce vulnerability.

9.5 Relationship—5: Integrated Planning and Policy

The environmentally friendly economy and long-term prosperity necessitate comprehensive and cohesive strategies for planning and policy formulation. Integrated planning and policy helps to collaboration among stakeholders from government, civic society and the commercial sector. Integrated planning and policy confirm that economic, social and environmental considerations are leading to more effective and balanced decision-making.

9.6 Relationship—6: Global Cooperation

Sustainable development and the green economy require international cooperation and collaboration. They are aware that global cooperation is necessary to address environmental issues that transcend national boundaries, such as contamination, loss of biodiversity, and climate change. To hasten the global adoption of environmentally friendly methods and solutions, international collaborations, information exchange, and dissemination of technology are crucial.

9.7 Relationship—7: Long-Term Perspective

Sustainable development and the green economy emphasize a long-term perspective, recognizing that actions taken today have implications for future generations. They prioritize intergenerational equity, considering the well-being and needs of future populations. The green economy supports communities' and ecosystems' resilience and long-term viability by encouraging sustainable activities and prudent resource management.

10 Impact of Green Economy on Employee Retention

The green economy can positively impact employee retention by providing various benefits and creating a more engaging and purposeful work environment. Organizations that embrace sustainability and create a supportive and fulfilling work environment are more likely to retain talented employees and build a committed workforce. The green economy's focus on sustainability, purpose-driven work, career growth, a positive work environment, and employee engagement can positively influence employee retention. Here are some ways in which the green economy can contribute to employee retention (Ahmad and Saxena 2023; Fazal et al. 2023; Jam and Jamal 2020).

10.1 Impact—1: Job Satisfaction

A great number of people are inspired by the constructive contribution to the preservation of the environment and sustainability. When workers discover that his/her work is congruent with his/her beliefs and that he/she are a part of a bigger purpose to promote sustainability, job satisfaction increases, which in turn leads to higher rates of employee retention.

10.2 Impact—2: Career Growth and Development

The green economy flourishes innovative concepts and the application of modern tools and methodologies. This opportunity for workers to acquire new knowledge, advance his/her careers and help to develop additional skills. Worker loyalty and employee retention may increase as a result of the emphasis on sustainability and the need for specialized employees in renewable energy and sustainable agriculture.

10.3 Impact—3: Competitive Compensation and Benefits

Renewable energy and sustainable technologies are experiencing a billow in demand for skilled labor. Now-a-days company is offering attractive salary packages that include assured, bonuses and other remuneration in order to attract and retain top employees or talent. Attractive benefits can keep good workers from leaving the job and help to attract new ones.

10.4 Impact—4: Positive Work Environment

The green economy focuses on positive work environment that ensure collaboration, innovation and employee welfare. Companies focused on sustainability are more likely to ensure a culture that fosters work-life balance, employee engagement and diversity. This will help to increase job satisfaction.

10.5 Impact—5: Recognition and Employee Engagement

Recognition and employee engagement can increase workplace satisfaction and loyalty. Green economy encourages employee participation by offering opportunity to submit ideas, participate in decision-making and take ownership of projects. Involving workers in sustainable activities boosts their engagement which leads to improved retention rates.

10.6 Impact—6: Workforce Stability

Workforce stability can generate more stable and sustainable employment prospects compared to other business that predominantly depend on non-renewable resources. The growth of sustainable infrastructure is anticipated to progress consistently, offering workers more job security and diminishing the likelihood of layoffs.

10.7 Impact—7: Workforce Health and Well-Being

The green economy focuses on sustainable and healthful work environment. This encompasses enhancing indoor air quality, advocating for sustainable transportation alternatives and maintaining physical activity. By emphasizing workers health and well-being, green economy businesses can foster a healthy workplace culture, enhancing employee satisfaction and employee retention.

11 Future Directions and Advancements

As the green economy continues to evolve, future research and advancements will play a crucial role in refining its implementation, effectiveness, and global impact. One key direction is the integration of digital technologies, such as AI, block chain and big data analytics, to enhance the measurement, tracking, and efficiency of green economic initiatives. These technologies can support real-time environmental monitoring, sustainable supply chain management, and carbon footprint assessments and make green policies more transparent and actionable (Sharma et al. 2024).

Another critical area is the expansion of green finance and investment mechanisms. The rise of green bonds, sustainability-linked loans, and ESG-driven investment strategies has demonstrated strong potential in mobilizing financial resources for sustainable projects. However, future research should focus on standardizing green finance regulations, improving risk assessment frameworks, and ensuring equitable access to sustainable funding, especially for emerging economies. The role of the circular economy in the green economic model is expected to grow. Future regulations should stress closed-loop production methods, resource efficiency, and waste reduction to help to develop sustainability practices in many different sectors. Governments and businesses must collaborate to implement policies that support green job creation, sustainable business models, and low-carbon innovations (Li et al. 2021).

Lastly, employee retention and workforce adaptation in the green economy should be explored further. As industries transition towards sustainable practices, it is essential to invest in green skills training, education programs, and workforce reskilling initiatives to ensure long-term employment stability. Addressing these aspects will drive a more inclusive, resilient, and innovation-driven green economy in the future.

12 Conclusion

The green economy is an innovative economic growth strategy combining social justice, environmental sustainability, and economic prosperity. It recognizes the urgency of climate change and social injustice. The ecological economy offers a more accessible and ecologically sensitive future by adopting sustainable behaviors, technology, and business structures. The green economy has several benefits. It increases resource efficiency, lowers waste and emissions, and advances technology (Hoque et al. 2023). The green economy boosts economic growth, job creation, and market competitiveness by investing in energy-efficiency improvements, sustainable agriculture, renewable energy, and clean technologies. It enhances the general welfare by reducing contamination and improving ecosystems that support human health. The green economy does, however, also have constraints and difficulties. Significant upfront investments, technical improvements, and alterations in consumer behavior are all necessary for the shift to a green economy. During this transition, some industries and communities may face short-term disruptions and job losses. Ensuring just and inclusive growth, where the benefits of the green economy are shared equitably and vulnerable groups are supported, is crucial. The green economy's relationship with sustainable development is closely intertwined. It offers a framework for combining social, environmental, and economic aspects for achieving SDGs (Islam and Ahmad 2017).

Innovation is a hallmark of the green economy. It drives technological advancements, promotes circular economy models and encourages social innovation and collaboration. The green economy fosters a culture of sustainability, job satisfaction, and employee engagement, which can contribute to higher employee retention rates. It offers numerous advantages, from creating green jobs and economic growth to reducing environmental impact and improving public health. However, challenges such as upfront investments, just transition and changing consumer behavior must be addressed (Loiseau et al. 2016). Through the advancement of sustainable initiatives and the emphasis on the welfare of employees, the green economy holds the promise of crafting a more environmentally conscious and thriving future for society as a whole.

References

Abdillah D, Panjaitan K, Stefanny N, Surbakti F (2022) The global competition in the digital society 5.0 era: the challenges of the younger generation. J Humanit Soc Sci Bus 1(3):75–80. https://doi.org/10.55047/jhssb.v1i3

Adarina R, Gazukina Y, Yankovskaya K (2019) Indicators of the "green economy" as a tool for monitoring the regional economy. IOP Conf Ser Earth Environ 395(1):121–137. https://doi.org/10.1088/1755-1315/395/1/012107

Ahmad S, Saxena C (2023) Mapping of insurance technology (Insur-Tech) research: a bibliometric analysis. In: Doctoral symposium on computational intelligence, vol 728, no 6, pp 715–726. https://doi.org/10.1007/978-981-99-3716-5_58

Ahmad S, Saxena C, Islam S, Karim R (2024) Impact of insure-tech on the premium performance of insurance business. SN Comput Sci 5(1):138–147. https://doi.org/10.1007/s42979-023-02462-0

Ahmad S, Karim R, Sultana N, Lima R (2025) InsurTech: digital transformation of the insurance industry. In: Financial landscape transformation: technological disruptions, 1st edn, vol 12, pp 287–299. https://doi.org/10.1108/978-1-83753-750-120251016

Alam S, Akter R, Khan S and Ahmad S (2025) FinTech: a new financial revolution. In: Financial landscape transformation: technological disruptions, 1st edn, vol 12, pp 301–316. https://doi.org/10.1108/978-1-83753-750-120251017

Aminudin N (2013) Corporate social responsibility and employee retention in green hotels. Proc Soc Behav Sci 105(4):763–771. https://doi.org/10.1016/j.sbspro.2013.11.079

Barbiroli G (2011) Economic consequences of the transition process toward green and sustainable economies: costs and advantages. Int J Sustain Dev World Ecol 18(1):17–27. https://doi.org/10.1080/13504509.2011.541592

Bozhkova V, Melnyk L, Derykolenko O, Yevdokimov Y, Dehtyarova I, Pasyevin O (2020) The system of indicators for alternative energy development in the context of the green economy. Int J Glob Environ Issues 19(1):70–89. https://doi.org/10.1504/IJGENVI.2020.114866

Brodziński Z, Kurowska K (2017) The issues of green economy innovativeness. In: International science conference rural development, vol 1, no 1, pp 931–937. https://doi.org/10.15544/RD.2017.144

Death C (2015) Four discourses of the green economy in the global South. Third World Q 36(12):2207–2224. https://doi.org/10.1080/01436597.2015.1068110

Fazal S, Ahmadi H, Sekhon H, Mortimer G, Sadiq M, Kharouf H, Abid M (2023) The role of green innovation and hope in employee retention. Bus Strateg Environ 32(1):220–239. https://doi.org/10.1002/bse.3126

Gasparatos A, Doll C, Esteban M, Ahmed A, Olang T (2017) Renewable energy and biodiversity: implications for transitioning to a green economy. Renew Sustain Energy Rev 70(12):161–184. https://doi.org/10.1016/j.rser.2016.08.030

Georgeson L, Maslin M, Poessinouw M (2017) The global green economy: a review of concepts, definitions, measurement methodologies and their interactions. Geo Geogr Environ 4(1):21–36. https://doi.org/10.1002/geo2.36

Hoque M, Ripon R, Ahmad S, Iqbal M, Kaur H, Kaur M (2023) Mini-review on blockchain technology for social media. AIP Conf Proc 8(2):2535. https://doi.org/10.1063/5.0113498

Islam Z, Ahmad S (2017) A trinomial probability model for occurrences of stock price change: evidence from Dhaka stock exchange. Sci J Appl Math Stat 5(1):24–30. https://doi.org/10.11648/j.sjams.20170501.14

Ivlev V, Ivleva M (2018) Philosophical foundations of the concept of green economy. Int Conf on Contem Edu Soci Sci and Eco Stud 2(1):869–873. https://doi.org/10.2991/cesses-18.2018.192

Jam M, Jamal N (2020) Impact of green human resources management practices on organizational sustainability and employee retention: an empirical study related to educational institutions. iRASD J Manag 2(1):38–48. https://doi.org/10.37896/YMER22.01/78

Kasztelan A (2017) Green growth, green economy and sustainable development: terminological and relational discourse. Prague Econ Pap 26(4):487–499. https://doi.org/10.18267/j.pep.626

Kenis A, Lievens M (2015) The limits of the green economy: from re-inventing capitalism to re-politicising the present 2(1):188–206. https://doi.org/10.4324/9781315769707

Kosoy N, Brown P, Bosselmann K, Duraiappah A, Mackey B, Martinez J, Thomson R (2012) Pillars for a flourishing earth: planetary boundaries, economic growth delusion and green economy. Curr Opin Environ Sustain 4(1):74–79. https://doi.org/10.1016/j.cosust.2012.02.002

Lavrinenko O, Ignatjeva S, Ohotina A, Rybalkin O, Lazdans D (2019) The role of green economy in sustainable development. Entrep Sustain Issues 6(3):1113–1126. https://doi.org/10.9770/jesi.2019.6.3(4)

Li L, Zhu B, Che X, Sun H, Tan M (2021) Examining the effect of green transformational leadership and environmental regulation through emission reduction policy on energy-intensive industry's employee turnover intention in China. Sustainability 13(12):6530. https://doi.org/10.3390/su13126530

Loiseau E, Saikku L, Antikainen R, Droste N, Hansjürgens B, Pitkänen K, Thomsen M (2016) Green economy and related concepts: an overview. J Clean Prod 139(2):361–371. https://doi.org/10.1016/j.jclepro.2016.08.024

McAfee K (2016) Green economy and carbon markets for conservation and development: a critical view. Int Environ Agreem Polit Law Econ 16(2):333–353. https://doi.org/10.1007/s10784-015-9295-4

Poltarykhin A, Alekseev E, Kudryavtsev V, Makhanova T, Voronkova O, Aydinov H (2018) Prospects for the development of the green economy of Russian Federation. Eur Res Stud 21(4):470–479. https://doi.org/10.35808/ersj/1135

Rana S, Das S, Shikder S, Ahmad S (2023) Industry 5.0: sustainable development in Bangladesh. Int J Eng Appl Sci Technol 8(1):106–111. https://doi.org/10.33564/IJEAST.2023.v08i01.018

Reganold J, Papendick R, Parr J (1990) Sustainable agriculture. Sci Am 262(6):112–121. https://doi.org/10.1038/scientificamerican0690-112

Sharma V, Gupta M, Taneja S (2024) Does FinTech adoption impact on the sustainability of small businesses: the mediating role of financial well-being. Global Knowl Mem Commun. https://doi.org/10.1108/GKMC-04-2024-0225

Spangenberg J (2016) The world we see shapes the world we create: how the underlying worldviews lead to different recommendations from environmental and ecological economics green economy example. Int J Sustain Dev 19(2):127–146. https://doi.org/10.1504/IJSD.2016.077208

Sulich A, Rutkowska M, Tutaj J (2019) The innovativeness of green sector enterprises. In: Knowledge-economy–society: challenges for contemporary economies–global, regional, network and organizational perspective, vol 17, no 3, pp 239–246

Sultana N, Rahman K, Lima P, Ahmad S (2023) Financial technology and competitive landscape in the banking industry of Bangladesh: an exploratory focus. In: International conference on emergent converging technologies and biomedical systems, vol 2, no 1, pp 225–241. https://doi.org/10.1007/978-981-99-8646-0_19

Talebzadehhosseini S, Scheinert S, Garibay I (2020) Global transitioning towards a green economy: analysing the evolution of the green product space of the two largest world economies. In: Proceedings of the eighth international conference on complex network and their applications, vol 882, no 2, pp 633–644. https://doi.org/10.1007/978-3-030-36683-4_51

Tarnawska K (2013) Eco innovations tools for the transition to a green economy. Econ Manag 18(4):735–743. https://doi.org/10.5755/j01.em.18.4.4485

Uddin M, Ferdous M, Rahaman A, Ahmad S (2023) Mapping of technostress research trends: a bibliometric analysis. In: 7th international conference on intelligent computing and control systems, vol 2, no 1, pp 938–943. https://doi.org/10.1109/ICICCS56967.2023.10142487

Vukovic N, Pobedinsky V, Mityagin S, Drozhzhin A, Mingaleva Z (2019) A study on green economy indicators and modeling: Russian context. Sustainability 11(17):4629–4647. https://doi.org/10.3390/su11174629

Zeya F, Sultana N, Rahman K, Ahmad S (2023) Digital transformation adoption in the insurance sector of Bangladesh: a quantitative study from the perspective of insurer. In: Global conference for advancement in technology, vol 2, no 1, pp 1–5. https://doi.org/10.1109/GCAT59970.2023.10353267

Zhang Z (2022) Evolution paths of green economy modes and their trend of hypercycle economy. Chin J Popul Resour Environ 20(1):1–11. https://doi.org/10.1016/j.cjpre.2022.03.001
Zhang L, Xu M, Chen H, Li Y, Chen S (2022) Globalization, green economy and environmental challenges: state of the art review for practical implications. Front Environ Sci 10(4):199–217. https://doi.org/10.3389/fenvs.2022.870271
Zhironkin S, Cehlár M (2022) Green economy and sustainable development: the outlook. Energies 15(3):1167–1181. https://doi.org/10.3390/en15031167

Green Insurance: Sustainable Development of Emerging Economy

Reshma Pervin Lima, Kazi Saifur Rahman, Nargis Sultana, and Shakil Ahmad

Abstract Insurance innovations and newness in the insurance industry have gained prominence recently, resulting in more effective and accessible insurance services to potential clients. The insurance sector continually changes to meet customers' needs in a challenging environment. The insurance sector can manage the risk of catastrophic environmental change. In addition to other effects, climate change emphasizes the scarcity of resources and the unpredictability of weather patterns, which drives the need for sustainable services and goods. Insurance products encompassing sustainable products' manufacture, use, and liability are classified as environmentally friendly and sustainable insurance products. Furthermore, they offer insurance against the ecological consequences of prospective climate change policies. In environmental protection, "Green Insurance" is typically used to refer to Environmental Pollution Liability Insurance or EPLI, and it serves the dual purposes of social and economic management. All general insurance business classes, such as unoccupied property insurance, insurance for empty buildings, company insurance, personal insurance, etc., are included in "Green Insurance." The green insurance sector promotes sustainable development, lowers the risks associated with industrial transformation, strengthens environmental protection capacities, and raises living standards for locals. This chapter aims to improve the prospects for the green insurance sector in emerging markets. This chapter demonstrates how environmental protection, social and economic development, and green insurance are related. It examines operating principles and development strategies for green insurance. The

R. P. Lima
Department of Accounting, Mawlana Bhashani Science and Technology University, Tangail, Bangladesh

K. S. Rahman
Department of Finance and Banking, Jatiya Kabi Kazi Nazrul Islam University, Mymensingh, Bangladesh

N. Sultana
Department of Finance and Banking, Comilla University, Comilla, Bangladesh

S. Ahmad (✉)
School of Business, University of Creative Technology Chittagong, Chittagong, Bangladesh
e-mail: s.a.srabon@gmail.com

V. Sharma et al. (eds.), *FinTech for Sustainable Finance and a Well-Blue Economy*, World Sustainability Series, https://doi.org/10.1007/978-3-031-92390-6_3

green insurance market is now necessary for the long-term expansion of the green economy in emerging countries.

Keywords Financial technology · Green insurance · Environmental protection · Sustainable development · Industrial transformation

1 Introduction

One of the biggest problems facing human society today is the effects of global climate change, which frequently results in natural disasters like temperature increases, sea level rise, hurricanes, and rainstorms that harm ecosystems, agricultural production, forest protection, and public health. Global climate change substantially threatens the economy's and society's healthy growth, generating considerable uncertainty (Global Risk Report 2020). Since the latter half of the twentieth century, there has occurred a distinct increase in the frequency of natural catastrophes globally and the associated economic losses, especially in economically disadvantaged areas such as East Asia, South Asia, the Southeast Asian region, and Africa. Consequently, it is imperative to enhance holistic disaster risk prevention to implement a strategy for natural risk mitigation and sustainable development, mitigate the effects of natural calamities, and establish a societal framework that coexists with hazards (Cremades et al. 2018). Although developing nations cannot utilize insurance processes to broaden and relocate climate change risk, employing insurance mechanisms to mitigate and transfer expenses from catastrophes has emerged as a crucial strategy for the global community in addressing climate change risk (Kong and Sun 2021).

Western nations have increasingly progressed toward material affluence since the Industrial Revolution, which brought about environmental catastrophe. In industrialized countries, a growing consciousness of environmental protection has emerged, and people gradually realize that, despite the earth's natural processes slowing the increase in pollution, they cannot make up for a load of environmental pollution brought on by human activity. Examples of this include the exponential growth of the population, rapid exhaustion of resources, and the acceleration of urbanization and industrialization (Eisenack et al. 2012). Green insurance offers insurance coverage for things like climate change adaptation, resource efficiency, and environmental betterment. Additionally, it provides services for environmental risk management. Global green finance projects are led by emerging economies, with most recipients being underdeveloped nations with stable political systems and high levels of credit risk (Zona et al. 2014). The insurance industry in developing countries has progressively improved green insurance products, offering risk control services for eco-friendly energy, transportation, buildings, eco-friendly technology, and eco-friendly forests as carbon sinks. An essential component of green financing is green insurance. Developing nations have improved their legal framework to direct the creation of green insurance and an extensive green insurance system (Belozyorov and Xie 2021). Human development should be predicated on environmental preservation through the advancement of technology and iterations in energy technology to facilitate industrial upgrading and improve the green assurance and governance framework.

Sustainable/greener products benefit society, the environment, and the economy throughout their life cycle, from the extraction of raw materials used in production to the product's final disposal. Like other types of green products, green insurance covers the development, manufacture, and use of these sustainable products and any liabilities connected to those activities. Additionally, they protect from the environmental effects of any future executive directors' decisions or inactions related to climate change. This notion of sustainable/green products also generally includes regulations where specific characteristics encourage sustainable or green behavior. As a result, a wide range of insurance products fall within the category of sustainable and green insurance products (Ahmad and Saxena 2022; Sustainable Products 2024).

This book chapter explores the concept of green insurance and its role in promoting sustainable development, particularly in emerging economies. It aims to comprehensively understand green insurance by discussing its key features, advantages, and available products. The chapter also highlights how green insurance contributes to environmental sustainability by mitigating risks associated with climate change and encouraging eco-friendly business practices. Additionally, it examines the role of green insurance in supporting environmental, social, and governance (ESG) objectives and aligning financial security with sustainability goals. By analyzing existing green insurance models and their potential impact, this chapter emphasizes the importance of integrating sustainable practices into the insurance sector. Ultimately, it encourages policymakers, insurers, and stakeholders to adopt and expand green insurance initiatives to foster a resilient and environmentally responsible economy.

In sum, this chapter has several reasons why it is crucial to contribute to the value of this research study. Firstly, by introducing top-level information about the financial environment of the world countries and their cultures, businesses become more competitive and hence try to identify new opportunities and accordingly come up with new attractive offers (Stricker et al. 2022). Second, it educates policy decisions because the findings help policymakers understand the nature and impact of international financial structure and cultures when enacting economic policies and placing regulatory measures. This understanding can result in policy development that more optimally addresses key social issues and is not hindered by traditional or strictly structured policy implementation. Third, the work encourages cross-cultural implications, which are critical given the growing diversity of global society. The study positively impacts society by reducing prejudice and increasing cultural sensitivity, promoting social relations and economic integration.

Last but not least, it helps in charting the strategic directions of the enterprise for businesses and investors involved. Adding financial and cultural considerations into strategic management leads to the generation of sustainable ideas and will also assist organizations in maintaining their supremacy in the international marketplace (Ahmad and Saxena 2023a). Thus, this study is indispensable for driving competitive advantage, effective policymaking, cultural harmony, and strategic foresight.

2 The Development of Emerging Country's Green Insurance

Green insurance refers to products and services that support environmentally friendly practices and cover green technologies, renewable energy projects, and climate-related risks. Developing green insurance in emerging countries can provide:

- Significant opportunities for promoting sustainable practices.
- Attracting investments in green technologies.
- Enhancing resilience to climate-related risks.

However, concerted efforts from multiple stakeholders are required to create an enabling environment and ensure the successful implementation of green insurance initiatives. Since the reform and opening up, especially in the past 20 years, it has not only been a period of rapid economic development in emerging countries but also, emerging countries' environmental problems have become increasingly prominent. Environment Pollution accidents are a global problem and the mainstay of emerging countries' sustainable and healthy economic development in the future. In this case, green insurance is used to control environmental pollution. Green insurance is ecological pollution liability insurance that restrains and transfers corporate environmental pollution utilizing market pollution risks and protects pollution victims' legitimate rights and interests. Green insurance has economic compensation and social management. The dual function of reasoning is that its development in the United States, Britain, and developed countries such as Germany has entered a mature stage; compared with other countries, emerging countries' green insurance started late. In the 1990s, insurance companies, That is to say, the initial green insurance products in emerging countries, were launched. However, the market size is tiny and suspended in the middle. At the same time, the problem of environmental pollution in emerging countries is gradually increasing. Heavy, emerging countries' green insurance has been promoted since 2007 after the vigorous promotion of relevant departments and enterprises. It entered a relatively rapid development stage, and significant development had been achieved by the end of 2015. The story of green insurance in emerging countries is essential to promoting sustainable development and mitigating the risks associated with climate change and environmental degradation (Barua and Aziz 2022; Liang 2017). Here are some key aspects of the development of green insurance in emerging countries.

2.1 Market Awareness

The first step in developing green insurance is to create awareness among insurance providers, businesses, and individuals about the benefits and importance of environmentally friendly practices. This can be done through educational campaigns, workshops, and collaboration with government agencies, NGOs, and international organizations.

2.2 Product Development

Insurance companies need to develop specialized green insurance products that cater to the specific needs of businesses and individuals involved in sustainable practices. This may include coverage for renewable energy installations, energy efficiency projects, sustainable agriculture, and green building construction.

2.3 Risk Assessment and Pricing

Green insurance requires a different approach to risk assessment and pricing. Insurance companies need to evaluate the unique risks associated with green technologies and climate change and develop models to quantify and price these risks accurately. This may involve collaboration with experts in climate science, renewable energy, and sustainability.

2.4 Partnerships and Collaboration

The development of green insurance in emerging countries often requires collaboration between insurance companies, government agencies, financial institutions, and other stakeholders. Partnerships can help share knowledge, leverage resources, and build a supportive ecosystem for sustainable development.

2.5 Regulatory Support

Governments play a crucial role in promoting the development of green insurance by establishing supportive regulatory frameworks. This may include providing incentives for insurance companies to offer green products, implementing mandatory coverage for specific environmentally sensitive sectors, and creating standards and certifications for green insurance products.

2.6 Capacity Building

Developing green insurance in emerging countries may require building the capacity of insurance companies, underwriters, and regulators. Training programs, knowledge-sharing platforms, and technical assistance from international organizations can help.

2.7 Public–Private Collaboration

The effectiveness of green insurance programs depends on public–private partnerships. The private sector may offer knowledge, new ideas, and market access, while the public sector can offer funding, regulatory incentives, and ways to split the risk.

2.8 Awareness Among Customers

It is essential to raise awareness among businesses and individuals about the availability and benefits of green insurance. This can be done through targeted marketing campaigns, partnerships with industry associations, and engagement with communities and local stakeholders.

3 The Features of Green Insurance in Emerging Countries

Green insurance encompasses several features that distinguish it from traditional insurance products. These features support environmentally friendly practices, provide coverage for green technologies and projects, and address the risks associated with climate change and environmental degradation (Stricker et al. 2022). Here are some key features of green insurance:

3.1 Coverage for Green Technologies

Green insurance offers coverage for various green technologies and environmentally friendly initiatives. This category includes green mobility solutions, energy-saving home appliances, environmentally friendly structures, and renewable energy projects like solar, wind, and hydrological plants. It may also cover sustainable agriculture practices, eco-friendly manufacturing processes, and waste management systems.

3.2 *Climate-Related Risk Coverage*

Green insurance addresses the risks associated with climate change and extreme weather events. It covers climate-related hazards such as floods, hurricanes, wildfires, and droughts. By offering financial protection against these risks, green insurance supports resilience and helps businesses and communities recover and rebuild sustainably after climate-related disasters.

3.3 *Sustainable Business Practices*

Green insurance encourages and rewards sustainable business practices. It may offer incentives or discounts to businesses that implement environmentally friendly measures, such as energy-efficient upgrades, waste reduction strategies, or sustainable supply chain practices. By promoting sustainable behavior, green insurance helps companies reduce their environmental impact and improve their long-term sustainability.

3.4 *Environmental Liability Coverage*

Green insurance may include coverage for environmental liabilities. This can involve protection against claims and costs associated with pollution, contamination, or environmental damage caused by insured activities. It helps businesses manage potential ecological risks and ensures that they have financial support in case of environmental incidents or legal disputes.

3.5 *Carbon Emission Reduction Initiatives*

Some green insurance products incorporate carbon emission reduction initiatives. They may provide coverage for carbon offset projects or offer risk management tools and financial incentives to support the development of emission reduction strategies. By incentivizing carbon reduction efforts, green insurance contributes to the fight against climate change and supports the transition to a low-carbon economy.

3.6 Risk Assessment and Advisory Services

Green insurance often includes specialized risk assessment and advisory services. Insurance companies may have experts who assess the unique risks associated with green technologies, renewable energy projects, or climate-related events. They can guide and recommend insured parties on risk prevention, mitigation measures, and sustainability best practices.

3.7 Partnerships and Expertise

Green insurance providers often collaborate with experts, environmental organizations, and industry stakeholders. They leverage partnerships to access specialized knowledge, stay up-to-date with emerging ecological risks and regulations, and develop innovative insurance solutions. By bringing together expertise from different domains, Green Insurance ensures that its products and services are well-suited to the evolving needs of businesses and individuals involved in sustainability.

As new environmental threats arise, green insurance adapts to help smooth the way to a more sustainable future. Policyholders gain real value from its features since it encourages sustainability, mitigates environmental hazards, and builds resilience to climate change (Ahmad and Saxena 2023a).

4 Current Green Insurance Products in the Markets

We know the following green and sustainable insurance products that are now on the market due to research conducted in developing nations using papers and websites from insurance companies (Bottani et al. 2014; Kral et al. 2020; Kunreuther 2016; Nobanee et al. 2021; Watabe et al. 2019).

4.1 Personal Lines

4.1.1 Green Property Reconstruction

This type of insurance covers repairs made with greener or more energy-efficient materials and newer, more efficient appliances and machines. Customers concerned about the environment may be eligible for premium discounts from time to time.

4.1.2 Reimbursement for Property Renewable Energy

This insurance protects a homeowner who uses a renewable energy source when the power goes out. As soon as the homeowner's green energy system is back globally, it may cover the following costs:

- Profit loss from selling extra energy to the local energy supplier.
- Extra expenses to purchase replacement electricity.
- Government or utility fees for permits, reconnections, or inspections.

4.1.3 Discount for Property Loss Mitigation Devices

Homeowners in disaster-prone locations are given premium credits if they employ storm-resistant construction methods or mitigation measures. Using window shutters to shield dwellings from powerful storms is one example.

4.1.4 Low Millage Rebate

Pay-as-you-drive vehicle insurance policies naturally encourage drivers to drive less, which reduces pollution and may slow the effects of global warming. These services offer customized vehicle insurance prices and savings based on how well and how much a consumer drives.

4.1.5 Discount for Low Emission Vehicles

Many insurance companies offer discounts for electric or hybrid passenger cars.

4.2 Commercial Lines

4.2.1 Green Commercial Fleets Upgrade

This product allows the company to replace existing vehicles with hybrid models to affirm the policy.

4.2.2 Policy for Renewable Energy Projects

These solutions protect businesses in the renewable energy sector, assisting them in risk management, lawsuit defense, and asset protection. These insurance services and products are made to cover a project's entire lifecycle, from conception through distribution. An illustration. Hydraulic, solar, wind, etc.

4.2.3 Policy for Property, Equipment, and Loss of Use in Renewable Energy

This kind of policy offers cost-of-replacement reimbursement for equipment with better-performing equivalents to stay up with the field of renewable energy's rapid technological advancement. Equipment that is in use, being built, or having just been acquired can be included in the policy. Examples—Green Roofs.

4.2.4 Policy for Green Building

In terms of this insurance, insurers assist clients in developing sustainably by assessing the plans and requirements for new buildings and offering advice on guaranteeing excellent construction and loss prevention. Following a covered loss, these products cover environmentally friendly building materials and construction, similar to the Personal Lines green property insurance.

4.2.5 Insurance for Energy Savings

Energy savings insurance policies may back up energy savings guarantees from energy service firms. Typically, policies last anywhere from five to ten years, and throughout that time, the insurer will pay out any difference between actual energy savings and a predetermined baseline.

4.2.6 Insurance for Emission Reduction Projects

Organizations capturing and storing significant amounts of carbon dioxide and other greenhouse gases are given insurance products and services. Before they can affect the environment, emission reduction efforts are implemented at major local sources of these pollutants, such as power plants.

4.2.7 Coverage for Green Buildings Against Negative Publicity

When a green building receives negative press, this reputation covering offers protection. Resources are allocated to engage crisis management specialists to address adverse media coverage, provide guidance and support to essential corporate executives, and deliver additional services to restore the organization's reputation.

4.2.8 Policy for Perishable Food Products

When produce is distributed from the grower to the retailer, these goods promote the use of tools that can lower the lost productivity and raise the overall quality of the crop. Technology tools continuously track the temperature, estimating how long it will stay fresh. Utilizing this data, product distribution can be optimized for quality, marketability, and minimal perishable waste.

4.2.9 Global Weather Insurance

The gaps left by conventional insurance coverage are filled by this product in general property damage policies. Insurers cover climate change and unpredictable weather. This could be useful for event planners who wish to safeguard their event if a specific meteorological variable, such as wind speed or precipitation, reaches a certain threshold within the coverage hours.

4.2.10 Political Risk Insurance

Financial protection is provided to stakeholders, including sponsors of projects, investors, and lenders, against risks stemming from governmental interference, embargoes, license cancellations, conflict, and violence from politics that may hinder the manufacturing, certification, and delivery of carbon credits.

4.3 Specialty Lines

4.3.1 Environmental Liability Insurance

This coverage impacts many different types of hazards and industries. Environmental duties stemming from statutes and judicial rulings are one example, as are essential business responsibilities and legal responsibility for contamination. The construction business is vulnerable to many hazards and activities that could cause losses for large corporations and smaller subcontractors.

4.3.2 Directors and Officers Insurance

It has been noticed that lawsuits against businesses thought to be causing climate change are increasing. Future legal action may arise even from a company's failure to disclose, evaluate, or adopt adaptation methods. Several insurance companies now offer directors' and officers' policies that include optional protection against lawsuits related to climate change.

4.3.3 Professional Liability Insurance

According to some insurers, sustainable practices, such as energy efficiency, are associated with reduced risk. Commissioning a building includes checking all subsystems' functionality, efficiency, and conformity, such as the plumbing, electrical, and HVAC systems. This helps the environment and reduces the likelihood of claims against professionals. Consequently, engineering and architecture companies that include building commissioning in their construction processes are awarded insurance premium credits.

4.3.4 Professional Liability Insurance for Building Raters

Many specialized experts who offer services related to energy efficiency frequently need more insurance coverage. A few qualified raters and residential energy surveyors have professional negligence, general culpability, and property insurance to protect themselves against accidents and possible litigation from company operations.

5 Benefits of Green Insurance Products

5.1 Direct Benefits

Offering environmentally friendly and sustainable insurance solutions gives insurers a competitive advantage (Ahmad and Saxena 2023b; Mansour 2023). Insurers can increase their part of the growing hybrid and electric vehicle market, for instance, by lowering their rates on these vehicles. The renewable energy sector has experienced meteoric growth thanks to government incentives. Because of its innovative and technically advanced nature, insurers can play a pivotal part in such endeavors. Those who get on the bandwagon for insurance early will be the first to reap the rewards of these new markets. Customers who care about sustainability and the environment may be more cautious drivers or have more safety features installed in their homes, which makes them better insurance candidates.

If an insurer actively offers cutting-edge green goods, it is considered environmentally conscious, corporately responsible, and forward-thinking. This can undoubtedly aid in its marketing and brand-building activities.

5.2 *Indirect Benefits*

Offering sustainable or green products has indirect advantages, such as promoting ecologically responsible behavior and mitigating risk for a new green initiative or technology (Sarkar 2013).

6 Green Insurance Role in Sustainable Development

Green insurance plays a significant role in promoting sustainable development by aligning insurance practices with environmental objectives. It involves integrating ecological considerations into insurance products, underwriting practices, risk assessment and claims management (Pfeifer and Langen 2021; Volz 2018). Here are some key roles of green insurance in sustainable development:

6.1 *Risk Management and Prevention*

Green insurance offers incentives and discounts for policyholders to adopt environmentally friendly practices and technologies. It covers the risks connected to sustainable projects, including eco-friendly transportation, green buildings, and renewable energy installations. Green insurance helps reduce environmental damage by reducing risks and encouraging sustainable practices.

6.2 *Transition to a Low-Carbon Economy*

Green insurance offers coverage for renewable energy initiatives, including solar and wind farms, to support the transition to a low-carbon economy. It facilitates the embrace of alternative energy sources and attracts investment by protecting these projects from potential risks, including breakdowns in infrastructure or natural disasters. Consequently, it reduces reliance on fossil fuels.

6.3 *Conservation and Biodiversity Protection*

Green insurance contributes to the conservation of natural resources and biodiversity protection. For example, it can cover eco-tourism ventures, protected areas, and conservation projects. This type of insurance helps manage the risks associated with preserving fragile ecosystems and encourages sustainable practices in the tourism sector.

6.4 Climate Change Adaptation

Green insurance supports climate change adaptation and resilience efforts by covering climate-related risks. This can include coverage for damages caused by extreme/adverse weather events, such as floods, hurricanes, or droughts. By offering financial protection, green insurance helps individuals, businesses, and communities recover from climate-related losses and encourages the implementation of climate-resilient measures.

6.5 Environmental Liability Coverage

Green insurance offers coverage for environmental liability, including pollution-related risks. It helps businesses manage the potential costs associated with environmental damage and contamination. By ensuring that companies are financially accountable for their environmental impacts, green insurance promotes responsible business practices and encourages the adoption of pollution prevention measures.

6.6 Sustainable Supply Chains

Green insurance promotes sustainable supply chains by covering environmental risks associated with production, transportation, and distribution processes. It also promotes ethical sourcing, waste management, and emission reduction as sustainable practices that companies should include in their supply chains.

7 ESG Features of Green Insurance

Green insurance supports sustainable development by integrating environmental considerations into insurance practices. It encourages risk management and prevention, facilitates the transition to a low-carbon economy, contributes to conservation efforts, supports climate change adaptation, addresses environmental liability, and promotes sustainable supply chains. By aligning insurance with sustainability objectives, green insurance helps drive positive environmental change while providing financial protection (Getter and Rowe 2006). ESG criteria have recently drawn the attention of the insurance sector and their supervisors, in addition to the financial aspects of climate change risks and their management. The necessity for financial supervisory bodies and monitored enterprises to be actively involved in activities that incorporate ESG risks into their business is a key lesson gained. This calls for a significant shift in institutional thinking. Supervisors must enhance their knowledge of ESG problems by providing information, guidance, and capacity building to achieve this involvement. Green insurance includes several ESG (environment, social, and governance) components that support ethical and responsible behavior (Fianu 2021; Ng 2021). Here are a few significant ESG features of green insurance.

7.1 Environmental Impact

Green insurance focuses on supporting environmentally friendly practices and projects. It protects initiatives and resources that promote environmental sustainability, including green energy projects, energy-efficient structures, ecologically friendly farming, and clean technologies. By endorsing these measures, green insurance aids in alleviating climate change, diminishing pollution, and preserving natural resources.

7.2 Climate Change Resilience

Green insurance addresses the risks associated with climate change. It covers climate-related events, such as extreme weather events, floods, and droughts. Green insurance supports resilience and adaptation efforts by providing financial protection against climate risks and promoting sustainable development in the face of a changing climate.

7.3 Social Impact

Green insurance also considers social issues by backing projects that improve society. For instance, it might cover projects that help people get along with each other, build communities, or ensure people can make a living in the long term. Green insurance can encourage and make investing in socially responsible projects easier, improving well-being and social justice.

7.4 Sustainable Investments

Green insurance companies often align their investment portfolios with ESG principles. They may invest in companies and projects demonstrating strong environmental and social performance, ethical business practices, and good governance. By incorporating sustainable investments, green insurers contribute to the growth of responsible financial markets and drive positive change through their investment decisions.

7.5 Risk Management

Green insurance focuses on proactive risk management of environmental and social risks. It encourages policyholders to adopt risk reduction measures and sustainable practices to mitigate potential losses. Insurers may offer incentives or discounts to businesses or individuals who implement environmentally friendly technologies or practices, thus promoting risk prevention and sustainable behavior.

7.6 Governance and Transparency

Green insurance companies typically demonstrate strong governance practices and transparency. They establish robust policies and procedures to ensure compliance with environmental regulations, ethical standards, and fair practices. Green insurers may also disclose their ESG performance and impact, allowing stakeholders to assess their commitment to sustainability and responsible business conduct.

7.7 Collaboration and Stakeholder Engagement

Green insurance fosters collaboration and engagement with various stakeholders. This includes partnerships with environmental organizations, government agencies, research institutions, and industry associations. By involving stakeholders in developing and implementing green insurance initiatives, insurers can leverage their expertise, gain insights, and enhance the effectiveness of their sustainability efforts (Table 1).

By incorporating these ESG aspects, green insurance provides financial protection, promotes sustainable practices, drives positive environmental and social change, and contributes to a more resilient and responsible economy.

Table 1 ESG features of green insurance

Environmental (E)	Climate change adaptation, biodiversity preservation, sustainable water and maritime resource use and protection, climate mitigation, shift to a circular economy, waste prevention and recycling, environmental pollution prevention and reduction, healthy ecosystems, and sustainable land use
Social (S)	Respect for established labor standards, observance of workplace health and safety protections, adequate compensation, equitable working conditions, opportunities for training and professional development, respect for trade union rights, and assurance of good product safety, including health protection
Governance (G)	Tax integrity, board management of sustainability, board compensation based on sustainability standards, assurances of employee rights, data protection safeguards, information release, and anti-corruption measures

Source ESG-Driven Approach (Khovrak 2020)

8 ESG Features of Green Insurance

Many people want green goods, meaning more people need green insurance options. The market of the future for green insurance looks good, thanks to government benefits and a shift in culture towards caring more about the environment. There are a lot of coverage shortfalls that insurers might fill as green technologies grow. These technologies have a lot of promise for people prepared to take advantage of them (Akomea et al. 2022). This is what the CERES study "From Risk to Opportunity: Insurer Strategies to Climate Change" suggests as possible green or sustainable products:

- According to Green Technology warranty and service agreements
- Insurance-related to carbon offsetting and trading programs
- Insurance about carbon trading schemes
- Climate-friendly agricultural insurance
- Green workers' compensation insurance with a discount
- Professional Responsibility for Energy Auditor

Mastering climate change and green technologies and developing new sustainable products takes time. Therefore, early investors in insurance will probably enjoy a fruitful harvest. Acquiring knowledge in this fresh and ever-changing field requires time and practice, so don't delay; act today before it's too late (Sultana et al. 2023a).

9 Future Directions and Advancements

This section will explore emerging trends, technological innovations, and policy developments shaping the future of green insurance. Key areas for future research include the integration of artificial intelligence and big data in assessing environmental risks, developing climate risk modeling for sustainable underwriting, and the role of blockchain in enhancing transparency in green insurance policies (Jangir et al. 2022). Additionally, the chapter will highlight recent advancements such as new green insurance products tailored for climate resilience, regulatory frameworks promoting sustainable insurance practices, and increasing global investments in eco-friendly risk management solutions. This section will provide valuable insights for researchers, policymakers, and industry stakeholders by addressing these aspects, fostering further exploration and innovation in green insurance for sustainable economic development.

10 Conclusion

The expansion of green insurance in the developing world has made up for the inefficiencies of corporate and governmental operations and altered the traditional development paradigm that relied on sacrificing the environment to achieve economic progress. Additionally, it can boost environmental protection capacities and offer financial support for reducing environmental pollution. When it comes to green finance policies, a growing economy like China's may set the standard. So, authorities need to speed up the creation of environmentally friendly insurance policies and make it easier for banks to offer such policies. Fintechs pose a systemic risk, but policymakers should push for their active participation in green insurance and other environmental protection measures (Sultana et al. 2023b). There needs to be a single centralized repository for ecological insurance that tracks national pollution levels, negative environmental impacts, energy efficiency, emission reduction, compensation costs related to pollution, and other relevant data. This database can then be used for research and development of insurance products, underwriting, processing of claims, and rate context, among other services. Green insurance requires a shift in both the supply and value chains. We can increase product coverage, broaden the reach of environmentally friendly insurance safety, and bring green insurance products to life through technological advancements in the financial sector. Insurance firms have met the demands of policyholders and the green insurance industry by developing scenario-based green insurance, introducing sustainable insurance with a range of risk protection features, and utilizing financial and digital technology. Green insurance is a win–win: it reduces risk and increases risk resilience while helping local governments, companies, and people deal with the aftermath of climate-related natural catastrophes (Yang and Zhang 2022).

Green insurance offers insurance applicants assurances of reimbursement and efficient risk management services, which support the sustainable development of environmental protection companies. This is achieved by integrating risk compensation services with technology and additional resource elements (Ahmad and Saxena 2023c). For developing countries like Bhutan, Bangladesh, and Nepal, these criteria can serve as a roadmap for generating green finance and achieving sustainable development.

References

Ahmad S, Saxena C (2022) Internet of things and blockchain technologies in the insurance sector. In: Proceeding of 3rd international conference on computing, analytics and networks, vol 2, no 1, pp 1–6. https://doi.org/10.1109/ICAN56228.2022.10007267

Ahmad S, Saxena C (2023) Artificial intelligence and blockchain technology in insurance business. In: Proceedings of the international conference on recent innovations in computing, vol 2, no 1, pp 61–71. https://doi.org/10.1007/978-981-99-0601-7_6

Ahmad S, Saxena C (2023b) Impacts of Covid-19 pandemic on life insurance business in India. AIP Conf Proc 2821(1):1–11. https://doi.org/10.1063/12.0019518

Ahmad S, Saxena C (2023) Mapping of insurance technology (Insur-Tech) research: a bibliometric analysis. In: Proceedings of fourth doctoral symposium computational intelligence, vol 726, no 9, pp 715–728. https://doi.org/10.1007/978-981-99-3716-5_58

Akomea I, Adeabah D, Ofosu D, Tenakwah E (2022) A review of studies on green finance of banks, research gaps and future directions. J Sustain Financ Invest 12(4):1241–1264. https://doi.org/10.1080/20430795.2020.1870202

Barua S, Aziz S (2022) Making green finance work for the sustainable energy transition in emerging economies. In: Energy-growth nexus of an era of globalization, pp 353–382. Elsevier. https://doi.org/10.1016/B978-0-12-824440-1.00014-X

Belozyorov S, Xie X (2021) China's green insurance system and functions. E3S Web Conf 311(12):01–08. https://doi.org/10.1051/e3sconf/202131103001

Bottani E, Ferretti G, Montanari R, Rinaldi M (2014) Analysis and optimization of inventory management policies for perishable food products: a simulation study. Int J Simul Process Model 9(2):16–32. https://doi.org/10.1504/IJSPM.2014.061429

Cremades R, Surminski S, Máñez Costa M, Hudson P, Shrivastava P, Gascoigne J (2018) Using the adaptive cycle in climate-risk insurance to design resilient futures. Nat Clim Change 8(1):4–7. https://doi.org/10.1038/s41558-017-0044-2

Eisenack K, Stecker R, Reckien D, Hoffmann E (2012) Adaptation to climate change in the transport sector: a review of actions and actors. Mitig Adapt Strategy Glob Change 17(5):451–469. https://doi.org/10.1007/s11027-011-9336-4

Fianu E (2021) Understanding environmental, social and governance (ESG) contributions in the downside systematic and systemic risk measurement of the insurance sector 3.9(2):1–42. https://doi.org/10.2139/ssrn.4091794

Getter K, Rowe D (2006) The role of extensive green roofs in sustainable development. HortScience 41(5):1276–1285. https://doi.org/10.21273/HORTSCI.41.5.1276

Global Risk Report (2020) Retrieved Feb 13, 2025 from https://www.weforum.org/publications/the-global-risks-report-2020/

Jangir K, Sharma V, Taneja S, Rupeika-Apoga R (2022) The moderating effect of perceived risk on users' continuance intention for FinTech services. J Risk Financ Manag 16(1):1–16. https://doi.org/10.3390/jrfm16010021

Khovrak I (2020) ESG-driven approach to managing insurance companies' sustainable development. Insur Mark Co 11(1):42–52. https://doi.org/10.21511/ins.11(1).2020.05

Kong F, Sun S (2021) Better understanding of the insurance mechanism in dealing with climate change risk with special reference to China. Int J Environ Res Public Health 18(6):2996. https://doi.org/10.3390/ijerph18062996

Kral P, Janoskova K, Lazaroiu G, Suler P (2020) Impact of selected socio-demographic characteristics on branded product preference in consumer markets. Manag Mark 15(4):570–586. https://doi.org/10.2478/mmcks-2020-0033

Kunreuther H (2016) Reducing losses from catastrophes: role of insurance and other policy tools. Environ Sci Policy Sustain Dev 58(1):30–37. https://doi.org/10.1080/00139157.2016.1112166

Liang J (2017) Analysis of green insurance development and product innovation in China. In: Proceedings of 5th symposium risk analysis and risk management in Western China, vol 7, no 2, pp 426–431. https://doi.org/10.2991/wrarm-17.2017.74

Mansour N (2023) Green technology innovation and financial services system: evidence from China. Businesses 3(1):98–113. https://doi.org/10.3390/businesses3010008
Ng A (2021) Green investing and financial services: ESG investing for a sustainable world. Palgrave Handb Glob Sustain 5(3):1–12. https://doi.org/10.1007/978-3-030-38948-2_104-1
Nobanee H, Alqubaisi G, Alhameli A, Alqubaisi H, Alhammadi N, Almasahli S, Wazir N (2021) Green and sustainable life insurance: a bibliometric review. J Risk Financ Manag 14(11):563–578. https://doi.org/10.3390/jrfm14110563
Pfeifer D, Langen V (2021) Insurance business and sustainable development. Risk Manag 5(1):1–15. https://doi.org/10.5772/intechopen.96389
Sustainable Products (2024) Retrieved Feb 13, 2025 from https://en.wikipedia.org/wiki/Sustainable_products
Sarkar A (2013) Promoting eco-innovations to leverage sustainable development of eco-industry and green growth. Eur J Sustain Dev 2(1):171–224. https://doi.org/10.14207/ejsd.2013.v2n1p171
Stricker L, Pugnetti C, Wagner J, Zeier Röschmann A (2022) Green insurance: a roadmap for executive management. J Risk Financ Manag 15(5):221–237. https://doi.org/10.3390/jrfm15050221
Sultana N, Lima R, Ahmad S (2023) Enabling technologies for human development in developing countries: a bibliometric analysis on Insur-Tech. In: Proceedings of international conference on computational intelligence, communication technology and networking, vol 1, no 1, pp 396–400. https://doi.org/10.1109/CICTN57981.2023.10140260
Sultana N, Rahman K, Islam K, Ahmad S (2023) An exploration of bancassurance: standpoints of banks, insurance companies and customers in Bangladesh. In: Proceedings of 4th international conference for emerging technology, vol 15, no 4, pp 1–7. https://doi.org/10.1109/INCET57972.2023.10169976
Volz U (2018) Fostering green finance for sustainable development in Asia. In: Routledge handbook of banking and finance in Asia, vol 11, no 4, pp 488–504. https://doi.org/10.2139/ssrn.3198680
Watabe A, Leaver J, Ishida H, Shafiei E (2019) Impact of low emissions vehicles on reducing greenhouse gas emissions in Japan. Energy Policy 130(6):227–242. https://doi.org/10.1016/j.enpol.2019.03.057
Yang R, Zhang R (2022) Environmental pollution liability insurance and corporate performance: evidence from China in the perspective of green development. Int J Environ Res Public Health 19(19):1–18. https://doi.org/10.3390/ijerph191912089
Zona R, Roll K, Law Z (2014) Sustainable/green insurance products. Casualty Actuar Soc E-Forum 1(1):1–8. https://www.casact.org/sites/default/files/database/forum_14wforum_zona_roll_law.pdf

Innovative FinTech Solutions for Ocean Resource Management

Ewelina Idziak and Vikas Sharma

Abstract The purpose is to explore how Fintech solutions can drive sustainability in ocean resource management. To support the goal, the examination of FinTech's solutions for critical challenges (overfishing, marine pollution, etc.) is presented. The purpose is completed by showcasing innovative technologies like blockchain, artificial intelligence, and decentralized finance. Dynamic ocean degradation is a fact, so there is an urgent need for sustainable policy, tools, and innovation implementation. On the other hand, there is a gap between the needs and technology application and the knowledge integration. Also, there is a lack of comprehensive studies showing "Good Practices" for effectiveness and scalability that encourage their use. The analysis employs a combination of comprehensive desk research, literature review, and comparative studies to establish an understanding of the innovative Fintech solutions for ocean resources. Data were gathered and also the study-case methodology was used to analyze examples for in-depth investigation. Visualization methods were used to make complex conclusions clear. The findings highlight the transformative and catalytic role of Fintechs in ocean resources. At the same time, knowledge alone is still not enough for efficient implementations, so the need for practical education is still pressing. A better understanding may come through widely spread "Good Practices," which help scale solutions in business practice. Practical implications are crucial for policymakers, financial institutions, technology developers, and businesses engaged in ocean resource management. Fintech services for the Blue Economy are in a growth phase, so we should expect them to improve. Implications grow in the following directions: regulatory frameworks, adoption and scaling, education, sustainable practices, and a long-term perspective.

E. Idziak
Department of Law and Economics, Uniwersytet Kazimierza Wielkiego, Bydgoszcz, Poland

Azerbaijan State University of Economics, Baku, Azerbaijan

Vilniaus Kolegija, Higher Education Institution, Vilnius, Lithuania

E. Idziak
e-mail: ewelina.idziak@ukw.edu.pl

V. Sharma (✉)
University School of Business, Chandigarh University, Mohali, India
e-mail: vikas.sharma62@yahoo.com

V. Sharma et al. (eds.), *FinTech for Sustainable Finance and a Well-Blue Economy*, World Sustainability Series, https://doi.org/10.1007/978-3-031-92390-6_4

Keywords Fintech · Sustainability · Blue economy · Ocean · Resource management

1 Introduction

Can FinTech be the missing link in saving our oceans? How can cutting-edge financial technologies, such as blockchain, artificial intelligence, and decentralized finance may help in ocean resource management challenges? FinTech is understood as all types of technological innovations in the finance sector. It could be computer programs and also other technology used to support (mainly) financial services. New technologies are understood as innovative technologies like blockchain, artificial intelligence, and decentralized finance. Nowadays, many approaches to the management of ocean resources are still in the development phase. The most popular approaches include, for example, limiting ocean access through permits, ocean zoning, restrictions on gear using or species harvesting, and imposing fish catch limits. It is emphasized by Science and Policy for People and Nature Organization that also important are community-based initiatives as well as informal or conventional administrations, or hybrid of techniques reliant on on local social-ecological contexts (Policy instrument 2024). Moreover, Levine, Richmond, and Lopez-Carr emphasize although marine resource management traditionally seems biophysically rooted but it has become clear that this field stems from political and cultural dimensions which link human lifestyles with spatial and territorial understanding in fundamental policy and decision-making frameworks. (Levine et al. 2015). Furthermore, support for ocean sustainability processes is also being sought in advanced technology, such as innovative FinTech solutions. This topic is a huge challenge, especially since scientific knowledge in this area is just emerging, while ocean pollution affects everything on the planet, as well as human beings.

To illuminate the importance of urgent actions to save the oceans it is enough to present selected data about ocean degradation:

- Fishery stock levels that are sustainable to biology declined by 2.3% from 2019 to 2021 resulting in a current fraction of 62.3% among assessed stocks. This fraction was 90% in 1974. Since the mid-1970s the proportion of stocks overfished shows a growing trend reaching 37.7% in 2021 from a base of 10% in 1974 (Food and Agriculture Organization of the United Nations 2024). Over the years, the percentage of overfished stocks has increased, while the proportion of under fished stocks has declined. This trend suggests that more fish stocks are being exploited beyond sustainable levels, which raises concerns about long-term fishery sustainability;
- Since 1960 more than 500 "dead zones" have appeared worldwide which today cover an area of 250,000 km^2 and continue expanding by 250,000 km^2 every ten years. Dead zones exist as underwater regions where sea creatures die due to deep water oxygen levels below survivability thresholds (NASA 2024). According to

recent reports these economic expenses from fishing operations combined with the losses in coastal tourism exceed tens of billions of dollars yearly (United Nations Development Programme 2015);

- The decline of ocean biodiversity exceeds the rate experienced since the beginning of human dominion (IPBES 2019). Marine habitats experienced a 30–50% loss of their vulnerable areas and recent decades show an escalating pattern of loss. (IPBES 2019). For example, coral reefs have reduced their extent by approximately 50% from 1870 until today while the fastest decline occurred in the last thirty years. As per present projections the future looks grim for coral reefs since rising temperatures may cause between 70 and 99% of Earth's corals to perish in upcoming decades (IPCC 2018).
- The issue of plastic pollution emerges from plastic waste making up 80% of sea pollution while roughly 8–10 million metric tons of plastic enter the ocean each year. According to UNESCO research plastic will surpass the number of fish found in the sea by 2050.The production of plastic items exceeded previous century levels during the last ten-year period. The Environmental Protection Agency confirms that one hundred percent of the plastics people have produced since the beginning remain in existence. The degradation process of plastic stretches between five hundred years and one thousand years. Although degradation occurs slowly from plastic over extended time periods it produces microplastics that remain intact rather than decomposing entirely. This problem requires proper plastic waste management in global regions especially underdeveloped countries responsible for ocean plastic sources and it influences future sea health (UNESCO 2022).

Among dozens of other instances these examples demonstrate the vast extent of harm. Restoring marine habitat suffers severely from multiple human-caused threats including climate change effects combined with pollution from warming seas, habitat destruction, overfishing along with unsustainable aquaculture practices in addition to marine construction projects, oil drilling operations, ocean acidification effects, coral bleaching incidents, high mercury levels and plastic contamination are all closely related to this problem. If they continue to increase or insufficient action is taken to prevent further damage, the well-being of this and future generations could be in peril. Dead zones are now a key stressor of marine ecosystems and rank with overfishing, habitat loss, and harmful algal blooms as global environmental problems. The world's 95% population living on 10% of the land mass alongside the land's total percentage against Earth's surface enables humans to seriously influence marine ecosystems. Modern human activity has resulted in 86.8% of global ocean areas sustaining adverse impacts (ENVIROTECH 2025). At the same time, human beings rely on the oceans, because of the oxygen to breathe, the food to eat, and the moderate climate to live in, so it is the right time to use high technology (blockchain, artificial intelligence, and decentralized finance) to help to act in a sustainable way to protect the oceans.

That's why it's crucial to explore and verify the latest research on how innovative FinTech solutions can help protect and manage ocean resources more effectively, which is presented in the next paragraph.

2 Review of Literature

FinTech is no longer just about banking. Innovative technologies may support also very narrow fields and answer to the needs that appear very rapidly in the new finances and business. Through a literature review, we aim to verify the extent to how scientific publications cover the topic of supporting ocean resources management by FinTech solutions. For this purpose, the Scopus database was explored in terms of the given search filters adequate for the chapter's topic. Scopus is one of the largest and most renowned bibliographic databases used to conduct scientific literature reviews. Its wide reach, covering thousands of peer-reviewed journals, books, and conference proceedings from various fields of science, makes it a valuable tool for researchers.

A couple of filters were tested and the effect of the three of them are presented in Table 1.

Table 1 Number of documents and frequency by year in Scopus database

No	Filter	Period	Number of documents	Visualization of publication frequency by year
1	TITLE-ABS-KEY: (fintech AND resources AND management)	2017–2025	184	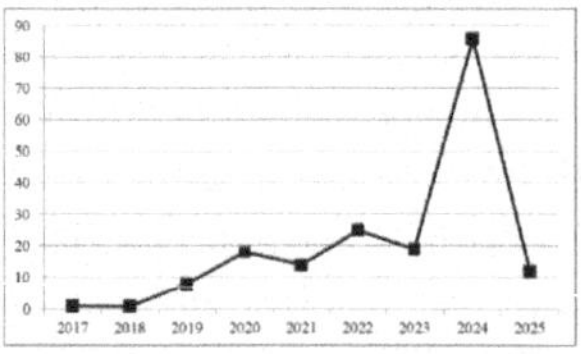
2	TITLE-ABS-KEY: (fintech AND ocean)	2020–2025	4	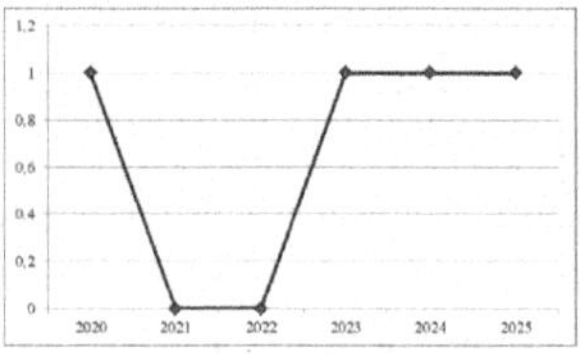
3	TITLE-ABS-KEY: (fintech AND resource AND management AND ocean)	2024	1	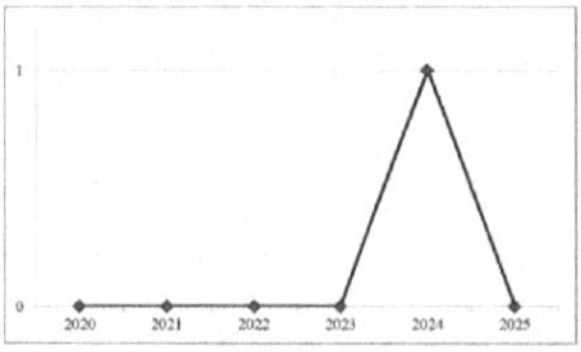

Source Created by author, based on Scopus database filtering

The data gathered in Table 1 highlight the different levels of research about FinTech's role in resource management in general, Fintech's role in wide aspects connected with oceans, and finally FinTech's role in ocean resources management. The first filter, "fintech AND resources AND management", retrieved 184 documents from 2017 to 2025, indicating a still not many, but starting to be a well-established and growing research area. The second filter, "fintech AND ocean", yielded only 4 documents from 2020 to 2025, indicating a much narrower research focus. The significant drop in the number of publications suggests that the intersection of FinTech and ocean-related topics still remains underexplored in the literature. Moreover, the third filter, "fintech AND resource AND management AND ocean", returned just 1 document from 2024, showing that the integration of FinTech in ocean resource management is an emerging and largely untapped research area. The presence of only one recent publication and as just mentioned paper from the conference with not exactly an adequate topic indicates that researchers have not even begun addressing this specific intersection. Given these important findings, it seems that future research should focus on exploring how FinTech can enhance sustainable ocean resource management, such as through blockchain-based traceability, AI-driven risk analysis, and blue finance initiatives. This is the significant gap that this book ambitiously tries to fill.

A deeper literature review helps to understand what kind of research and in which directions are already presented in scientific papers on the crossroads of FinTech and ocean resources management. Firstly, focusing on the very narrow area of those 4 selected documents about the connection of FinTech and oceans. It is noted that only one paper from 2024 has already been cited two times and one paper only once, what also suggests a very early stage of publications about this topic. Thanh Ha's paper presents the first research about FinTech and blue economy connections. The research period spanned from September 14th, 2020, until 11th August 2023. The investigation demonstrates how FinTech relates to blue bond movement volatility. This study showcase the advancement of Fintech demonstrated a reciprocal relationship with the growth of the blue economy. The blue economy index functions as a net shock receiver during the period extending from late 2021 through the second part of 2022 yet the sample's fintech indexes demonstrate mostly net shock transmitting responsibilities. The escalating Ukraine–Russia tension poses an ongoing threat to the future development of sustainable blue economic activities. The development of Fintech stands essential for advancing the blue economy as established in research (Thanh Ha 2025). The next paper about emerging an current supervisory outline for cybernetic currencies in Mauritius was classified by Scopus filter to the "ocean" category because of the location of Mauritius in the Indian Ocean. Balgobin and Seeam present the background of Mauritius's role to the African countries as a FinTech hub, because of the rapid development of technology, but so far Mauritius has not so far executed any precise guidelines to discourse the use of Blockchain, bitcoins or cryptocurrencies in its authority. The Board of Investment Mauritius ("BOI") operates the Regulatory Sandbox Licensing Scheme which stands as the primary legal instrument applicable to this domain. Through this scheme investors gain access to conduct business activities in regulatory gaps that exist between existing Mauritian law or its inadequate provisions. This research paper introduces different regulatory structures of the United States and Canada and China as tools for comparative analysis to explore regulatory systems that remain absent from African and Indian Ocean nations. (Balgobin and Seeam 2020). One more paper is worth consideration. Hassan presents Somalia's example of banking the unbanked population. The Somali economy's strong ties to the Indian Ocean can serve as an excellent case study because it permits the study of how organizations create market-ready solutions for consumers and conducts product competition analyses and identifies how market data can forecast future product positions and helps develop new product strategies. Premier Wallet—the first digital wallet in Somalia. The narrative demonstrates how Premier Wallet changed the banking market while following its inception path. Premier Bank enabled customers to purchase goods and collect cash and shop online and make top-ups through its mobile wallet system even without opening an account at Premier Bank. The financial inclusion process got finalized because of this system which reached Somali citizens who were not in the banking system. The Wallet Send feature emerges as the disruptive addition to the study which challenged the widespread Hawala system in Somalia. Mobile clients received access to money transfer through 110 countries which allowed them to deposit funds directly into family or friend mobile wallets and nearby cash withdrawal points. Premier Wallet faced challenges in market acceptance due to Somalia's low 9% internet penetration combined with the lack of national IDs and intense competition from WAAFI which received support from Hormuud Telecom. The case explores strategic moves made by Afrah to establish Premier Wallet as Somalia's leading mobile money solution for consumers (Hassan 2023). As it is seen, the topic of FinTech combined with the topic of oceans has been very poorly presented in the scientific literature so far.

The topic connected with FinTech and resources management is much wider presented in the Scopus database. During the last 9 years, 184 documents connected to this topic appear. It is seen that there is a growing trend. It is still the beginning of 2025, and during the first month, 12 documents have been identified as related to FinTech and resource management. It seems that the interest in the topic of the crossroad of FinTech and resource management is growing in a logarithmic way, which is presented in Fig. 1.

The trend in Fig. 1 shows an exponential increase in the "Y" variable (number of documents in Scopus base about FinTech and resource management) over time, particularly from 2023 to 2024, where the growth is very steep for the number of scientific documents about FinTech and resource management. The exponential trendline equation suggests that the data follows a rapid growth pattern with a relatively strong fit ($R^2 = 0.8436$) indicating that the model explains approximately 84.36% of the variance in the data. This sharp rise in 2024 is a strong signal about a significant increase in activity, probably because of technological advancements, regulatory changes, or increased market adoption. Therefore, it is worth analyzing the most influential papers in this topic, because it may be expected that the still undervalued category of issues at the intersection of FinTech and oceans will evolve in this direction. The most influential papers were chosen by the highest number of citations and adequacy for the topic. Scopus database filtering presents that for the filter:

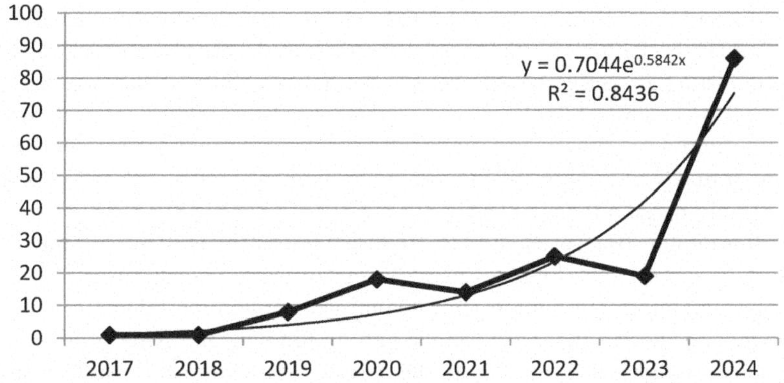

Fig. 1 Trend of scientific publication about FinTech and resource management. *Source* Created by author, based on Scopus database

TITLE-ABS-KEY (fintech AND resources AND management) the highest participation for the topic by subject area have such disciplines as: Economics, Econometrics and Finance (17.8%), Environmental Science 14.8%), Social Sciences (14.0%), Business, Management (13.1%), Computer Science (12%), Engineering (9.2%), Decision Sciences (5.4%) and others. The most active researchers are from countries: China, Pakistan, India, Malaysia, Indonesia, Saudi Africa, United States, United Kingdom, South Korea and Bangladesh.

The most cited (252 citation) research paper selected in the area of FinTech and resource management is about FinTech development as a way for a transition towards a low-carbon economy. Tao, Su, Naqvi, and Rizvi stated the research question: "Whether the Fintech development is helping economies towards a smooth transition towards a lower level of carbon and greenhouse gas emissions?" They concluded that, The financial technology sector has transformed the financial industry by reshaping its overall structures since the 2000s. Multiple tangible benefits demonstrate that Fintech systems provide increased efficiency together with competitive advantages and improved risk-sharing abilities. The same factor has made possible the effective distribution of resources. Moreover, The banking sector institutions have gained efficiency and competitive advantages because of the changes. It must also be taken into contemplation that this system can potentially offer benefits across multiple sectors. Studies show that when controlling factors are integrated Fintech development leads to decreased emissions of greenhouse gases. These research findings demonstrate durability through both 2SLS and GMM estimations that manage the endogeneity between Fintech development (Tao et al. 2022).

The second most popular paper (174 citations) demonstrates how blockchain technology partnered with predictive energy trading based on machine learning enables resource management improvement within FinTech applications. The authors show that the system can effectively predict both short-term and prolonged energy demand with accuracy through the implementation of deep learning models including

RNN and LSTM and Bi-directional LSTM. The implementation of a permissioned blockchain (Hyperledger Fabric) ensures security, transparency, and accountability while supporting incentive-based energy transactions. Future improvements may focus on hybrid machine learning models and incorporating additional environmental factors like humidity, temperature, and wind speed to further enhance predictive accuracy, which may bring innovativeness in many branches, also with the perspective for blue economy (Jamil et al. 2021).

One more interesting research (110 citations) boons the perception of an unified resolution to simplify the logistics bankrolling for capital-constrained SMEs in E-commerce marketing. The authors showed that a blockchain-based platform can help e-commerce businesses that have trouble getting funding. By using smart contracts with a special system, they made sure the process was secure and worked the same way every time. They also combined blockchain with a multi-agent system, which made the whole process more independent, clear, and fair, helping build trust. In the future, researchers could look into ways to make contracts work faster and better, making it easier for e-commerce companies to get the financial help they need. It seems that this solution might be applicable in other branches too (Li et al. 2020).

In summary, these selected studies present the transformative potential of FinTech and blockchain technologies in different sectors. The first paper emphasizes how FinTech can contribute to a low-carbon economy. The second paper shows how a blockchain-enabled predictive energy trading platform. The third research highlights the role of blockchain in facilitating financing for small e-commerce businesses. All together, these examples present how widely FinTech may be used, and heading to the topic of this chapter it is also important to highlight how the innovative FinTech solutions for ocean resource management may be applied. The next part of this chapter emphasizes the role of FinTech for ocean resource management.

3 FinTech Solutions for Ocean Resource Management

As one can see, business and the needs of real finance are ahead of scientific publications, which are missing in the field of FinTech and ocean resource management. Business life takes its own course and applies innovative technologies quicker than it is reflected in scientific publications that were reviewed in the previous subchapter. Ocean resource management is such an example. It is therefore worth learning from practical examples and "Good Practices" presented in this subsection.

Firstly, it is important to establish what ocean resource management means for this chapter. There are many definitions of ocean resource management. Also could be seen the evaluation of this concept from rigid definitions at the beginning of the creation of this concept to the more dynamic ones observed today. One of the first formal, international documents that constituted the importance of ocean protection (among others) was The United Nations Convention on the Law of the Sea (UNCLOS), adopted in 1982. It is a comprehensive international treaty that defines nations' rights and responsibilities regarding the world's oceans. It governs maritime boundaries, resource management, and environmental protection (UNCLOS 1982). The United Nations Convention on the Law of the Sea (UNCLOS 1982) does not explicitly define "ocean resource management" as a standalone concept, but establishes the legal framework for how marine resources should be managed, conserved, and utilized. It is emphasized that ocean resource management should protect and balance sovereign rights while maintaining sustainability and international cooperation. Lavine, Richmond, and Lopez-Carr present in their paper the definition of marine resources management, which is also possible to use for ocean purposes. They precise the management of marine resources as a "politically and culturally driven process, shaped by human lives and perceptions, where notions of both space and place shape policies and decision-making in fundamental ways" (Levine et al. 2015). More up-to-date approaches emphasize that oceans are dynamic environments. Maxell and other authors highlight that the authors demonstrate that most spatial marine management approaches utilize fixed boundaries to enclose inherently mobile marine elements and users. The success of stationary marine resources with these management tools does not guarantee effectiveness for dynamic ocean management because resources and users require adaptive spatial and temporal management. The authors propose dynamic ocean management as an alternative approach which applies quick spatial and temporal adjustments to oceanic adaptation and user integration based on biological, oceanographic, economic data streams. Dynamic management allows managers to adjust space–time boundaries of managed areas while achieving an improved balance between ecological needs and economic interests (Maxell et al. 2015). This approach opens the possibility for the integration of new technology to support more effective, dynamic management. The concept is to skip from static to real-time management by gathering huge data sets for better decision-making. Different types of data may be integrated into dynamic management processes (existing databases, biological, remotely-sensed, socio-economic, and user distribution data for advanced analysis or modeling). Nowadays, it is possible with innovative technology. The visualization of the dynamic ocean resources management concept is presented in Fig. 2.

In the case of ocean resources management, there could be different types of data sets, for example: existing databases, from remote sensing equipment, animal tracking, fisheries observer data, data from space and many others according to the needs. New technology like blockchain, artificial intelligence, and decentralized finance might be used to support the decision-making processes.

3.1 FinTech Practices for Ocean Resource Management

Nowadays, FinTech as a technology changes very quickly and is very flexible in creating new connections or solutions. The most popular categorization for Fintechs is done by the type of activity, for example, payments and transfers, digital banking and neobanks, lending and credit, investments and asset management, insurance, regulatory, blockchain and decentralized finance, embedded finance, and

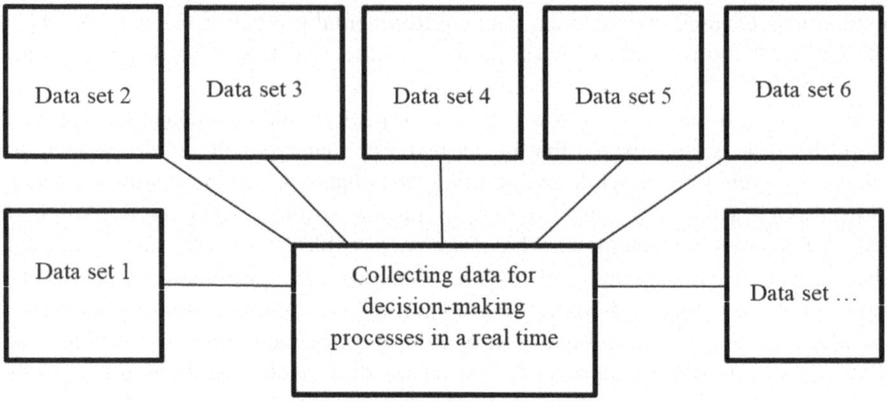

Fig. 2 Dynamic ocean resources management. *Source* Created by author

B2B fintech solutions. It could be observed that there are still new ways to technology use. According to the Cambridge Centre for Alternative Finance, 14 market segments could be explored interactively, by chasing the connection and new entities that appeared on the market. FinTech entities are divided into the following segments: digital lending, digital payments, enterprise tech provisioning, crypto asset exchange, digital capital raising, wealth tech, insurtech, regtech, digital banks, digital custody, digital identity, alternative credit analytics, digital saving, consensus services (University of Cambridge 2025). They monitor 4576 FinTech entities, where 4147 are noted as active and 429 as inactive. Table 2 presents eight examples of spectacular FinTech solutions for ocean resources management.

Those selected examples show that the adoption of FinTech in ocean resources management has a strong power to transform the ocean's industries and in consequence have an impact on sustainability, regulatory efficiency, and financial accessibility.

Examples presented in Table 2 may be categorized into six groups, selected by the type of FinTech solution. This is not an exhaustive list, but only practical examples that can be used to find further innovative applications of FinTech for managing ocean resources in the future. To make conclusions more synthetic it is helpful to group the following highlights:

- Blockchain and DeFi—help to increase transparency and support sustainable financing. The examples of ocean industries improve traceability in seafood supply chains and financial transparency in marine conservation funding. It helps to make financial flows monitoring and combat illegal fishing;
- RegTech and AI—help to fight illegal fishing by AI and satellite tracking by real-time monitoring that supports regulatory efficiency and at the same time ocean conservation;

Table 2 Spectacular FinTech solutions for ocean resources management

Type of FinTech solution	Subject	Project explanation	Effect	Amount of investments	Expected financial effect	Project's name and website
Blockchain and DeFi	Sustainable fishing	Blockchain-based system rewarding fishers with tokens for reporting catch data. Fiscoin incentivizes supply chain stakeholders to share data from the point of harvest to the point of consumption	Increased seafood traceability and sustainability	Undisclosed	Potential multimillion-dollar impact by reducing illegal fishing	Fishcoin, fishcoin.co
Blockchain and DeFi	Blue bonds for marine conservation	A sovereign bond issued to finance ocean conservation and sustainable fisheries. Bonds as such are not FinTech solutions, but blockchain improves transparency	$15 M raised for sustainable marine projects	$15 M	Supports long-term economic and environmental sustainability	Seychelles Blue Bond, worldbank.org
RegTech and AI	Anti-illegal fishing	Uses AI and satellite data to monitor and detect illegal fishing activities globally	Reduced illegal fishing, and improved regulatory compliance. Funded by Google, Oceana, and SkyTruth	Undisclosed	Increased transparency in fisheries	Global fishing watch, globalfishingwatch.org

(continued)

Table 2 (continued)

Type of FinTech solution	Subject	Project explanation	Effect	Amount of investments	Expected financial effect	Project's name and website
InsurTech	Climate risk insurance for coastal areas	AI-powered insurance provides automatic payouts to fishermen affected by extreme weather	Faster financial relief for affected communities	Undisclosed	Improved financial resilience for coastal businesses	Blue marble, bluemarblemicro.com
WealthTech	Blue economy investments	Provides funding and mentorship to startups in the sustainable ocean economy	Facilitates the growth of marine conservation startups	€145 M+	Growth in ocean-based sustainable businesses	BlueInvest (EU Initiative), blueinvest-community.converve.io
LendTech	Microloans for fishers	Provides mobile-based microloans for fishermen to finance boats and equipment	Increased financial inclusion in coastal communities	$250 M+ in total funding (for various applications)	Expansion of financial access for small-scale fisheries	M-Kopa, m-kopa.com
Payments and PayTech	Digital payments for fishermen	Enables unbanked coastal communities to receive digital payments securely	Financial inclusion for remote ocean-dependent workers. The first fintech company in Myanmar	$200 M+ (Ant Financial-backed)	Growth in financial transactions in remote areas	Wave Money, wavemoney.com.mm
WealthTech and carbon markets	Ocean-based carbon credits	Tokenizes carbon credits from ocean conservation projects to trade on DeFi platforms	Monetization of marine carbon sequestration	$4 M seed funding	Expanding voluntary carbon market for blue economy	Toucan Protocol, toucan.earth

Source Created by author

- InsurTech—supports financial resilience against climate risks. It helps to mitigate financial losses by faster and automated financial relief to coastal communities affected by extreme weather events;
- WelthTech—helps to provide funding to ocean-related startups and support sustainable ocean businesses;
- Lend Tech and PayTech—help financial inclusion for small fishers. They enable access to micro-loans and digital payments for small-scale fishers and unbanked coastal communities and reduce financial barriers;
- WelthTech and Carbon Market Monetization—provide a new revenue stream for ocean conservation and contribute to climate action and sustainable finance.

Those case studies show the effective adoption of FinTech in different ocean resource management, which confirms strong possibilities for a transformative impact on oceans, their resources, and stability, but also possibilities to support regulatory efficiency and financial accessibility for so far excluded society. It seems that the examination of those examples proves that FinTech technologies are able to provide innovative ways to finance marine conservation, combat illegal activities, and ensure a more resilient blue economy. It is expected that the rapid changes in the interest of FinTech and the rapid development of innovative technologies will play a crucial role in the fight for ocean sustainability.

3.2 Fishcoin Case-Study

Fishcoin is a spectacular example of blockchain technology supporting a very difficult area of seafood supply chain transparency. The industry faces two major problems consisting of 89% overexploitation or complete depletion of wild marine fish stocks and supply chain losses reaching 60% of all harvested seafood with many fishing areas still operating outside legal regulations (IUU) theft producing more than 700 kg of stolen seafood every second (Fishcoin 2025). The harvest of seafood occurs mostly through small-scale producers operating within developing nations, which results in absent data, subsequent management issues, and sustainability problems. The solution for those problems comes from Fishcoin. The Eachmile serves as developer for this program. Eachmile received selection from the US State Department to extend mFish (application) software development for improving artisanal fisher livelihoods and generating valuable data that strengthens global fisheries management (Eachmile 2025).

Fishcoin represents a blockchain system supporting data ecosystem operations with stablecoin token features that motivate users to provide seafood supply chain information. It might be used for all kinds of seafood. Technically it is based on Ethereum. By definition, Fishcoin is a blockchain-based traceability and incentive system designed to improve sustainability and transparency in the seafood supply chain. It rewards fishers and other supply chain stakeholders with digital tokens for reporting accurate catch data. This system helps combat illegal fishing, enhances seafood traceability, and incentivizes responsible fishing practices.

The Fiscoin implements Trace Protocol as an open-source platform that creates a scalable data-sharing scheme with reward mechanisms to drive sustainability projects through digital transformation. Different ecosystems can join this decentralization platform to share data and data incentives proportionally based on shared data. Due to their role in verifying the Legal Reported and Regulated status of products these digital vouchers known as tokens provide instant mobile airtime awards and/or government e-voucher payments or direct payments for stakeholders based on country-specific rules which drive sustainability practices above government standards (Fishcoin 2025).

The reward structure developed for data acquisition coupled with the interoperable platform enables third-party tools manufacturers (including IoT sensors) to join the system along with dApp developers to receive appropriate compensation. The system enables seafood buyers to obtain more productive information about product security and quality standards. The Trace Protocol functions as a crucial worldwide solution that needs various established partnerships to reach its main goal which includes informed seafood buyers who make smarter choices and governments can handle fisheries with current information coupled with payment mechanisms for data-sharing participants (Fishcoin 2025).

The visualization of the process and the comment help to understand step-by-step how the Fishcoin works. Step 1 is presented in the Fig. 3.

Step 1 is the beginning of the process. Tagging procedures begin at this particular moment. Multiple tagged components are linked to both a fisherman and his fishing vessel. Fishcoin tokens could be issued to the fisherman once he completes tagging work as well as records data transfers then releases the catch. The fisherman would receive tokens valued at US$0.15c to perform this action while earning more than double the amount the fisher would have received for the juvenile coral trout. The Fishcoin tokens enable him to increase his Internet access with mobile top-ups while allowing research centers or NGO analysts to understand the management needs of

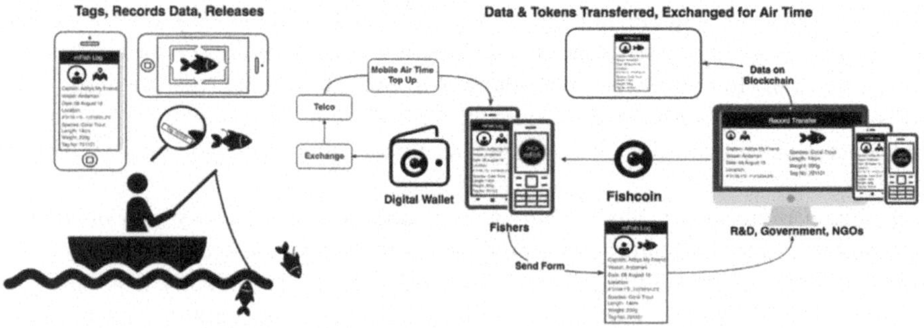

Fig. 3 Step 1. Fisher tags a fish, records data, and releases a fish. Once the record is sent the fisher receives Fishcoin tokens that can be exchanged for air time top-up. *Source* Douglas (2018)

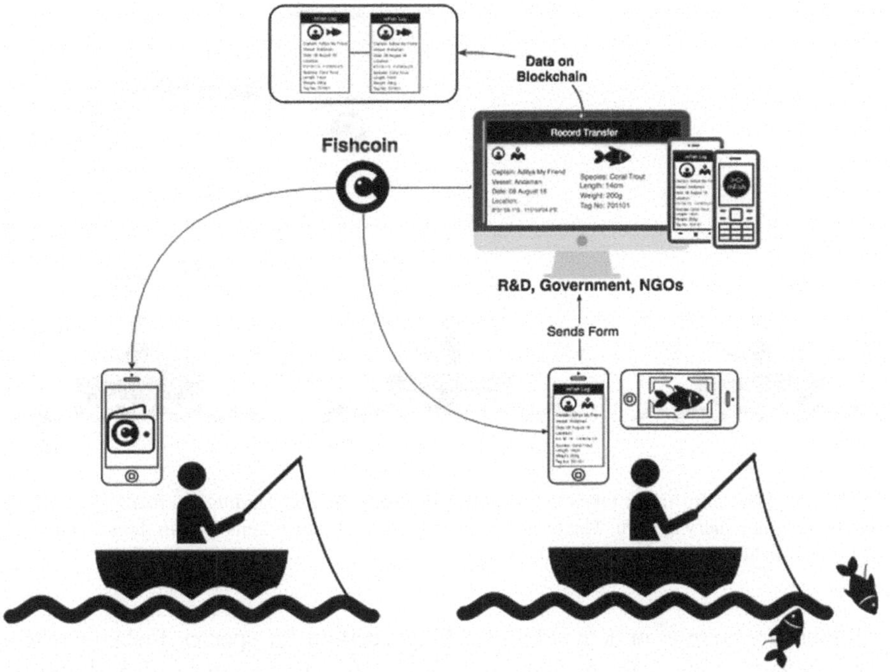

Fig. 4 Step 2. The tagged fish is recaptured, the tag number and data are recorded, and then transferred. The second fisher receives tokens along with the first fisher who originally tagged the fish. *Source* Douglas (2018)

his fisheries that support both his financial source and his family's protein intake (Douglas 2018). Step 2 is visualized in Fig. 4.

Step 2 happens when the second fisherman who catches that same fish receives an additional reward through Fishcoin tokens equivalent to US$0.20c for both observing and releasing the matured fish. The second fisherman receives tokens at the same moment, the first fisherman could also receive more tokens thus further rewarding him to be a part of the database, presents in Fig. 4 (Douglas 2018). Then comes the next explanation in Fig. 5.

Step 3 takes place when a third fisherman succeeds in catching this fish. The fish went into the market because it reached a fit size for sale and the fisher then recorded all data while bringing it aboard his boat. As the third fisher gets his tokens the other two fishers simultaneously acquire double the number of tokens. The fish value undergoes a substantial increase from US$0.08c to $8.00 during the period when first and second fishers receive their tokens together with second and first fishers receiving the transferred tokens which amount to a dollar or two. The first and second fishers obtain additional tokens while the third fisher potentially obtains greater value from selling the fish. The system enables maintenance of big fertile females both for stock recovery and as a fish supply in the fishery (Douglas 2018).

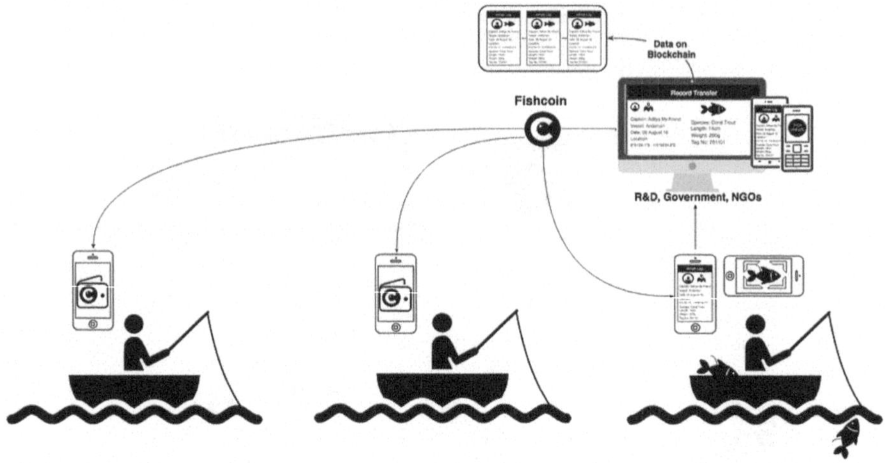

Fig. 5 Step 3The third fisher takes the record of information after landing the fish as he receives tokens from the fishery system. The payment system bestows new tokens to both the second fisher and the first fisher. *Source* Douglas (2018)

If there are no rules or demand for tracking a certain fish species, Fishcoin tokens probably won't come from the first buyer. But if the fish is part of a Fishery Improvement Project (FIP), the first buyer might get tokens if they can sell it to a market that pays extra for sustainability or rewards traceability. The benefits come since the cost is low and it improves food security and transparency, so governments or organizations could fund the tokens until fishers see the value of protecting resources. Over time, fishers would self-regulate, earning access rights or quotas and adopting better fishing practices.

Fishcoin is an effective and practical example of a FinTech solution that uses blockchain to solve real-world problems in the seafood industry. It creates a transparent and sustainable supply chain by incentivizing data collection with digital tokens. This innovative approach not only combats illegal fishing but also improves resource management, making it a powerful tool for environmental and economic sustainability.

Considering all these factors, it's also important to focus on the quality and real value of the solutions being implemented. Sharma, Jangir, Gupta, and Rupeika-Apoga demonstrated that high-quality FinTech payment services attract more users. Their research suggests that people are more influenced by how beneficial they find these services rather than just how easy they are to use. This challenges the common assumption that ease of use is the most crucial factor in technology adoption and emphasizes the need for FinTech companies to prioritize reliability and real value to keep users engaged (Sharma et al. 2024a).

Another key aspect, especially from a global perspective, is the impact of FinTech adoption on sustainability, considering businesses of different sizes and scalability. For instance, Sharma, Gupta, and Taneja's study found that FinTech plays a vital role in helping small businesses remain sustainable. A major factor in this is financial well-being—businesses that manage their finances effectively have a better chance of long-term success. Their research highlights the importance of financial stability for small businesses to fully benefit from FinTech and secure their future growth (Sharma et al. 2024b).

4 Conclusion

This chapter has explored various technological advancements (blockchain, artificial intelligence, decentralized finance, and digital payment systems) that present a strong possibility to support ocean resource management. The main purposes are transparency, regulatory efficiency, and financial accessibility in marine conservation efforts. The proof that it works was presented by selected case studies. Moreover, the gap between technological advancements and their real-world applications underscores the necessity for interdisciplinary collaboration between financial institutions, policymakers, marine scientists, and technology developers, so this book is a trial to fill the gap. The main goals for the future that come from the literature review and case-study examination are: the development of standardized regulatory frameworks, investment in education and capacity-building programs, further innovation and implementation of FinTech innovations. There is hope for fostering economic resilience in ocean-dependent industries by leveraging FinTech effectively. Presented practical example—Fishcoin is a smart FinTech solution that uses blockchain to make the seafood industry more transparent and sustainable. By rewarding fishers for tracking their catches, it helps fight illegal fishing and improve resource management.

References

Balgobin P, Seeam A (2020) Developing an effective regulatory framework for virtual currencies in Mauritius. In: ACM international conference proceeding series 2nd international conference on intelligent and innovative computing applications, ICONIC 2020, Article number 3415126. https://doi.org/10.1145/3415088.3415126

Douglas A (2018) Fishcoin use case: incentivising mark, release, re-capture & conservation. Medium. https://medium.com/fishcoin/fishcoin-use-case-incentivising-mark-release-re-capture-conservation-eeac104a731e. Accessed 10 Feb 2025

Eachmile. https://eachmile.co/results. Accessed 10 Feb 2025

ENVIROTECH. How much of our oceans have we damaged? https://www.envirotech-online.com/news/water-wastewater/9/breaking-news/how-much-of-our-oceans-have-we-damaged/46941. Accessed 10 Feb 2025

Fishcoin. https://fishcoin.co/. Accessed 10 Feb 2025

Food and Agriculture Organization of the United Nations (2024) The state of world fisheries and aquaculture—blue transformation in action. Rome. https://doi.org/10.4060/cd0683en. Accessed 10 Feb 2025

Hassan MM (2023) Premier wallet: banking the unbanked population in Somalia. Emerald Emerg Mark Case Stud 13(4):1–16. https://doi.org/10.1108/EEMCS-01-2023-0030

IPBES (2019) IPBES global assessment summary for policymakers. https://www.ipbes.net/news/ipbes-global-assessment-summary-policymakers-pdf. Accessed 10 Feb 2025

IPCC (2018) Summary for policymakers. In: Masson-Delmotte V, et al (eds) Global warming of 1.5 °C. An IPCC special report on the impacts of global warming of 1.5 °C above pre-industrial levels and related global greenhouse gas emission pathways, in the context of strengthening the global response to the threat of climate change, sustainable development, and efforts to eradicate poverty. World Meteorological Organization, Geneva, Switzerland, 32 pp

Jamil F, Iqbal N, Imran AS, Kim D (2021) Peer-to-Peer energy trading mechanism based on blockchain and machine learning for sustainable electrical power supply in smart grid. IEEE Access, 9, pp 39193–39217, 9358144. https://doi.org/10.1109/ACCESS.2021.3060457. Accessed 10 Feb 2025

Levine AS, Richmond L, Lopez-Carr D (2015) Marine resource management: culture, livelihoods, and governance. Appl Geogr 59(2015):56–59. https://doi.org/10.1016/j.apgeog.2015.01.016. Accessed 10 Feb 2025

Li M, Shao S, Ye Q, Xu G, Huang GQ (2020) Blockchain-enabled logistics finance execution platform for capital-constrained E-commerce retail. Robot Comput Integr Manuf 65:101962. https://doi.org/10.1016/j.rcim.2020.101962. Accessed 11 Feb 2025

Maxell SM, Hazen EL, Lawison RL et al (2015) Dynamic ocean management: defining and conceptualizing real-time management of the ocean. Marine Policy 58:42–50. https://doi.org/10.1016/j.marpol.2015.03.014

NASA (2024) Coastal dead zones id: 30479. https://svs.gsfc.nasa.gov/30479/. Accessed 10 Feb 2025

Science and Policy for People and Nature, Marine resource management, Policy instrument. https://www.ipbes.net/policy-support/tools-instruments/marine-resource-management (2024). Accessed 10 Feb 2025

Sharma V, Jangir K, Gupta M, Rupeika-Apoga R (2024a) Does service quality matter in FinTech payment services? An integrated SERVQUAL and TAM approach. Int J Inf Manage Data Insights 4(2):100252

Sharma V, Gupta M, Taneja S (2024b) Does FinTech adoption impact on sustainability of small businesses: mediating role of financial well-being. Glob Knowl Mem Commun

Tao R, Su C-W, Naqvi B, Rizvi SKA (2022) Can Fintech development pave the way for a transition towards a low-carbon economy: a global perspective. Technol Forecast Soc Chang 174:121278. https://doi.org/10.1016/j.techfore.2021.121278

Thanh Ha L (2025) Gauging the dynamic interlinkage level between Fintech and the global blue economy performance. J Econ Stud 52(1):55–71. https://doi.org/10.1108/JES-10-2023-0613

UNESCO (2022) Ocean plastic pollution an overview: data and statistics. Ocean Literacy Portal. https://oceanliteracy.unesco.org/plastic-pollution-ocean/. Accessed 10 Feb 2025

United Nations Development Programme (2015) Issue brief—ocean hypoxia dead zones. https://www.undp.org/publications/issue-brief-ocean-hypoxia-dead-zones. Accessed 10 Feb 2025

United Nations, United Nations Convention on the Law of the Sea (UNCLOS) (1982). [Online]. https://www.un.org/depts/los/convention_agreements/texts/unclos/unclos_e.pdf

University of Cambridge, Cambridge Centre for Alternative Finance, Atlas (2025). https://ccaf.io/atlas/visualisation/graph

Green Investment and Fintech: Financing the Blue Economy

Harshad.L. Jadhav, Suresh N. Maind, Priyanka P. Pandey, and Gaurav G. Pawar

Abstract The interplay between green investment and financial technology (fintech) is emerging as a transformative illustration. Green investment's linkage with financial technologies has presented a new era of sustainability warming various aspects of it. The blue economy or ocean-based development is a recent and revolutionary idea which is catalyzed as an important axis of green concept. Blue economy, includes industries such as sustainable fisheries, renewable marine energy, eco-tourism etc. necessitates s'nificant and targeted funding to achieve innovation, minimizing the environmental issues, and provide a rigorous economic growth. The traditional financing methods have often struggled to align with the unique challenges and opportunities of blue economy, including high risks, undefined gains, regulatory complexities, and the requirement of sustainability measurements. Fintech is revolutionizing the blue economy's access to cash through its ability to provide creative financial solutions. By means of peer-to-peer lending platforms, blockchain-enabled transparency, green bonds, and AI-powered risk assessment models, fintech is simplifying investments that meet Environmental, Social, and Governance (ESG) standards. Digital financial system integration increases accountability in project execution, promotes inclusivity, and speeds up resource mobilization. Furthermore, by fostering data-driven decision-making, the collaboration of fintech and green investment helps close the gap between environmental conservation efforts and financial stakeholders. This chapter explores the evolving nexus of green investment and fintech in catalyzing sustainable development within the blue economy. The area of the study has been taken the coastal states of India. The secondary data of financing systems and financial technology devoted towards the ocean economy has been taken and discussed. By analyzing the correlation between green investment and fintech the research

Harshad.L. Jadhav (✉) · P. P. Pandey
Department of Economics and Research Centre, Karmaveer Bhaurao Patil College, Vashi, Navi Mumbai, Maharashtra 400703, India
e-mail: harshad.jadhav9921@gmail.com

S. N. Maind
Mumbai School of Economics and Public Policy, University of Mumbai, Mumbai, India

G. G. Pawar
Department of Economics, Abasaheb Marathe Arts and New Commerce, Science College, Rajapur, Ratnagiri, Maharashtra 416702, India

V. Sharma et al. (eds.), *FinTech for Sustainable Finance and a Well-Blue Economy*, World Sustainability Series, https://doi.org/10.1007/978-3-031-92390-6_5

presents an explanatory study for equitable, efficient, and impactful investment in ocean sustainability.

Keywords Green investment · Fintech · Ocean economy · Fishery industry · Marine energy · Eco-tourism

1 Introduction

The current era is related with various dimensions of sustainability as integrated growth, sustainable environment, social sustainability and improved governance. The blue economy—being an important section of sustainable development which seeks the promotion and preservance of marine resource. It also categorized among various subparts like sustainable fisheries marine renewable energy, marine tourism coastal infrastructure etc., the development of each sub parts is quite required for the development of a string green economy (Paun et al. 2021). Till date the traditional structure of financing the economies is struggling through various challenge like high risk, uncertain returns, regulatory barriers etc., the same applies with the blue or marine economy (Silver et al. 2021).

In the growing period of sustainability, the yellow symbol of innovation along with blue of marine economy creates a strong structure of green economy. Here the innovative tools of finances, wide spectrums of green investment are providing a strong base to the capital flows for the development of blue economy. Green investment stands for such innovative financial systems and process which helps in development of environmental sustainability, such as green bonds, sustainable equity funds, and impact investing (OECD 2022). On the other hand, modern days financial technologies such as fintech—the strongest growing phase includes digital payment systems, blockchain, artificial intelligence (AI), and peer-to-peer lending platforms is completely devoted towards the inclusion and transparency (Chen et al. 2020). The integration of these two factors provides a strong base and growing opportunities to development of a strong blue economy or marine sustainability.

A new framework that makes use of support vector machine techniques to robustly adapt and optimize business strategies, hence ensuring alignment with sustainability pretensions needs to develop. By utilizing RL methods, social enterprises can simulate functional scripts, gain knowledge from feedback received from the real world, and iterate to improve their models to improve both their social value and their financial performance. The potential for RL to address difficult decision-making challenges that are vital in social entrepreneurship. It also guides perpetration and offers practical perspectives on the subject. RL may be used to push further successful and sustainable business practices, which will ultimately contribute to the larger goal of producing a positive social impact (Rathod et al. 2024).

2 Background of the Blue Economy, Green Investments, and Fintech

The blue economy is not only just an economic model; but also, is a sustainability-driven aspect that acknowledges the value of oceans as economic, social, and environmental assets (World Bank 2023a). It aims to balance economic growth, sustainable development with marine conservation, ensuring that ocean resources are used efficiently, equitably, and sustainably. The United Nations' Sustainable Development Goal (SDG) 14 ("Life Below Water") underscores and reminds the necessity of sustainable marine resource management and leadership to combat challenges such as climate change, overfishing, habitat destruction, and pollution (United Nations 2023). Despite its tremendous potential, the blue economy faces multiple regulatory and financial barriers. Unlike traditional land-based economic activities, ocean-based industries often involve higher capital risks, long investment horizons, and uncertainties due to environmental factors and regulatory changes. Moreover, many investors lack the expertise to evaluate the financial viability of ocean-based ventures, further limiting capital and funding access (Nakamura and Elder 2021). This highlights the need for targeted financial solutions that align investor confidence and belief with sustainable ocean development.

2.1 Green Investment: Connecting Channel for Sustainable Blue Growth

Green investment and finance work as the major financing channel to provide financial assistance to the blue economy, it manages the flow of capital through various mechanisms such as green bonds, blue bonds, sustainable investment funds, and public–private partnerships (PPPs) (Bennett et al. 2021). By directing financial flows toward environmentally responsible projects, green investment ensures that businesses and organizations in the marine sector adopt sustainability best practices while achieving economic returns. Blue bonds, in particular, have gained traction and interest as a financing tool specifically designed and developed for marine conservation and sustainable ocean industries. Governments, corporations and global financial institutions are also incentivizing green investments through policies such as carbon credits and pricing, subsidies for sustainable businesses, and risk mitigation tools and technologies (OECD 2022). These incentives reduce and manage the financial risks associated with investing in the blue economy and attract institutional investors seeking long-term value creation aligned with ESG goals and SDC goals.

Investing in training for the fishing community is crucial for achieving Sustainable Development Goals (SDGs), particularly those related to poverty alleviation, sustainable livelihoods, and responsible consumption. Enhanced training equips fishermen with modern techniques and sustainable practices, which can lead to better fish yields while preserving marine ecosystems. Empowering the community through education fosters resilience against climate change impacts, ensuring long-term environmental sustainability and economic stability. Additionally, well-trained fishermen can contribute to food security and nutrition by improving the quality and quantity of seafood available. Overall, targeted investment in training not only uplifts the fishermen's socio-economic status but also aligns with global efforts to create a balanced and sustainable future for all (Bhave et al. 2022).

2.2 Fintech: Tool to Develop Smooth Flow of Sustainable Funds

The rise and acceptance of financial technology (fintech) has introduced new opportunities for sustainable financing, making investment in the blue economy more accessible, efficient, and transparent (Zhang et al. 2022). Fintech solutions such as blockchain-based smart contracts, AI-driven risk assessments, peer-to-peer lending, and digital asset securitization and tokenization are disrupting traditional financial models by improving capital formation and flow efficiency and reducing transaction costs (Chen et al. 2020). One of the most promising fintech applications in ocean sustainability is blockchain technology, which improve and enhances transparency in marine supply chains, fisheries management, and sustainable trade certification. Blockchain-based platforms enable real-time tracking of fishery resources, ensuring transparent compliance with sustainability standards while preventing illegal, unreported, and unregulated (IUU) fishing (Paun et al. 2021). Similarly, AI-driven predictive analytics help investors assess and understand environmental risks, enabling data-driven investment decisions in blue economy projects.

2.3 A Sustainable Compound of Green Investment and Fintech for Growth of Blue Economy

The integration of green investment and fintech is reshaping ocean finance by introducing and implementing digital-first, sustainability-driven investment models. Fintech platforms facilitate crowdfunding for marine conservation projects, democratizing access to capital for small-scale coastal enterprises and local fisheries (Bennett et al. 2021). Moreover, AI-powered fintech solutions are enhancing ESG reporting standards, allowing and enabling investors to evaluate environmental impact metrics with greater accuracy (Zhang et al. 2022). As digital financial ecosystems evolve, fintech's role in de-risking ocean investments and operations will become increasingly prominent. Innovations such as tokenized blue bonds and AI-driven environmental scoring models can attract and enhance mainstream institutional investors while ensuring alignment with sustainability principles (Silver et al. 2021). However, challenges such as cybersecurity threats, regulatory uncertainties, and data privacy concerns must be addressed and arrested to fully realize fintech's potential in advancing the blue economy.

3 Significance of Green Investment and Transformative Role of Financial Technologies

The growing issues and complexities related with resources depletion, climate change and economic viabilities has increased the importance of green investment. The green investment has a large number of compounds, it effects all aspects of human and environment, and some major focus areas are promotion of renewable energy, green infra, and optimum utilization of natural and manmade resources. Among the various compounds green investment also inherit the growing aspects of ocean economy. Ocen economy is not a new area it is always related with the evolution of human but to remove the complexities and challenges lies with the traditional way of working modern days paradigm of green investment plays a crucial role in development of blue economy. It has widened the landscape of sustainability which creates a new model to explore the large areas of blue economy (Bennett et al. 2021). Seemingly the development of sustainable financing sources helps in development of ocean economy but still it dwells with some challenges high capital requirements, long payback periods, and uncertain regulatory landscapes (Chen et al. 2020). To smoothen or to remove these barriers, financial technologies play important role with modifying the investment farmwork, easy and efficient capital allocation, visible and accessible for growing areas of blue economy.

Green investment is the base for sustainable economic development as it includes inflow of financial resources into environmentally responsible projects. The marine industries require regular flow of capital, that is transforming the conventional financial frameworks lacking to create strong base due to high risks and the lack of standardized valuation metrics for sustainability impacts (OECD 2022). Green and blue bonds are the major sustainable financing mechanism. Blue bonds support ocean-based spectrums, and green bonds are channelized for climate protection or specially for renewable energy developments (Paun et al. 2021). These bonds offer long-term returns and easy accessibility for environmental projects. Public–private partnerships (PPPs) also support in mitigating the risk for private investors in marine sustainability (Nakamura and Elder 2021).

Green investment in the blue economy increases employment, strengthens coastal resilience, and encourages innovations. The marine renewable has lesser down the pressure on conventional sources of energy (Bennett et al. 2021). Green-funded sustainable aquaculture helps combat overfishing and preserves marine resources. While discussing about significance of fintech in transforming blue investment by easy capital access, hedging, and viability. Traditional financing often excludes small businesses due to strict credit rules, but fintech platforms enable peer-to-peer lending, decentralized finance, and microfinancing for sustainability-focused entrepreneurs

(Zhang et al. 2022). The developing financial technologies such as blockchain, AI, machine learning has smoothened the working of financial assistance. They provide tracking systems for supply chains, corban trading, risk calculations and also creates accessibility towards quick and automated ESG investment options portfolios (Chen et al. 2020).

4 Theoretical Framework—Previous Studies

4.1 Economic Globalization and Blue Economy in the Indian Ocean

Various studies explore the effect of economic globalization and green finance on the development of the ocean economy. The study represents the ways of global sustainable financial mechanism like green bonds and green investment funds, have played the significant role in developing blue resource and ocean sustainability. In this study an econometric analysis has been utilized to acknowledge the role of green finance in improving blue economy base, job creation, preservation and development of aquaculture. It also projects the challenges and obstacles of developed blue economy (Beeharry and Demir 2024).

4.2 Sustainable Development and Financial Mechanisms: Green Hydrogen and Blue Bonds

Some studies analyze the role of financial mechanisms, particularly green hydrogen initiatives and blue bonds, in sustainable ocean development. The study identifies the various financial instruments encouraging ocean sustainability, marine renewable energy, and sustainable fisheries. Through the analysis of multiple case studies it concludes that blue bonds have successfully attracted private investments for marine economy development (Pozhilova 2024).

4.3 Comprehensive Literature Review of Blue Finance

Some studies present a detailed review on previous studies done on blue capitals; it focuses on the effect of different financial instruments on ocean projects. The study represents that sustainable finance have an important role in developing sustainable economic growth. The authors represent different financial models, including government subsidies, impact investments, and private–public partnerships, that have successfully enhanced the blue economy projects. The review suggests that while blue finance has potential, inconsistent policies and a lack of standardized financial instruments hinder its full-scale implementation (Witantri et al. 2024).

4.4 Marine Spatial Planning in Algeria: Challenges and Opportunities

Some studies highlighted the need for strategic financial planning, incorporating international blue bonds and government-backed sustainable funds to support marine industries. They encompass the Algeria's marine spatial planning and its reliance on sustainable finance for blue economy development. The study identifies the importance of financial capacity in addressing climate change and preserving marine biodiversity. It concludes that even with the availability of financial resources, regulatory obstacles and investment risk factors hinder progress in Algeria's blue economy (Chahtou et al. 2025).

4.5 Greening Indian Financial Markets for Sustainable Development

Some studies explored the growth of Indian financial market for sustainability focusing on development of blue economy. It presents sustainable finance instruments such as blue bonds and sustainable investment frameworks focusing on developed blue economy along with preservance of aquaculture resources. The authors analyze regulatory frameworks, government incentives, and private sector contributions in financing blue economy projects. The study emphasizes the role of financial institutions in supporting ocean-based industries like fisheries, coastal tourism, and renewable energy while reducing carbon footprints. Challenges such as policy fragmentation and investor hesitancy due to perceived risks in blue economy investments are also discussed. The paper concludes that a well-structured financial ecosystem backed by policy coherence is crucial for achieving sustainable blue economic growth in India (Panda et al. 2024).

5 Case Studies: Global Best Practices

5.1 Poseidon Projects from Perenco (UK)—Blockchain for Carbon Credits

Poseidon project is a pioneer project of Perenco UK-based fintech firm leveraging blockchain technology to track and trade carbon credits, directly benefiting marine conservation projects. More than 150 firms are using this platform to assess their corban neutrality. The company's platform integrates carbon offsetting into everyday transactions, allowing consumers and businesses to fund ocean conservation efforts with micro-contributions. It supports more than 30 marine conservation projects. In 2023, Poseidon reported facilitating the offset of over 1.8 million metric tons of CO_2, significantly benefiting marine biodiversity and coral reef conservation.

5.2 Sea Ahead (USA)—Investment in Blue Tech Startups

Sea Ahead is a Boston-based fintech accelerator that funds and mentors' startups focused on sustainable ocean-based industries. They support blue tech startups. Blueswell is a startup program offered by them. It provides venture capital and fintech solutions for companies developing aquaculture technology, marine biodiversity tracking, and clean energy solutions. It has invested more than 100 million in more than 20 blue tech startups focusing on aquaculture and marine technology. One of Sea Ahead's notable investments is in Sway, a startup that develops biodegradable seaweed-based packaging to replace single-use plastics. By 2023, Sway successfully secured $4 million in funding through Sea Ahead, allowing the company to scale operations and reduce marine plastic pollution.

5.3 Blue Finance (Philippines)—Marine Conservation Bonds

Blue Finance is a fintech organization that develops financial instruments such as blue bonds to fund marine conservation projects. It partners with governments and NGOs to secure sustainable financing for marine protected areas. It has been declared as global funds for coral reef. In 2021, Blue Finance issued the first "Blue Bond" worth $15 million to fund the restoration of coral reefs and mangrove ecosystems in the Philippines. The initiative supported the protection of 1.5 million hectares of marine areas, boosting local ecotourism revenue by 20% within two years.

5.4 PROBLUE Program by World Bank Group

Focusing on SDG 14 world bank has initiated PROBLUE multi doner trust fund. It major moto is healthy ocean, healthy economies and healthy communities. It supports fishery and aquaculture, warns against marine pollution, focuses on sustainable economic development through ocean tourism, energy, marine transport and marine infrastructure. PROBLUE financing activities has reached up to 144 million US Dollars. Around $18 million of PROBLUE RETF (Recipient Executed Trust Funds) financing supports 1.4 billion of world bank lending.

6 Research Framework—A Correlative Analysis

The chapter focuses on the wide spectrum of sustainable or green finance. Blue finance and fintech are the part of sustainable finance. As the sustainable finance will grow the blue economy and fintech will defiantly grow. The blue economy works as intersection between various aspects of green economy as renewable energy, tourism, environmental conservation in terms of ocean resources. Here by the focus on sustainability automatically includes blue or ocean development. While talking about fintech which is also a part of sustainability, and it works as a medium for the smooth conduction of the process related with green finance. The correlation study is just a way to prove the concept that development of blue economy is positively related with sustainable parameters.

6.1 Structure on Costal States in India

India is rich country in terms of natural resources, where at one side is covered by mountains and one side by desserts, the two part of our country is seashore only. It has a longest coastal area of approx. 7517 km. The ocean economy contributes about 4% of total GDP and it supports around 95% trade of the country (TERI 2021). The structure of blue economy in India includes 9 coastal states, 4 union territories and around 1380 islands. The strong structure of trade in India is provided with 12 major pots and 187 small and medium size pots. The Indian ports contribute in about 633.80 million tons of cargo in 2019. The blue economy contributes in Indian economy in form of employment, trade, food products, tourism, and marine infra. A detailed study of coastal states has presented below.

6.1.1 Gujarat

Gujarat has the longest coastline in India (1600 km) and it has major ports. The state includes total 40% of India's total shipment. The major ports situated here are—Mundra, Kandla, and Pipavav along with many small ports. It is ancient door of ship transport in India. Mundra Port alone handled 144 million metric tons of cargo in 2023. Gujarat is also home to India's largest shipbreaking industry in Alang, which contributes significantly to maritime recycling. But resource depletion issues like coastal erosion, ocean pollution and industrial pollution has increased the challenges for the blue economy development.

6.1.2 Maharashtra

Maharashtra, having 720 km-long coastlines, is a center for fisheries production, coastal tourism, and offshore oil exploration. The state's Konkan region is famous for variety of sea-foods, producing a revenue of around ₹30,000 crore annually. Major ports as Nhava Sheva (Jawaharlal Nehru Port) holds around 55% of India's container traffic. Mumbai and Alibaug attract millions of tourists annually, contributing to marine-based tourism and employment. It also suffers from pollution issues affecting the blue economy.

6.1.3 Goa

Goa is completely devoted towards the tourism. It has a coastline of 101 km, contributes at a large level in marine-based tourism. The tourism sector contributes around 15% to Goa's GDP, with nearly 8 million tourists visiting the state annually for its beautiful beaches and water sports. The state is also exploring sustainable blue economy initiatives such as marine conservation projects and artificial reef restoration. The major challenges it facing are—over-tourism, plastic pollution, and beach erosion hindering the blue economy.

6.1.4 Karnataka

Karnataka with 300 km coastline is a hub for fisheries, marine biodiversity research, and port activities. The state generates nearly ₹5000 crore annually from marine fisheries, employing over 500,000 people. The ports of Mangalore and Karwar facilitate coastal trade and fish exports. Karnataka has some major marine research institutions like NITK Surathkal, focuses on the development of marine biotechnology and sustainable fisheries.

6.1.5 Kerala

Kerala, has 590 km-long coastlines, poses a developed fisheries and aquaculture industry which provides jobs around 50 lakh people. The state is a leader in seafood exports, especially in shrimp and tuna, and generates more than ₹5000 crore yearly. Kochi port is a key maritime trade hub, and the state government is investing in blue economy startups that focus on seaweed farming, marine biotechnology, and deep-sea fishing. However, frequent cyclones, coastal erosion, and marine pollution threaten Kerala's blue economy sustainability.

6.1.6 Tamil Nadu

Tamil Nadu, with a 1076 km-long coastline, is a leader in shipbuilding, marine renewable energy, and fisheries. The state has the second-highest marine fish production in India, with exports exceeding ₹6000 crore annually. Tamil Nadu also has the highest offshore wind energy potential in India (33 GW), with plans to develop 5 GW of offshore wind power by 2030. Major ports like Chennai, Tuticorin, and Ennore drive trade, but marine pollution and overfishing remain critical challenges.

6.1.7 Andhra Pradesh

Andhra Pradesh, with a 972 km coastline, leads India's seafood export industry, particularly in shrimp farming, contributing ₹10,000 crore annually. The state's ports at Visakhapatnam and Kakinada facilitate major trade routes. It also has a strong aquaculture sector, producing 30% of India's shrimp exports. However, mangrove destruction and pollution from aquaculture farms pose sustainability concerns.

6.1.8 Odisha

Odisha's 480 km-long coastline supports fisheries, shrimp farming, and marine ecotourism. The state earns ₹4500 crore from seafood exports annually, particularly from Paradip and Dhamra ports. Odisha is also a leader in mangrove conservation, with projects in the Bhitarkanika National Park protecting coastal biodiversity. However, frequent cyclones and rising sea levels threaten long-term blue economy sustainability.

6.1.9 West Bengal

West Bengal, is one of the largest states with 210 km coastline, is famous for inland and coastal fisheries sector. The Sundarbans region, a UNESCO World Heritage Site, plays a key role in marine biodiversity conservation and blue carbon omission. West

Bengal accounts for 50% of its total fish production from coastal waters. The state also focuses on inland water transport, with the Kolkata port facilitating trade with Bangladesh and Southeast Asia.

7 Fintech Structure—A Data Study

Fintech (financial technology) is the major aspect sustainable finance providing smoother assistance to financial integration, efficient investment measures and virtual financial supports to develop coastal environment, ocean industries, and marine conservations. Here we represent a developing strong structure of fintech. Fintech mainly include the following layers to support blue economy globally (Fig. 1).

7.1 Digital Payments and Financial Inclusion

Microfinance institutions and digital bank solutions are providing a huge fund ratio to fishery and small marine industries. The structure of transaction in fishery sector is also lead by UPI payment systems like other sectors. The Pradhan Mantri Matsya Sampada Yojana (PMMSY) is focusing on the financial inclusion and digitization of fishery industries. In terms of digitization fintech firms takes the prominent position. The various publics sector, private sector and leading commercial banks are providing fintech assistance to marine industries.

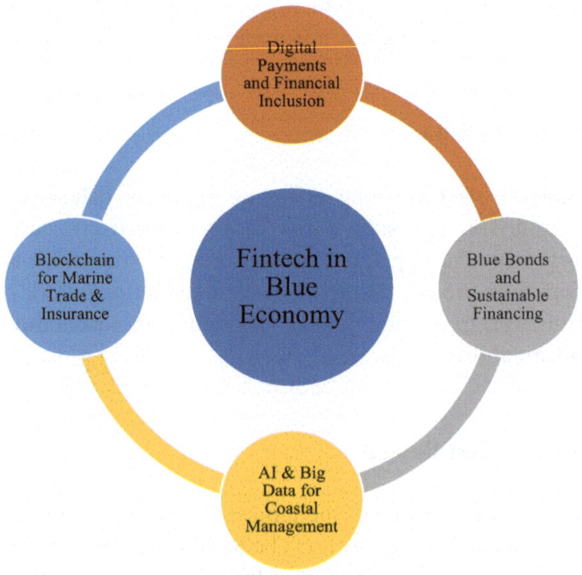

Fig. 1 Layers of fintech in blue economy. *Source* By authors

7.2 Blue Bonds and Sustainable Financing

Indian firms raised nearly $7.5 billion funds with route of sustainable finance in 2021 compared to $4.8 billion in 2019. Axis, the leading bank of private sector in cooperation with International Finance Corporation issued a blue investment of worth $500 million for the development of blue finance market. The RBI's Green and Sustainable Finance Framework allocates ₹20,000 crore ($2.4 billion) for fintech-backed marine conservation projects. Banks like State Bank of India (SBI), ICICI bank and fintech startups such as Blue Finance India are focusing on marine insurance products for fishermen to tackle down the risk factors.

7.3 Blockchain for Marine Trade and Insurance

Blockchain-based seafood traceability has been implemented in Visakhapatnam and Kochi ports, reducing fraud and improving seafood export quality. It has developed a research potential to enlarge the sea food production, to identify market access and reputation, to ensure quality assurance etc. Such the technologies are creating a strong foundation for the development of ocean-based economy.

7.4 Artificial Intelligence and Big Data Modelling for Coastal Management

AI and Big Data modelling helps in maintenance, compilation, assessment of data related to blue economy. The emerging systems of remote sensing, predictive analysis and other innovations are helping in ocean resource management, pollution control sustainable fishing practices and under water explorations.

8 Blue Economy—New Era to Focus

The blue economy is an oldest route of trading and communication is going in a face of transformation to increase its productivity. Marine or ocean resources are available in abundant quantity in India. But the traditional way of working or environmental challenges has forced to focus on revival of blue economy. Now it is emerging as a major sector in Indian the futuristic outlook of blue economy in India includes the given structures (Fig. 2).

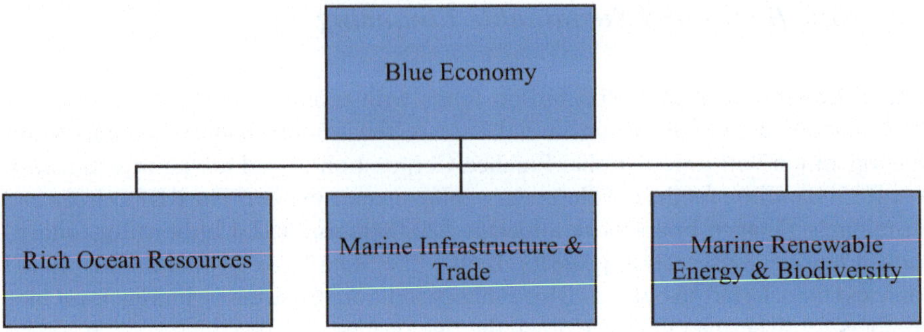

Fig. 2 Structure of blue economy in India. *Source* By authors

8.1 Rich Ocean Resources

The blue economy in India inherits the rich ocean resources which includes the fishery, sea foods, aquaculture etc. India is the second-largest fish producer globally, and in marine fisheries its contribution around $16 billion to the economy. Andhra Pradesh, Gujarat, and West Bengal lead in shrimp exports, and in 2023, India exported seafood of worth $7.2 billion. The Pradhan Mantri Matsya Sampada Yojana (PMMSY) aims to double aquaculture productivity by 2025 through financial incentives and modern fisheries technology.

8.2 Marine Infrastructure and Trade

Blue economy supports Indian economy through marine infra and trade. It works as the backbone for the trade supports. 95% of India's trade is carried by the major ports like Mundra, Chennai, and Visakhapatnam which carries around 1.5 billion metric tons of cargo annually. The Sagarmala Project has funded $96 billion for coastal special economic zones, modernized ports, and digitalized logistics. By 2030, India's shipping sector aims to be carbon–neutral, reducing emissions by 30% through green infrastructure (Ministry of Ports 2023).

8.3 Marine Renewable Energy and Biodiversity

India has an estimated is third largest renewable energy producer globally. Among the various renewable energy sources, offshore wind energy holds a prominent position with the potential of 70 GW. Gujarat and Tamil Nadu have the leading in offshore wind energy projects. The government targets 110 GW or more of offshore wind capacity by 2050. Mangrove afforestation campaigns, especially in Odisha and West

Bengal, aim to preserve and increase coastal biodiversity against climate change, reducing economic losses from cyclones.

9 Corelative Study on Green Investment, Fintech and Blue Economy—Rising Paradigm of Sustainability

The presentation of correlation study among green investment, fintech and blue economy is limited due to unavailability of authentic data. The green funds which include green bonds and a limited supply of blue bonds have been included. Total green fund structure has been also taken in study. The most important inclusion fintech details are restricted due to the unavailability of data. Some authentic details published in RBI reports has been included in study, it represents the top fintech companies and their working areas. The idea about blue economy has been taken mainly as the development of fishery industry in production and exports. The relativity among these three do not come directly. We assume that as government is focusing on development of marine sustainability, one part of green financing is devoted for the development of blue economy. So here we have taken the total of Green Financing and green bonds. Top fintech firms and the total fund raised by them over the approx. ten year has been included. The following chart details included in the study (Tables 1, 2, 3, 4, 5, 6, 7, 8, 9 and Figs. 3, 4, 5).

The above study represents that there is strong relativity between green funds or bonds and blue economy which is led by fishery industry's production and export. The correlation analysis reveals a strong link between green finance and fisheries growth. Green bond issuance and total green investments (0.99 correlation) show a direct impact on fish production and export funding. Additionally, fish production and government release funds (0.93 correlation) indicate substantial financial support driving fisheries expansion and sustainability. The other related concepts can be presented with the help of Table 5 where government budget shows a significant

Table 1 Green bond issuance in the economy

Years	Green bond issuance in approx. (USD billion)
2014	0.5
2015	1.5
2016	3
2017	5
2018	7
2019	9.5
2020	12
2021	14.5
2022	17.5

Source RBI Bulletin 2024

Table 2 Estimated green funds trends

Year	Total green investments (₹ Billion)	Key developments areas
2014	150	Introduction of India's first green bond
2015	180	Renewable Energy Push (175 GW target by 2022)
2016	220	SEBI issues guidelines for Green Bonds
2017	270	Surge in corporate green bond issuance
2018	350	Green Climate Fund (GCF) approvals
2019	420	Government-backed ESG investments rise
2020	500	COVID-19 slows investment, but green recovery plans emerge
2021	600	Record green bond issuance in India
2022	720	India commits to Net Zero by 2070 at COP26
2023	850	Sovereign Green Bonds launched
2024	950+	Continued expansion of climate finance, Blue Economy 2.0

Source Google search of various authentic records

Table 3 Various online platforms in India

Rank	Company	Founded	Total amount raised ($ Million)	Description
1	Paytm	2010	1000	App-based wallet for consumer payments
2	BharatPe	2017	154.3	QR code-based payment app
3	Policybazaar	2008	150	Online insurance comparison platform
4	CRED	2018	145.6	Rewards-based platform for credit card bill payments
5	KhataBook	2016	140.6	Digital ledger account book
6	Acko	2017	101.6	Tech-enabled automotive insurance
7	ZestMoney	2015	30.47	Online platform for point-of-sale financing
8	Lendingkart	2014	87.87	Online platform providing working capital for SMEs
9	InCred	2016	85.9	Alternative lending platform focusing on SME, consumer and personal, home and education loans
10	Pine Labs	1998	85	PoS software solutions for offline retailers
11	Billdesk	2000	84.8	Payment gateway
12	Digit Insurance	2016	84.35	Insurance platform for individuals

Source RBI report, Nov 2020

Table 4 Central assistance for fisheries sector in India

Sr. no	Year	Fish production (in lakh tonnes)			Release fund (in lakhs)
		Marine	Inland	Total	
1	2010–11	32.50	49.81	82.31	29161.24
2	2011–12	33.72	52.94	86.66	30585.88
3	2012–13	33.21	57.19	90.40	29832.77
4	2013–14	34.43	61.36	95.79	31812.95
5	2014–15	35.69	66.91	102.62	35335.98
6	2015–16	36.00	71.62	107.62	41680.94
7	2016–17	36.25	78.06	114.31	42411.18
8	2017–16	37.56	89.48	127.04	42077.97
9	2018–19	38.53	97.20	135.73	67552.37
10	2019–20	37.27	104.37	141.64	64026.84
11	2020–21	34.76	112.49	147.25	88384.76
12	2021–22	41.27	121.21	162.48	136114.81

Source Government of India. Handbook on Fisheries Statistics, 2018–2019

Table 5 Budget allocated to various blue economy sectors through the different ministries

Sr. no	Ministry	Budget 2015–16 (In Rs. Crore)	Budget 2019–20 (In Rs. Crore)	Budget 2020–21 (In Rs. Crore)
1	Agriculture and allied activities (department of fisheries)	410	805	825
2	Environment, forests and climate change (national coastal management programme)	100	95	103
3	Earth sciences	508	1700	2070
4	Shipping and ports	1440	1900	1800
5	Tourism	1570	2190	2500
Total		4028	7590	7300

Source Blue Economy: India's Pathway to Sustainable, Secure, and Resilient Economy

increase in allocation on sustainable project. And as the maximum funding of green bonds goes with the renewable energy production and offshore wind energy is an important source of renewables and an integral part of blue economy. The study of Table 5 represents it as the green funding increases, the funding support to blue economy also rises. Tourism or eco-tourism is also the part of sustainable funds where marine tourism is an important segment of eco-tourism. Sustainable funds for development of reliable and resilience economy encompass the blue economy development.

Table 6 Trends and growth of fish production in India

Year	Fish production (in lakh tonnes)			Annual average growth rate (%)		
	Marine	Inland	Total	Marine	Inland	All India
2000–01	28.11	28.45	56.56	−1.44	0.78	−0.33
2001–02	28.3	31.26	59.56	0.68	9.88	5.3
2002–03	29.9	32.1	62	5.65	2.69	4.1
2003–04	29.41	34.58	63.99	−1.64	7.73	3.21
2004–05	27.79	35.26	63.05	−5.51	1.97	−1.47
2005–06	28.16	37.56	65.72	1.33	6.52	4.23
2006–07	30.24	38.45	68.69	7.39	2.37	4.52
2007–08	29.2	42.07	71.27	−3.44	9.41	3.76
2008–09	29.78	46.38	76.16	1.99	10.24	6.86
2009–10	31.04	48.94	79.98	4.23	5.52	5.02
2010–11	32.5	49.81	82.31	4.7	1.78	2.91
2011–12	33.72	52.94	86.66	3.75	6.28	5.28
2012–13	33.21	57.19	90.4	−1.51	8.03	4.32
2013–14	34.43	61.36	95.79	3.67	7.29	5.96
2014–15	35.69	66.91	102.6	3.66	9.04	7.11
2015–16	36	71.62	107.62	0.87	7.04	4.89
2016–17	36.25	78.06	114.31	0.7	8.99	6.21
2017–18	36.88	89.02	125.9	1.73	14.05	10.14

Source Government of India. Handbook on Fisheries Statistics, 2018. 2019. Ministry of Fisheries, Animal Husbandry and Dairying, Department of Fisheries, New Delhi. September, http://dof.gov.in/sites/default/files/2020-08/HandbookonFS2018.pdf

Table 7 Growth of shipping and port facilities in India

Year	Major ports		Non-major ports	
	Cargo (in MT)	Growth (in %)	Cargo (in MT)	Growth (in %)
2013–14	555.5	2	417.20	7.5
2014–15	581.3	5	470.6	12.8
2015–16	606.5	4	465.9	−1
2016–17	648.4	7	485.2	4.14
2017–18	679.4	5	528.5	8.9
2018–19	699.1	2.9	582.5	10.22
2019–20	704.8	0.8	–	–

Source Ministry of Shipping. EXIM Bank Research

Table 8 Correlation analysis

Green bond issuance (USD billion)	Total green investments (₹ billion)	Total fish production (lakh tonnes)	Fish export (lakh tonnes)	Release fund (lakhs)
0.5	150	82.31	32.5	29161.24
1.5	180	86.66	33.72	30585.88
3	220	90.4	33.21	29832.77
5	270	95.79	34.43	31812.95
7	350	102.62	35.69	35335.98
9.5	420	107.62	36	41680.94
12	500	114.31	36.25	42411.18
14.5	600	127.04	37.56	42077.97
17.5	720	135.73	38.53	67552.37
19.5	850	141.64	37.27	64026.84

Source Ministry of shipping. EXIM bank research

Table 9 Correlation matrix

Green Bond Issuance	1.00	0.99	1.00	0.95	0.92
Total Green Investment	0.99	1.00	1.00	0.93	0.94
Total Fish Production	1.00	1.00	1.00	0.95	0.93
Fish Export	0.95	0.92	0.95	1.00	0.86
Release Fund	0.92	0.94	0.93	0.86	1.00
	Green Bond Issuance	Total Green Investment	Total Fish Production	Fish Export	Release Fund

0.99-1.00	0.95-0.99	0.90-0.95	0.85-0.90

Source- by authors

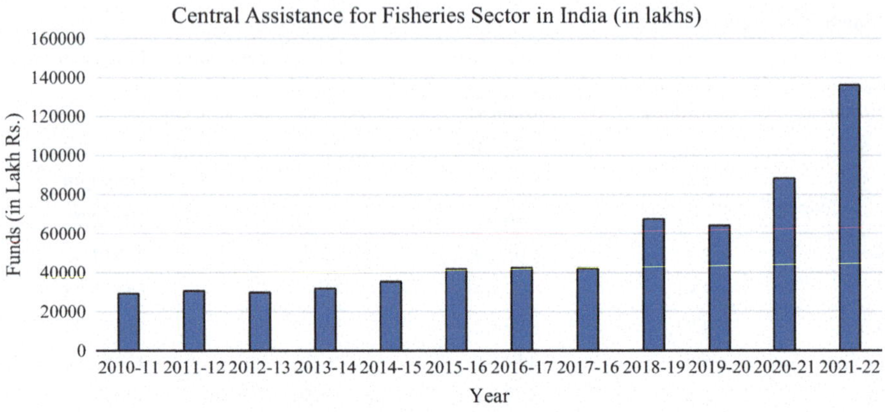

Fig. 3 Central Assistance for Fisheries Sector in India. *Source* Government of India. Handbook on Fisheries Statistics, 2018. 2019. Ministry of Fisheries, Animal Husbandry and Dairying, Department of Fisheries, New Delhi

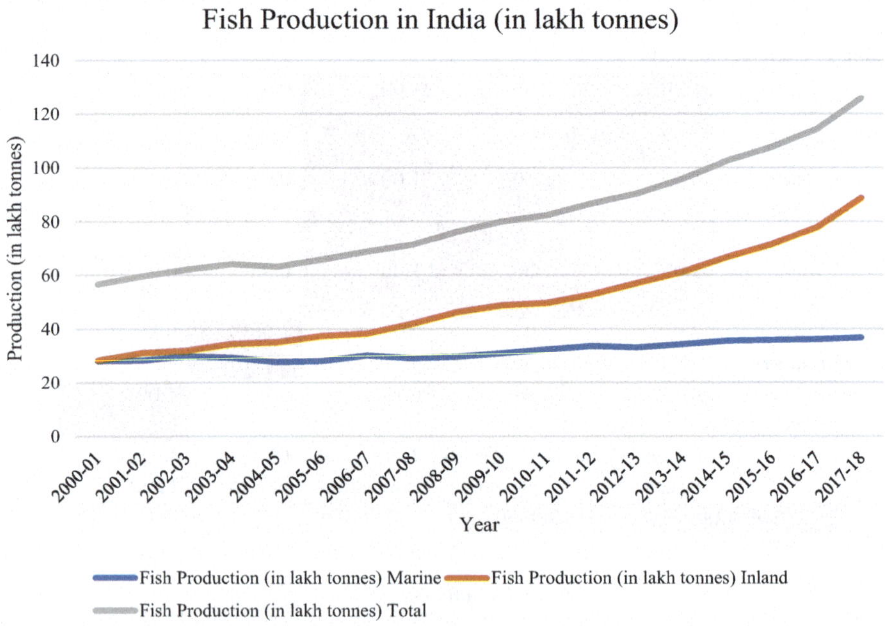

Fig. 4 Trends and growth of fish production in India. *Source* Government of India. Handbook on Fisheries Statistics, 2018. 2019. Ministry of Fisheries, Animal Husbandry and Dairying, Department of Fisheries, New Delhi. September, http://dof.gov.in/sites/default/files/2020-08/HandbookonFS 2018.pdf

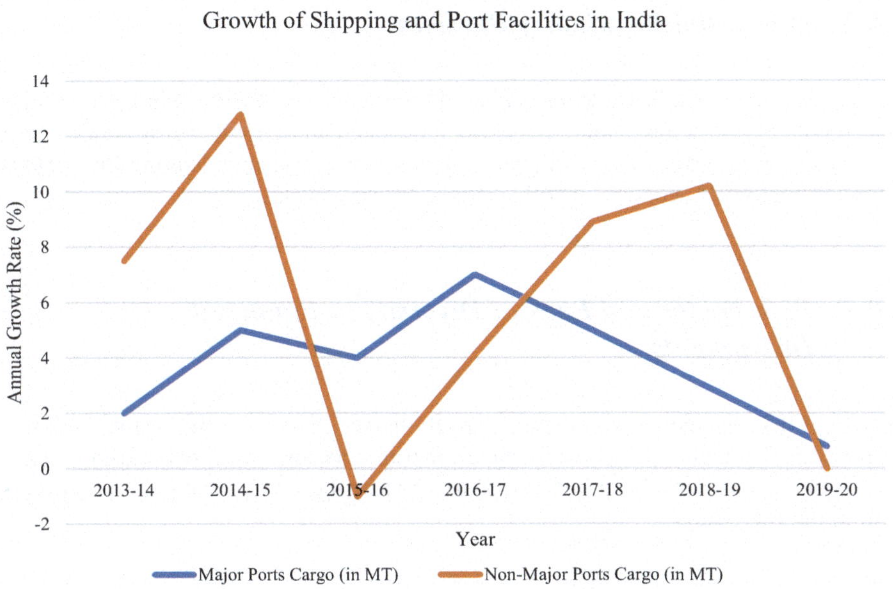

Fig. 5 Growth of shipping and port facilities in India. *Source* Ministry of shipping. EXIM bank research

10 Discussion and Challenges

10.1 Less Investment or Less Focus on Blue Economy

Despite the ocean's critical role in supporting over 4% of GDP and contribution around 95% of total trade, investments in blue economy are very less. The blue economy in India still works on the traditional measures and receives a less and unorganized sources of fundings. Even having a potential to grow in different vicinity, it is still lacking to have the central focus of the economy.

10.2 Unorganized Data and Lack of Authencity Measurement

The criteria of a strong business models are one of the major challenges faced by blue economy in India. Fishery the largest aspect of blue economy is a part of primary sector. The whole primary sector is identified as the unorganized sector, similarly fishery sector also effects from same. There is no regulatory measure to establish a strong business model.

10.3 Policy and Regulatory Barriers

India is lacking towards the development of blue economy in terms of proper policies and regulatory structure. Some concerns have been started but still it is not a part of discussion in terms of central issues of India. While acknowledging the various growing sectors of economy it still stands at far way back.

10.4 Risk Factor and Lack of Diversity in Financial Instruments

The high risk associated with investments in marine industry is the major challenge before development. Innovative financial instruments, such as blue bonds and debt-for-nature swaps, have been introduced to mitigate these risks, but their adoption is still in the early stages.

11 Conclusion

The above study explores the combination of green investment and financial technologies in development of blue economy in India. Green finance instruments like green bonds and blue bonds are play an important role in channelizing the funds for development of different spectrums of marine economy. These investments align with global sustainability goals, ensuring economic viability while preserving marine ecosystems. At the same time, financial technology (fintech) is revolutionizing capital allocation, enhancing transparency, and improving risk assessment in blue economy projects. Innovations such as blockchain-based tracking for marine resources, AI-driven sustainability analytics, and digital investment platforms have streamlined funding for sustainable ocean initiatives. In the study we have explored the correlation between green investment and growing aspects of blue economy, which is positive and highly correlated. The lack of data on fintech funds for blue economy has restricted our study in quantitative terms, but through the qualitative analysis it could be analyzed that in the development of blue economy, the sources of funding technologies of green investment, plays a pivotal role. The major challenges in development of financial assistance for blue economy has been also discussed. The major challenges faced in growth of financial mechanism is due the lack of strong regulatory parameters, and it creates a risk to investment. Government supports and strong framework can provide assurance to risk mitigations. Simultaneously the large range of insurance factors must be also increased to develop a strong and risk averting foundation.

References

Axis bank's investment in blue economy (2024). https://www.axisbank.com/about-us/press-releases/ifc-partners-with-axis-bank-to-provide-a-500-million-climate-loan-in-india

Beeharry Z, Demir Y (2024) Economic globalisation and the islands of the Indian Ocean: an econometric analysis. Island Stud J. https://islandstudiesjournal.org/article/125908.pdf

Bennett NJ, Blythe J, White CS, Campero C (2021) Blue economy and ocean finance: a pathway to sustainability or a road to unsustainable ocean use? Mar Policy 128:104468. https://doi.org/10.1016/j.marpol.2021.104468

Bhave AV, Waware AK, Pawar GG, Mengal SG, Solkar NAR (2022) Training needs of fisherwomen in South Konkan region of Maharashtra state. J Surv Fish Sci 9(1):740–745. https://www.researchgate.net/publication/376797257_Training_Needs_Of_Fisherwomen_In_South_Konkan_Region_Of_Maharashtra_State

Chahtou S, Guerfi M, Mezouar K (2025) Marine spatial planning in Algeria: challenges and opportunities. J Coast Res. https://meridian.allenpress.com/jcr/article/113/SI/120/505141

Chen Z, Li Y, Wu Y, Luo J (2020) The role of fintech in sustainable finance: a comprehensive review. Sustainability 12(23):9986. https://doi.org/10.3390/su12239986

European Union Publications (2020) Sustainable finance and the blue economy. https://publications.europa.eu/resource/cellar/e8c10a7d-2a1f-11eb-9d7e-01aa75ed71a1.0002.01/DOC_1

Green Finance LAC (n.d.) Our initiatives: blue economy. https://greenfinancelac.org/our-initiatives/blue-economy

Green Money (n.d.) Unlocking climate tech capital for blue economy ventures. https://greenmoney.com/unlocking-climate-tech-capital-for-blue-economy-ventures

KPMG India (2024) Unlocking India's blue economy: a fintech-driven future. https://kpmg.com/in

KPMG India (2024) Unlocking India's blue economy: a pathway to sustainable growth and prosperity. https://kpmg.com/in/en/blogs/2024/10/unlocking-indias-blue-economy-a-pathway-to-sustainable-growth-and-prosperity.html

Medium (n.d.) Innovative financing strategies for the blue economy: the role of blue bonds. https://medium.com/global-climate-solutions/innovative-financing-strategies-for-the-blue-economy-the-role-of-blue-bonds-3f6bcb8a5317

Ministry of Fisheries, Government of India (2023a) Annual report on blue economy and digital finance integration. https://dof.gov.in

Ministry of Fisheries, Government of India (2023b). India's fisheries sector: annual report 2022–23. https://dof.gov.in

Ministry of Ports, Shipping, and Waterways (2023) Sagarmala programme: annual review. https://sagarmala.gov.in

Nakamura H, Elder M (2021) The financial challenges of scaling up blue economy initiatives. J Sustain Finan Investment 11(2):230–246. https://doi.org/10.1080/20430795.2020.1837074

OECD (2022) The role of green finance in sustainable ocean economies. OECD blue economy report. https://www.oecd.org/environment/blue-economy-finance.pdf

Panda P, Chari L, Smark C (2024) Greening Indian financial markets for sustainable development. Account Bus Finan J. https://www.academia.edu/download/117622557/viewcontent.pdf

Paun C, Thomaidis N, Boulanger M (2021) Financing sustainable marine projects: the impact of green bonds and fintech. J Clean Prod 310:127450. https://doi.org/10.1016/j.jclepro.2021.127450

Pozhilova N (2024) The concept of sustainable development and financial mechanisms: green hydrogen and blue bonds. Reliab Theory Appl. https://cyberleninka.ru/article/n/the-concept-of-sustainable-development-and-financial-mechanisms-green-hydrogen-and-blue-bonds

Rathod US, Thomas SA, Maruti FS, Kadam B, Jadhav HL, Bdair MJ (2024) Reinforcement learning for sustainable business model development in social entrepreneurship. In: Second international conference computational and characterization techniques in engineering & sciences (IC3TES), Lucknow, India, pp 1–5. https://doi.org/10.1109/IC3TES62412.2024.10877619

Reserve Bank of India (RBI) (2023a) Blue finance and fintech innovations. https://www.rbi.org.in

Reserve Bank of India (RBI) (2023b) Green and sustainable finance framework report. https://www.rbi.org.in

Reuters (2025) Comment: the dangers of ditching Europe's green deal. https://www.reuters.com/sustainability/boards-policy-regulation/comment-dangers-ditching-europes-green-deal-2025-02-17

Rowan NJ (2023) The role of digital technologies in supporting and improving fishery and aquaculture across the supply chain—Quo Vadis? Aquacult Fish 8:365–374

Silver JJ, Gray NJ, Campbell LM, Fairbanks LW, Gruby RL (2021) The contested blue economy: conflict and cooperation in ocean governance. Front Mar Sci 8:613253. https://doi.org/10.3389/fmars.2021.613253

The Energy and Resources Institute (TERI) (2021) Blue economy: an ocean of livelihood opportunities in India. https://www.teriin.org/article/blue-economy-ocean-livelihood-opportunities-india

The Energy and Resources Institute (TERI) (2022) Fintech and sustainable finance in India's blue economy. https://www.teriin.org

United Nations (2023) Exploring the potential of the blue economy. https://www.un.org/en/desa/exploring-potential-blue-economy

Witantri G, Giovanni A, Kurniasari E (2024) Comprehensive literature review of blue finance. In: Proceedings of the international conference on economics and business. https://conference.unib.ac.id/index.php/BICEMBA/article/view/57

World Bank (2023a) India's digital finance for sustainable marine trade. https://www.worldbank.org

World Bank (2023b) India's blue finance and fintech revolution. https://www.worldbank.org

Zhang L, Wang S, Liu H (2022) The digital transformation of sustainable finance: the role of fintech in green investment. Financ Res Lett 47:102556. https://doi.org/10.1016/j.frl.2022.102556

Harnessing Digital Finance for Sustainability and Equity in Marine and Seaside Economies

Shalini Singh, Pooja Singh, Samriddhi Pandey, and Shubhanker Yadav

Abstract The integration of digital finance into marine and seaside economies offers transformative potential to foster sustainability, equity, and economic growth. This chapter explores the role of financial technology (FinTech) in addressing the unique challenges faced by coastal and marine industries, including resource depletion, economic instability, and socio-environmental disparities. It highlights how innovations like blockchain, artificial intelligence (AI), and mobile financial platforms can enhance financial inclusion for coastal communities, streamline investment in sustainable marine projects, and promote transparent governance. Through case studies and data-driven analysis, the chapter investigates global best practices where digital finance has enabled resource optimisation, resilience in the face of climate change, and support for blue economy initiatives. By examining the interplay of technology, finance, and sustainability, this chapter aims to provide actionable insights into leveraging digital finance for building resilient and equitable marine ecosystems, contributing to a sustainable blue economy.

Keywords Digital finance · Sustainable blue economy · Marine economies · Seaside economies · Financial technology (FinTech) · Blockchain · Artificial intelligence (AI) · Sustainable development · Blue economy governance

---

S. Singh (✉)
Assistant Professor, School of Commerce, Finance and Accountancy, CHRIST (Deemed to Be University), Bengaluru, India
e-mail: shalini.dec1226@gmail.com

P. Singh (✉)
Freelancer, Researcher, Gujarat, India
e-mail: psngh87@gmail.com

S. Pandey
Student, BCOM (Applied Finance and Analytics), School of Commerce, Finance and Accountancy, CHRIST (Deemed to Be University), Bengaluru, India

S. Yadav
Assistant Professor, Jaipuria Institute of Management, Lucknow, India
e-mail: shubhanker141@gmail.com

V. Sharma et al. (eds.), *FinTech for Sustainable Finance and a Well-Blue Economy*, World Sustainability Series, https://doi.org/10.1007/978-3-031-92390-6_6

1 Introduction

The oceans encompass three-quarters of the Earth's surface, hold 97% of the planet's water, constitute 99% of its habitable area and are acknowledged for their capacity to produce novel bioproducts and materials utilised in diverse sectors. They also offer vital ecosystem services necessary for the sustainability of coastal communities and their related value chains (Pace et al. 2023). However, modern times have witnessed a dramatic shift in humankind's relationship with the oceans, seas and marine resources and how we use and exploit their resources. These resources are finite and are experiencing escalating and frequently cumulative human impacts like overfishing and plastic pollution, and as a result, the unhealthy and non-resilient oceans are incapable of sustaining economic growth (Food and Agriculture Organization of the United Nations 2020; Gao et al. 2022; Nham and Ha Thanh 2023). Acknowledging the ocean's fundamental importance to human survival on Earth SDG 14 of the UN Sustainable Development Goals (SDGs) underscores the sustainable use of marine resources, including the efficient management of fisheries, aquaculture, and tourism (1). "Blue Economy" has recently become strongly linked to marine resources and industrialised economies. The blue economy refers to the sustainable utilisation of ocean resources for economic development while safeguarding marine ecosystems. World Bank; United Nations Department of Economic and Social Affairs (2017) has broadly defined the Blue Economy as the sustainable use of ocean resources to promote economic growth, enhance livelihoods, provide employment, and maintain ocean ecosystem health. The need to balance sustainable development's economic, social, and environmental aspects concerning seas is a fundamental element of the blue economy (Bax et al. 2022; Haas et al. 2021). However, the term Blue Economy is still ambiguous; terms like marine economy, ocean economy that has been interchangeably used for blue economy, but an analysis of these terms conducted by Martínez-Vázquez et al. (2021) found that sustainability and governance are common elements at the in blue economy and marine economy.

In contrast, the marine and ocean economies' common elements are industry development and location. However, this study uses the marine economy, ocean economy, and blue economy interchangeably as the Blue Economy directly pertains to oceanic resources and encompasses the sustainable and responsible utilisation of marine resources for economic endeavours (Bari 2017; Lee et al. 2020) and also the economic activities of the BE encompass all facets of marine, maritime, and coastal zones that exert a direct or indirect influence on the economy (Lee et al. 2021). *World Bank provides a working definition as follows: "Blue Economy is a marine-based economic development that leads to improved human wellbeing and social equity, while significantly reducing environmental risks and ecological scarcities."* (Smith-Godfrey 2016).

Human use of marine resources is seeing rapid growth. It is on the verge of a new epoch of industrialisation and marine resource exploitation, presenting enhanced innovation, growth, and job possibilities. As the blue economy expands, marine enterprises and the use of oceanic space, resources, and ecosystems must be ecologically sustainable; economic activities must correspond with the long-term carrying capacity of ocean ecosystems and must also be cognisant of regional differences, conditions, and resource requirements (Kraemer 2017; Smith-Godfrey 2016). The conservation of the environment has emerged as a primary focus among stakeholders, encompassing the business and financial sectors. The financial industry is vital in efficiently distributing financial resources in society (Hossain et al. 2024). Many recent studies by Hossain et al. (2024); Li and Xu (2023); Song et al. (2024); Wang et al. (2025) have shown that the development of digital finance benefits the green economy. In light of this, finance has emerged as a crucial tool for promoting sustainability, equity, and resilience within the blue economy (Sumaila et al. 2020). Digital finance, enabled by advancements such as blockchain, artificial intelligence (AI), mobile financial platforms, and big data analytics, is revolutionising the administration and utilisation of maritime resources. It aids in addressing significant challenges such as financial marginalisation, governance inefficiencies, and environmental deterioration in the blue economy. To enhance the comprehensiveness of FinTech for sustainable development, (Hasan et al. 2024) analysed dimensions where FinTech can impact and advance sustainable development. The analysis discovered three dimensions based on the Sustainable Development Goals. The first component addresses the importance of access to finance, encompassing inclusive finance, innovative financial services, inexpensive lending options, and related financial matters. The second dimension discussed the impact of innovative financial services on the economy, specifically regarding poverty alleviation, income inequality reduction, and economic development. At the same time, the third component emphasised the enhancement of environmental quality, namely the reduction of carbon emissions, as a pivotal environmental concern influencing sustainable development. Advancements in Fintech have generated novel prospects for financial inclusion, sustainable investments, energy efficiency, and improved financial literacy (Phukan and Kumar 2024). However, high costs and lack of digital infrastructure pose significant challenges to Fintech-based sustainable development (Hasan et al. 2024). Still, easy access to financial resources is crucial for enhancing green innovation and implementing green growth strategies in nations experiencing severe environmental degradation and misallocating financial resources (Song et al. 2024). As the term Blue Economy originates from the Green Economy, extending sustainable principles to marine resources and the ocean specifically, this chapter aims to contribute to the literature about the role of digital finance in addressing the unique challenges coastal and marine economies face.

2 Research Objective

1. To analyze the role of digital finance in fostering sustainability and equity within marine and seaside economies.

2. To examine public perception and stakeholder sentiment regarding digital financial initiatives for sustainable coastal development using sentiment analysis across social media, news articles, and other textual data.
3. To evaluate the socio-economic impacts and acceptance of digital finance initiatives in promoting financial inclusion and sustainable growth in marine-dependent communities.
4. To identify challenges and opportunities in leveraging digital finance for sustainable blue economy governance.

2.1 *Why This Study?*

While many studies like (Rajan Kumar et al. 2024; Shan et al. 2023; Thanh Ha 2025), etc., provide essential insights into the link between financial development and sustainable marine economic growth, there is a lack of research employing sentiment analysis to explore the interplay among digital finance, sustainability, and equality in marine and coastal economies. This study utilises sentiment analysis across social media, news articles, and diverse textual data to evaluate public perception and stakeholder sentiments regarding digital finance initiatives to enhance sustainability and equality in coastal regions. This approach would yield a more comprehensive understanding of this initiative's socio-economic impacts and acceptance among the public.

This study is divided into six main sections: The introduction summarises the blue or marine economy and digital finance, emphasising the importance of the study and the research questions it seeks to explore. The second section presents the literature review. The third section provides an overview of the research methodology employed. The fourth section discusses the analysis and interpretation of the collected data. The manuscript concludes with two sections that discuss the study's findings, their broader implications, and final remarks. The sections delineate potential avenues for future research and identify the limitations faced during the study.

3 Review of Literature

3.1 *Marine Economies Challenges*

Marine economies, encompassing enterprises dependent on coastal and oceanic ecosystems, are vital for global livelihoods and food security. These economies face various challenges, including resource depletion, climatic vulnerability, and socio-economic inequalities, all of which are thoroughly documented in the extant literature (Gao et al. 2022; Narwal et al. 2024; Österblom et al. 2023).

3.1.1 Resource Depletion

The unsustainable exploitation of marine resources, particularly fisheries, is a significant challenge for marine economies. Studies like Andrews et al. (2021) and Eadie and Hoisington (2011) demonstrate that overfishing has led to a worldwide decline in fish stocks, with over 30% of assessed fish populations categorised as overfished (FAO 2020). This decline endangers food security, particularly in developing coastal nations where fish is a primary protein source. Furthermore, the degradation of marine habitats, such as coral reefs and mangroves, exacerbates the issue (Eadie and Hoisington 2011). The seas are frequently portrayed in two opposing ways: as potential areas for growth, expansion, and advancement and as fragile ecosystems necessitating conservation. The "Blue Economy" concept has arisen in the last ten years due to tensions between these two ideas (Nham et al. 2023). Du et al. (2024) underscored the economic importance of these ecosystems, illustrating that their deterioration leads to reduced fisheries yield and increased vulnerability to coastal hazards. Despite numerous laws and international agreements, notably UN Sustainable Development Goal 14—Life Below Water, challenges in implementation and enforcement persist, particularly in regions with limited governmental capacity. Further climate change poses another challenge to marine economies.

3.1.2 Climate Susceptibilities

The oceans are crucial in regulating the Earth's temperature and are recognised for their potential in mitigating climate change. Concurrently, they have historically served as repositories for anthropogenic waste and have been profoundly affected by climate change, as demonstrated by phenomena such as ocean acidification, elevated temperatures, rising sea levels, etc. (Zhang et al. 2023), [67]. Hence, climate change is one of the main concerns regarding the future of the marine economy. According to NASA, global sea levels have risen by more than 10 cm from 1993 to 2024 (Masterson et al. 2024). The influence of increased water temperatures on global coastal systems includes marine heatwaves, ocean acidification, deoxygenation, changes in salinity, and rising sea levels that disrupt the marine environment and the livelihoods reliant on it (Daly et al. 2021). Climate change poses a substantial threat to marine life, encompassing coral reefs and fisheries, hence impacting marine ecosystems, economies, and communities, especially those dependent on natural resources, as also reported by Bacciu et al. (2021) wherein the study investigated the risk due to climate in the Mediterranean blue economy; Vrontisi et al. (2022) studied the impact of climate change on the blue economy of European island. Hoegh-Guldberg et al. (2019) demonstrate that temperature anomalies have precipitated coral bleaching episodes, severely impacting coral reef ecosystems vital for tourism and fisheries. Furthermore, small island developing states (SIDS) are especially susceptible to these alterations, with increased threats of environmental degradation, economic volatility, and relocation (Mycoo 2018; Petzold and Magnan

2019; Senaratne 2024). The research emphasises the necessity for adaptive management solutions, encompassing ecosystem-based approaches and using traditional knowledge to enhance resilience in marine-dependent communities.

3.1.3 Socio-Economic Inequities

The Blue Economy was envisioned to achieve sustainable wealth for a nation or region, ensuring the welfare of all citizens and humankind while protecting the environment, especially the marine ecosystem (Bari 2017). The blue economy was designed to prioritise social equality but is incessantly unheeded (Österblom et al. 2023); neo-liberal corporations aim to dominate the blue economy discourse. The increase in marine industrial developments and activities has unequally distributed ocean obligations and benefits, worsening economic inequities and power imbalances (Bennett et al. 2021; Croft et al. 2024; Engen et al. 2021). A study by Croft et al. (2024) stated that Indigenous peoples, women, small-scale fishermen, low-income workers, racialised minorities, coastal communities, and distant populations with insufficient employment and infrastructure are in danger of environmental degradation and unsustainable development. Black Americans, Black Canadians, and Indigenous peoples dwell near factories and absorb more seafood and environmental contaminants. Similarly, Bennett (2019) identified disparities in access to marine resources, poverty, and conflict as significant concerns in many coastal communities, with unequal access to marine resources acting as a primary catalyst for poverty and violence. Women often occupy vital positions in post-harvest activities and small-scale aquaculture and face systemic marginalisation and restricted access to financial and technological resources (Rao et al. 2024). Moreover, industrial corporations often monopolise the economic benefits derived from marine resources to the detriment of artisanal fishermen and small-scale operators. Addressing these discrepancies requires robust governance frameworks prioritising equity, community engagement, and capacity building (Schweinberg and Raspotnik 2024).

3.2 Digital Finance and Sustainability

Digital finance refers to financial services provided by mobile phones, computer networks or cards connected to a reliable payment system (Ozili 2018). Digital finance integrates digital technology and financial services, encompassing both financial services offered by emerging internet companies and traditional financial services that have undergone digital transformation. In other words, digital finance transitions offline financial services and transactions to the Internet. This evolution reduces the accessibility barriers to financial services and enhances efficiency (Fu et al. 2023). An emerging consensus recognises that finance and technology profoundly impact the efficiency and sustainability of an economy (Xin et al. 2024). Li and Xu (2023) investigated the impact of digital finance on urban green economic efficiency from a financial standpoint. They concluded that digital finance is positively connected with urban green economic efficiency. Yuan (2022) examined the influence of digital finance on economic development from 2000 to 2012. The study concluded that environmental pollution and economic growth are crucial for economic development. A comprehensive analysis revealed that digital finance has significantly contributed to this endeavour to foster coordinated and sustainable development of China's regional economy and environment. [68] analysed how digital finance can more effectively facilitate the green transition of the economy by examining the impact of digital finance on the green total factor energy efficiency. Wang et al. (2022) reported that digital finance enhances the effectiveness of tourist resource allocation by stimulating demand for tourism consumption and optimising the supply of tourism resources. Digital finance has the potential to enhance ecological resilience; as evidenced by Jiang and Jiang (2024), the broad reach and substantial use of digital banking can greatly enhance ecological resilience.

Furthermore, digital finance also assists in protecting resources and clarifying the complex relationship between resource curse and digital finance. [69] emphasised that digital inclusive finance can significantly lessen the effects of the resource curse in areas with a wealth of natural resources. However, Bakry et al. (2024) study concluded a U-shaped relationship between digital finance and CO_2 omission, implying that digital finance would take time to materialise. Until then, other measures like using renewable energy, etc., could be adopted.

A few more studies related to (a) digital finance and sustainability and (b) digital finance and socio-economic equity.

	Author/s	Research objectives	Findings	Analysis tool
Digital finance and sustainability	Than Ha (2025)	To understand the relationship between Fintech development and the blue economy	The study reported a two-way connectedness when analysing the relationship between the fintech index and blue bond volatility. It stated that the development of Fintech can positively impact the blue economy	Cutting-edge model-free connectedness approach
	Shan et al. (2023)	To understand the impact of lending to the blue economy and digitalisation on the banking sector	The study reported that blue lending and digitalisation boost bank performance, lower risks, and support investment in sustainable and blue economies	Quantitative Aanlysis

(continued)

(continued)

	Author/s	Research objectives	Findings	Analysis tool
	Nham and Ha (2023)	To assess the impact of financial development on marine living resources	The finding underscored the significance of finance in achieving a sustainable blue economy. The development of financial markets is more critical than that of financial institutions	Panel-corrected standard errors model (PCSE) and a feasible generalized least square (FGLS) estimates model
	Rajan Kumar et al. (2024)	To investigate the relationship among smart technology, business sustainability, and environmental sustainability in the context of small and medium-sized organisations (SMEs)	The study indicated that smart technologies improve business sustainability, whereas enhanced business practices indirectly facilitate their effect on environmental sustainability	Structural equation modelling
	Song et al. (2024)	To examine digital finance development's impact on Chinese Provinces' green growth	The advancement of digital finance positively influences green development; however, the impact is asymmetric, primarily demonstrating regional differences and threshold effects	Panel data analysis
	Salman and Ismael (2023)	To evaluate if digital financial inclusion (DFI) supports Egypt's CO_2 reduction initiatives	The findings indicated that in a developing nation such as Egypt, characterised by low DFI (<0.5), digital financial services serve as a complement to conventional finance, and while the use of digital financial services may reduce carbon emissions in the long term, it does not have the same effect in the short term	Autoregressive Distributive Lag (ARDL) model

(continued)

(continued)

	Author/s	Research objectives	Findings	Analysis tool
Digital finance and socio-economic equity	Shang (2023)	To investigate the impact of digital finance on educational equity	The research indicated that digital finance enhances educational equity by reducing credit constraints for the rural population and diminishing the income disparity between urban and rural areas	Panel data analysis
	Suhrab et al. (2024)	To examine the link between income equality and digital finance inclusion	The findings indicated that adopting technology innovation and infrastructure development could mitigate income inequality in BRICS nations by facilitating faster, more secure, and cost-effective transactions while improving access to markets and essential services	Statistical model—fixed effects regression and feasible generalized least squares (FGLS)
	Showkat et al. (2024)	To investigate the impact of adopting digital financial services on improving women's financial independence	The results indicate a substantial and positive correlation between using digital financial services and enhancing women's financial decision-making capabilities. This research advances the discourse on fostering gender equality and economic empowerment	Quantitative analysis using structural equation modelling
	Museba et al. (2021)	To examine the effects of fintech, mobile money, and digital financial services in Uganda, as well as assess their social implications via financial inclusion within the Ugandan market	The findings indicated a positive effect on sustainable development objectives, including Gender Equality (SDG 5), Decent Work and Economic Growth (SDG 8), enhanced financial inclusion via mobile money, and the reduction of inequalities (SDG 10)	Qualitative analysis—case study approach

(continued)

(continued)

	Author/s	Research objectives	Findings	Analysis tool
	Al-Smadi (2023)	To examine the relationship between digital finance and financial inclusion in Middle East and North Africa	The findings indicated that digital finance improves access to, availability of, and utilisation of financial services in the MENA area, hence elevating the overall degree of financial inclusion	System generalised method of moments panel method

Despite the growing body of literature on digital finance and sustainability and on socio-economic equity, there is a paucity of research utilising sentiment analysis to examine its role in fostering socio-economic equity, particularly within coastal and marine economies. Recent research predominantly focuses on quantitative economic models or legal frameworks, overlooking public perception and stakeholder opinions about digital financial initiatives. Understanding public sentiment may provide significant insights into digital finance effectiveness, challenges, and acceptance in achieving sustainability and equity.

4 Research Methodology

This study aims to analyze public opinion on online comments related to leveraging digital finance for sustainability and equity in marine and seaside economies. Web scraping techniques were utilized to gather textual data, and sentiment analysis was conducted using Python-based Natural Language Processing (NLP) tools. The methodological framework ensures data accuracy, relevance, and compliance with ethical considerations. This study follows a quantitative research approach, as it involves the numerical analysis of sentiment scores derived from textual data. It is also an empirical study, as it does not manipulate variables but rather analyzes pre-existing data collected from online platforms.

4.1 Data Collection Strategies

Web Scraping Process

- Web scratching was performed to extricate comments from online stages utilizing Python libraries such as BeautifulSoup.
- The scratched information incorporates user comments.

- Websites were chosen based on their significance to the research theme, guaranteeing a assorted dataset.
- Preprocessing steps included evacuation of copy sections, extraordinary characters,, and non-English writings.

4.2 Testing Technique

- A random sampling method was utilized to choose comments from diverse online gatherings and social media stages.
- Selection criteria guaranteed that the comments were important, freely accessible, and had adequate printed substance for examination.

4.3 Data Analysis Methods

Sentiment Analysis

- Assumption investigation was performed utilizing Characteristic Dialect Preparing (NLP) models.
- The essential Python library utilized for opinion examination was NLTK.
- Comments were categorized into positive, impartial, and negative opinion classes based on opinion scores.

Impact Estimation

- To survey the real-life affect, estimation patterns were connected with outside occasions, such as financial shifts, arrangement changes, or community responses in marine and ocean side economies.

Reliability and Validity

- Unwavering quality:
  Different scratching runs were conducted to guarantee consistency in information collection.
- Legitimacy:
  The dataset was approved by cross-referencing with third-party estimation datasets to guarantee exactness.
- Execution Assessment:
  The precision of assumption classification was tried employing a labeled dataset with known estimation scores.

Moral Contemplations

- As it were freely accessible information was collected, guaranteeing compliance with moral rules.
- No Personally identifiable data (PII) was put away or analyzed.

- The study respects website terms of benefit.

Limitations

- Web scratching restrictions and CAPTCHAs may have constrained information collection.
- Opinion investigation models may confuse mockery or context-specific expressions.
- The ponder does not account for subjective inclinations show in comment balance frameworks.

This methodology guarantees an efficient and moral approach to analyzing the effect of online estimation on real-life occasions, especially within the setting of marine and ocean side economies and computerized back supportability.

4.4 Interpretation

4.4.1 Sentiment Analysis of Tweets on Blue Economy and Its Effect

The blue economy alludes to the sustainable utilize of sea assets for financial development, moved forward vocations, and marine biological system wellbeing. It incorporates businesses such as fisheries, tourism, renewable vitality, and marine biotechnology.

With the rise of digitalization, fake insights (AI), and green back, discourses approximately the blue economy have ended up progressively pertinent. Understanding open opinion towards these improvements is significant, because it impacts policymaking, speculation, and mechanical selection.

This report presents an opinion examination of tweets related to the blue economy, categorized into positive, impartial, and negative estimations. The experiences drawn from this examination offer assistance us evaluate the open discernment, rising trends, and potential challenges within the blue economy segment.

4.4.2 Sentiment Analysis Methodology

The sentiment analysis was conducted using **Natural Language Processing (NLP)** techniques, particularly polarity scoring. Each tweet was assigned a **sentiment score between −1 and +1**, where:

- **Negative Sentiment (−1 to <0)**: Expresses criticism, concern, or skepticism.
- **Neutral Sentiment (0)**: Lacks strong emotional tone, often factual or descriptive.
- **Positive Sentiment (>0 to 1)**: Shows support, optimism, or enthusiasm.

4.4.3 Outline of Analysed Datasets

Three datasets containing tweets related to the blue economy were analysed:

Record 1—Contains common talks on advanced fund and marine vitality.

Record 2—Centres on the effect of digitalization on the blue economy.

Record 3—Talks about AI, green back, and exchange within the blue economy.

Each dataset was assessed independently to recognize patterns, key themes, and the by and large effect on the blue economy.

4.4.4 Sentiment Distribution and Key Perceptions

Neutral Sentiment (Larger part in All Datasets)

The larger part of tweets in all datasets were unbiased, showing that numerous discourses are educational, expository, or specialized. Such tweets incorporate inquire about discoveries, arrangement declarations, or common perceptions approximately the blue economy.

The tall predominance of impartial tweets suggests that discourses are frequently knowledge-driven instead of sincerely charged.

Cases of Unbiased Discourses

"The selection of AI in marine asset administration is increasing."

"Blue back models point to coordinated maintainability with financial growth."

"Renewable marine vitality has the potential to back coastal economies."

Effect on the Blue Economy

Positive:

Energizes information sharing and educated decision-making.

Negative:

Needs solid engagement, meaning open intrigued may not be effectively driving approach alter.

4.4.5 Positive Opinion (Higher in Record 2)

Positive assumption tweets reflect eagerness, trust, and optimism toward improvements within the blue economy.

File 2 had the most elevated positive assumption, demonstrating that digitalization within the blue economy is by and large seen as useful.

Steady discourses centred on mechanical headways, arrangement changes, and maintainability activities.

Key Topics in Positive Assumption

Mechanical Progressions: Computerized back and AI are moving forward effectiveness in marine businesses. Savvy arrangements are optimizing fisheries administration and diminishing squander.

Supportability and Green Back:

Green fund is making a difference move towards an naturally feasible blue economy. Renewable sea vitality sources are picking up footing as a climate-friendly elective.

Arrangement and Government Back:

Numerous governments are expanding venture in sea preservation and feasible marine businesses.

Public–private organizations are advancing capable marine asset utilization.

Effect on the Blue Economy.

Positive:

Empowers speculation, development, and arrangement headways.

Negative:

Can make over-expectations, possibly driving to disillusionment on the off chance that guaranteed improvements drop brief.

4.4.6 Negative Estimation (Higher in Record 3)

Outline

Negative assumption tweets speak to concerns, reactions, or skepticism approximately blue economy advancements.

Record 3 had the most noteworthy number of negative tweets, recommending that talks around AI and green fund are more questionable.

Common concerns spun around moral, financial, and natural dangers.

Key Topics in Negative Estimation

- Natural Concerns:
  Over-exploitation of marine assets in spite of supportability claims.
  Negative affect of mechanical exercises on marine biodiversity.
- Skepticism Toward AI and Advanced Back:
  Fear of AI supplanting human occupations in marine businesses.

Concerns approximately the monopolization of blue economy assets by enormous tech firms.

- Financial and Economic Risks:

  Potential abuse of green fund components for profit-driven thought processes.

  Disparity in get to to blue economy benefits, favouring expansive enterprises over small-scale anglers and coastal communities.

Impact on the Blue Economy

Positive:

Distinguishes basic ranges requiring approach enhancements.

Negative:

In the event that concerns are not tended to, open skepticism might moderate down selection of advantageous advancements.

Opinion Patterns and Their Suggestions

- Opinion Sort Perceptions Suggestions on Blue Economy
- Unbiased Lion's share of tweets Demonstrates an progressing scholastic and policy-driven talk, but needs passionate engagement.
- Positive Higher in Record 2 Proposes bolster for digitalization and green fund within the blue economy.
- Negative Higher in Record 3 Highlights concerns almost AI, budgetary disparity, and natural dangers.

4.5 Key Takeaways

- Developing Intrigued in Digitalization:
- The positive gathering of computerized fund and AI proposes expanding open bolster for technology-driven arrangements within the blue economy.
- Concerns Approximately Moral and Natural Issues:
- Whereas AI and green fund offer openings, open skepticism must be tended to through straightforward approaches and moral shields.
- Require for More noteworthy Open Engagement:
- The tall number of unbiased tweets proposes that open engagement in blue economy discourses ought to increment.

Mindfulness campaigns and open gatherings might bridge the hole between policymakers, industry pioneers, and the common population.

4.6 Suggestions for Policymakers and Partners

1. Reinforce Open Communication and Mindfulness

   Increment endeavours to teach the open on AI and advanced back within the blue economy.

   Advance straightforwardness in how unused innovations and budgetary models affect marine businesses.
2. Address Natural and Financial Concerns
   - Guarantee that AI and advanced back models are created in a way that does not abuse marine assets or neighbourhood communities.
   - Energize maintainable speculation methodologies that adjust with long-term marine preservation objectives.
3. Cultivate Public–Private Organizations

   Governments ought to collaborate with private firms to advance impartial get to to blue economy benefits.

   Create comprehensive budgetary models that bolster small-scale fisheries and coastal economies.

5 Conclusion

- The assumption examination uncovers a complex but advancing open discernment of the blue economy.
- Positive discourses highlight the potential of digitalization, AI, and green fund.
- Negative sentiments indicate challenges that ought to be proactively overseen.
- Unbiased opinion proposes a solid scholarly and specialized talk, but more noteworthy open engagement is required.

By leveraging positive patterns, tending to concerns, and cultivating comprehensive arrangements, the blue economy can proceed to create in a economical and impartial way.

References

Al-Smadi MO (2023) Examining the relationship between digital finance and financial inclusion: evidence from MENA countries. Borsa Istanbul Rev 23(2):464–472. https://doi.org/10.1016/j.bir.2022.11.016

Andrews N, Bennett NJ, Le Billon P, Green SJ et al (2021) Oil, fisheries and coastal communities: a review of impacts on the environment, livelihoods, space and governance. Energy Res Soc Sci 75:102009. https://doi.org/10.1016/j.erss.2021.102009

Bacciu V, Hatzaki M, Karali A, Cauchy A, Giannakopoulos C, Spano D, Briche E (2021) Investigating the climate-related risk of forest fires for Mediterranean Islands' blue economy. Sustainability 13(18):10004. https://doi.org/10.3390/su131810004

Bakry W, Nghiem XH, Bhatti MI, Al-Mohamad S, Cui L (2024) Digital finance and sustainable development: evidence from developing nations. Sci Prog 107(3):00368504241278823. https://doi.org/10.1177/00368504241278823

Bari A (2017) Our oceans and the blue economy: opportunities and challenges. Proc Eng 194:5–11. https://doi.org/10.1016/j.proeng.2017.08.109

Bax N, Novaglio C, Maxwell KH, Meyers K, McCann et al (2022) Ocean resource use: building the coastal blue economy. Rev Fish Biol Fish 32:1–19. https://doi.org/10.1007/s11160-021-09636-0

Bennett NJ (2019) In political seas: engaging with political ecology in the ocean and coastal environment. Coast Manag 47(1):67–87. https://doi.org/10.1080/08920753.2019.1540905

Bennett NJ, Blythe J, White CS, Campero C (2021) Blue growth and blue justice: ten risks and solutions for the ocean economy. Mar Policy 125:104387. https://doi.org/10.1016/j.marpol.2020.104387

Croft F, Breakey H, Voyer M, Cisneros-Montemayor A (2024) Rethinking blue economy governance–A blue economy equity model as an approach to operationalise equity. Environ Sci Policy 155:103710. https://doi.org/10.1016/j.envsci.2024.103710

Daly J, Knott C, Keogh P, Singh GG (2021) Changing climates in a blue economy: assessing the climate-responsiveness of Canadian fisheries and oceans policy. Mar Policy 131:104623. https://doi.org/10.1016/j.marpol.2021.104623

Du X, Wu S, Zhou J (2024) How does digital finance affect rural capital misallocation? Evidence from China. Appl Econ 1–14. https://doi.org/10.1080/00036846.2024.2323019

Eadie L, Hoisington C (2011) Stocking up: securing our marine economy. Centre for Policy Development, Sydney. https://cpd.org.au/wp-content/uploads/2011/09/stocking-up_final_for-web.pdf

Engen S, Hausner VH, Gurney GG et al (2021) Blue justice: a survey for eliciting perceptions of environmental justice among coastal planners and small-scale fishers in Northern Norway. PLoS ONE 16(5):e0251467. https://doi.org/10.1371/journal.pone.0251467

Food and Agriculture Organization of the United Nations (2020) The state of world fisheries and aquaculture 2020: sustainability in action. FAO, Rome

Fu Z, Zhou Y, Li W, Zhong K (2023) Impact of digital finance on energy efficiency: empirical findings from China. Environ Sci Pollut Res 30(2):2813–2835. https://doi.org/10.1007/s11356-022-22320-5

Gao LH, Ning J, Yan A, Yin QR (2022) A study on the marine ecological security assessment of Guangdong-Hong Kong-Macao Great Bay Area. Mar Pollut Bull 176:113416. https://doi.org/10.1016/j.marpolbul.2022.113416

Haas B, Mackay M, Novaglio C, Fullbrook L, Murunga M et al (2021) The future of ocean governance. Rev Fish Biol Fish 32:1–18. https://doi.org/10.1007/s11160-020-09631-x

Hasan M, Hoque A, Abedin MZ, Gasbarro D (2024) FinTech and sustainable development: a systematic thematic analysis using human-and machine-generated processing. Int Rev Financ Anal 95:103473. https://doi.org/10.1016/j.irfa.2024.103473

Hoegh-Guldberg O, Pendleton L, Kaup A (2019) People and the changing nature of coral reefs. Reg Stud Mar Sci 30:100699. https://doi.org/10.1016/j.rsma.2019.100699

Hossain MR, Rao A, Sharma GD, Dev D, Kharbanda A (2024) Empowering energy transition: green innovation, digital finance, and the path to sustainable prosperity through green finance initiatives. Energy Econ 136:107736. https://doi.org/10.1016/j.eneco.2024.107736

Jiang W, Jiang N (2024) From sustainable development towards resilience: does digital finance matter in enhancing ecological resilience? Sustain Dev. https://doi.org/10.1002/sd.3257

Kraemer RA (2017) A sustainable ocean economy, innovation and growth: a G20 initiative

Lee KH, Noh J, Khim JS (2020) The blue economy and the United Nations' sustainable development goals: challenges and opportunities. Environ Int 137:105528. https://doi.org/10.1016/j.envint.2020.105528

Lee KH, Noh J, Lee J, Khim JS (2021) Blue economy and the total environment: mapping the interface. Environ Int 157:106796. https://doi.org/10.1016/j.envint.2021.106796

Li H, Xu R (2023) How does digital finance affect the efficiency of urban green economies? Evidence from China. Finan Res Lett 58:104595. https://doi.org/10.1016/j.frl.2023.104595

Martínez-Vázquez RM, Milán-García J, de Pablo VJ (2021) Challenges of the blue economy: evidence and research trends. Environ Sci Eur 33(1):1–17. https://doi.org/10.1186/s12302021005021

Masterson V, Hall S, North M (2024) Sea level rise: everything you need to know. World Economic Forum. https://www.weforum.org/stories/2024/09/rising-sea-levels-global-threat/

Museba TJ, Ranganai E, Gianfrate G (2021) Customer perception of adoption and use of digital financial services and mobile money services in Uganda. J Enterp Commun People Places Glob Econ 15(2):177–203. https://doi.org/10.1108/JEC-07-2020-0127

Mycoo MA (2018) Beyond 1.5 °C: vulnerabilities and adaptation strategies for Caribbean Small Island developing states. Reg Environ Change 18(8):2341–2353. https://doi.org/10.1007/s10113-017-1248-8

Narwal S, Kaur M, Yadav DS, Bast F (2024) Sustainable blue economy: opportunities and challenges. J Biosci 49(1):18. https://doi.org/10.1007/s12038-023-00375-x

Nham NTH, Ha Thanh L (2023) The role of financial development in improving marine living resources towards sustainable blue economy. J Sea Res 195:102417. https://doi.org/10.1016/j.seares.2023.102417

Nham NTH, Hoa TTM, Ha LT (2023) Influences of digitalisation on sustaining marine minerals: a path toward sustainable blue economy. Ocean Coast Manag 239:106589. https://doi.org/10.1016/j.ocecoaman.2023.106589

Österblom H, Wabnitz CC, Tladi D, Allison EH, Arnaud-Haond S et al (2023) Towards ocean equity. In: The blue compendium: from knowledge to action for a sustainable ocean economy. Springer International Publishing, Cham, pp 485–521. https://doi.org/10.1007/978-3-031-16277-0_13

Ozili PK (2018) Impact of digital finance on financial inclusion and stability. Borsa Istanbul Rev 18(4):329–340. https://doi.org/10.1016/j.bir.2017.12.003

Pace LA, Saritas O, Deidun A (2023) Exploring future research and innovation directions for a sustainable blue economy. Mar Policy 148:105433. https://doi.org/10.1016/j.marpol.2022.105433

Petzold J, Magnan AK (2019) Climate change: thinking small islands beyond Small Island Developing States (SIDS). Clim Change 152(1):145–165. https://doi.org/10.1007/s10584-018-2363-3

Phukan PK, Kumar PP (2024) The intersection of fintech and sustainability: a catalyst for positive change. In: The adoption of fintech. Productivity Press, pp 98–113. https://doi.org/10.4324/9781032644165-7

Rao N, Hooper L, Gray H (2024) A systematic review of the impact of post-harvest aquatic food processing technology on gender equality and social justice. Nature Food 5(9):731–741. https://doi.org/10.1038/s43016-024-01034-6

Rajan Kumar VK, Saunila M, Rantala T, Ukko J (2024) The interplay between smart technologies, business sustainability, and environmental sustainability: an empirical analysis of SMEs. Corp Soc Responsib Environ Manag. https://doi.org/10.1002/csr.2966

Salman D, Ismael D (2023) The effect of digital financial inclusion on the green economy: the case of Egypt. J Econ Dev 25(2):120–133. https://doi.org/10.1108/JED-05-2022-0087

Schweinberg MJ, Raspotnik A (2024) Slaying the toothless tiger: social equity as the legal element of the blue economy. J Environ Dev 10704965241238290

Senaratne M (2024) Achieving SDGs in Small Island Developing States: financing the blue economy through collaboration. https://www.orfonline.org/public/uploads/posts/pdf/20240417105927.pdf

Shan S, Mirza N, Umar M, Hasnaoui A (table) The nexus of sustainable development, blue financing, digitalisation, and financial intermediation. Technol Forecast Soc Chang 195:122772. https://doi.org/10.1016/j.techfore.2023.122772

Shang R (2023) Divide or dividend: how digital finance impacts educational equality. Financ Res Lett 55:103858. https://doi.org/10.1016/j.frl.2023.103858

Showkat M, Nagina R, Nori U, Baba MA, Shah MA (2024) Empowering women in the digital age: can digital financial services fulfil the promise of financial autonomy and gender equality in the attainment of Sustainable Development Goal 5? Cogent Econ Finan 12(1):2342459. https://doi.org/10.1080/23322039.2024.2342459

Smith-Godfrey S (2016) Defining the blue economy. Marit Aff J Natl Marit Found India 12(1):58–64. https://doi.org/10.1080/09733159.2016.1175

Song Y, Gong Y, Song Y (2024) The impact of digital financial development on the green economy: an analysis based on a volatility perspective. J Clean Prod 434:140051. https://doi.org/10.1016/j.jclepro.2023.140051

Suhrab M, Chen P, Ullah A (2024) Digital financial inclusion and income inequality nexus: can technology innovation and infrastructure development help in achieving sustainable development goals? Technol Soc 76:102411. https://doi.org/10.1016/j.techsoc.2023.102411

Sumaila UR, Walsh M, Hoareau K, Cox A et al (2020) Ocean finance: financing the transition to a sustainable ocean economy. World Resources Institute. https://medblueconomyplatform.org/wp-content/uploads/2020/11/file-library-85582906edbb79cca982.pdf

Thanh Ha L (2025) Gauging the dynamic interlinkage level between fintech and the global blue economy performance. J Econ Stud 52(1):55–71. https://doi.org/10.1108/JES-10-2023-0613

Vrontisi Z, Charalampidis I, Lehr U, Meyer M et al (2022) Macroeconomic impacts of climate change on the blue economy sectors of southern European islands. Clim Change 170(3):27. https://doi.org/10.1007/s10584-022-03310-5

Wang Q, Yang L, Yue Z (2022) Research on development of digital finance in improving efficiency of tourism resource allocation. Resour Environ Sustain 8:100054. https://doi.org/10.1016/j.resenv.2022.100054

Wang Y, Cui L, Zhou J (2025) The impact of green finance and digital economy on regional carbon emission reduction. Int Rev Econ Financ 97:103748. https://doi.org/10.1016/j.iref.2024.103748

World Bank; United Nations Department of Economic and Social Affairs (2017) The potential of the blue economy: increasing long-term benefits of the sustainable use of marine resources for small island developing states and coastal least developed countries. World Bank, Washington, DC. http://hdl.handle.net/10986/26843. Accessed 24 Nov 2024

Xin C, Fan S, Guo Z (2024) Can digital finance promote inclusive growth to meet sustainable development in China? A machine learning approach. Environ Dev Sustain 26(10):26647–26677. https://doi.org/10.1007/s10668-023-03748-2

Yuan F (2022) Research on the impact evaluation of digital finance on the synergy between economic development and ecological environment. J Environ Public Health 2022:1714609. https://doi.org/10.1155/2022/1714609

Zhang W, Chen F, Liu E, Zhang Y, Li F (2023) How does digital inclusive finance promote the financial performance of Chinese cultural enterprises? Pac Basin Finan J 82:102146. https://doi.org/10.1016/j.pacfin.2023.102146

Blockchain and AI for Advancing Sustainability in the Blue Economy: Innovating Marine Resource Management and Environmental Stewardship

C. V. Suresh Babu, M. S. Asmaa Begum, and Simeon Karafolas

Abstract This study explores how blockchain and artificial intelligence (AI) can be merged to support the advancement of sustainability in the blue economy, specifically in managing marine resources and safeguarding the environment. The ultimate aim is to establish how the technologies enhance transparency, traceability, and efficiency in fighting challenges such as overfishing, marine pollution, and climate change. The target audiences are researchers, policymakers, business experts, and environmental groups. The method employed is a systematic literature review, case study examination, and critical analysis of technological applications in marine conservation. Results indicate that blockchain enhances data integrity and trust, and AI maximizes resource allocation and predictive analytics for ecosystem monitoring. The research identifies the potential of smart contracts, decentralized finance, and AI-based analytics to transform sustainability efforts. The implications indicate an international cooperation and regulatory structures to enhance these technologies' advantages.

Keywords Artificial intelligence · Blockchain · Blue economy · Climate change · Decentralized finance · Marine conservation · Smart contracts · Sustainability · Transparency

C. V. Suresh Babu (✉) · M. S. Asmaa Begum
Department of Information Technology, Hindustan Institute of Technology and Science, Bay Range Campus, Padur, Chennai, India
e-mail: pt.cvsuresh@hindustanuniv.ac.in

S. Karafolas
Department of Accounting and Finance, University of Western Macedonia, 50100 Koila Kozanis, Greece

V. Sharma et al. (eds.), *FinTech for Sustainable Finance and a Well-Blue Economy*, World Sustainability Series, https://doi.org/10.1007/978-3-031-92390-6_7

1 Introduction

Blockchain and artificial intelligence (AI) are widely recognized as technologies with the potential to fundamentally transform and innovate across multiple industries and sectors. While the term blockchain began to captivate the global tech community in 2008 through the launch of the Bitcoin cryptocurrency by Nakamoto (2008), the fundamental principles and concepts it is based on have been in practice since the 1980s (Tripathi et al. 2023). Specifically, the first attempt to create a decentralized and cryptographic digital money came out as early as 1983 in the seminal paper titled "Blind signatures for untraceable payments" (Chaum 1983).

Emerging technologies such as blockchain and artificial intelligence (AI) are being harnessed to promote sustainability within the blue economy. Blockchain technology brings transparency to seafood supply chains, which serves to hold fisheries accountable and dissuade illegal, unreported, and unregulated (IUU) fishing. Blockchain makes it possible for stakeholders to track the source of seafood, track supply chain activity, and boost regulatory compliance through decentralized and unalterable recording.

AI enhances sustainability by providing real-time monitoring and predictive analytics on marine ecosystems. AI-driven systems process massive amounts of data from satellites, underwater drones, and IoT sensors and deliver insightful information about ocean conditions. These observations aid in the identification of environmental threats like oil spills, coral bleaching, and poaching, enabling rapid intervention and conservation action. In sea transport, AI-optimized shipping routes improve fuel efficiency, lowering greenhouse gas emissions and the environmental footprint of global trade. Further, AI-based models assist climate resilience by forecasting changes in marine ecosystems, enabling policymakers to design forward-looking conservation policies. Experts urge caution, highlighting the complex challenges and mixed opinions regarding the integration of AI with blockchain technology (Sam 2024).

The convergence of AI and blockchain provides a disruptive solution to sustainable ocean management. By improving transparency, efficiency, and foretelling, these technologies enable governments and industries to reconcile economic growth with environmental stewardship. With escalating global dependence on the ocean's resources, leveraging innovation is key to maintaining the long-term stability and health of marine ecosystems as well as promoting sustainable economic growth.

1.1 Why Sustainability in Ocean Resource Use

Sustainability is the golden key upon which the Blue Economy must be constructed. We can meet present requirements without sacrificing future generations' right to benefit from the marine ecosystem by managing ocean resources in a sustainable manner.

Healthy oceans not only promise the planet ecological balance, but billions of human beings depend on them for nutrition, climate regulation, and economic opportunities.

But habitat loss, pollution, and overfishing threaten the marine ecosystems severely. Strategic management measures are required to limit the threats from fishing operations. These involve sustainable fishing, promoting renewable marine energy, and investment in pollution control measures. Protection of marine diversity through the preservation of mangroves and coral reefs is vital to the health of the oceans and to creating resilience to environmental and economic shocks. Long-term environmental sustainability will be allowed by the ocean blue economy. sustainable management of the ocean so that coastal economies can flourish.

1.2 Blue Economy Challenges Today

Nonetheless, despite its vast potential, the Blue Economy itself has numerous challenges that hinder its sustainable development. IUU and overfishing drain the stocks and undermine conservation efforts (Chaum 1983). Plastic and oil spills, which further destabilize ocean systems and marine life, are also increasing. Climate change aggravates this by enhancing ocean acidification, sea level rise, and acceleration of biodiversity loss (IPCC 2021). One of the key limits to sustainability is the secrecy and lack of accountability in managing marine resources. Ineffective regulatory enforcement and risky control mechanisms enhance unsustainable extraction. Solutions to such issues rely on new, technology-driven involvements that enhance good governance and protection programs.

1.3 Future Technologies Contribution to Redressing Blue Economy Issues

Blockchain and artificial intelligence are two of the emerging technologies that are revolutionizing the Blue Economy. Blockchain makes it possible to have open and transparent accounting that can ensure seafood origin traceability, prevent illegal fishing, and enhance supply chain transparency, increasing consumers' trust and facilitating sustainable sourcing. On the other hand, AI facilitates real-time monitoring of the oceans, predictive forecasting of resource consumption, and automated conservation processes. All of this empowers the decision-makers with useful information to enhance the effectiveness of resources, put regulations in place, and become sustainable so that a blue economy can become more powerful.

1.4 Objectives of the Chapter

- To give a rounded view of the blue economy and its critical components.
- To highlight the importance of sustainability in the management and application of ocean resources.
- To identify and analyze the current challenges obstructing sustainable development in the blue economy.
- To explore the role of emerging technologies, such as blockchain and AI, in addressing these challenges.
- To provide an enabling framework by which technology might promote innovation within the management of marine resources, including environmental protection.

2 Theoretical Background

2.1 Definition and Scope of Blue Economy

Blue Economy is integrated with the principles for the beneficial utilization of the ocean's resources for sustainable purposes with a blend of economic prosperity, environmental concerns, and human welfare. Historically, it has considered the traditional sectors: fisheries, maritime transport, and coastal tourism; however, for the World Bank (2017) (Satoshi 2022), new sectors such as marine biotechnology, deep-sea mining, and offshore renewable energy are now being included. This concept does not only represent the mere exploitation of marine resources but also the fact that oceans provide critical functions within the services for ecosystems in regulating climate, sequestering carbon, and providing protection for coastlines (OECD 2016).

The Blue Economy, through integrated approaches, pays much attention to the management of marine resources and biodiversity conservation along with social inclusion. It is aimed at bringing sustainable development where the harm inflicted on marine ecosystems is minimized while the long-term benefits are maximized for communities and industries. Therefore, this requires partnerships among governments, industries, and research institutions to ensure there is a proper balance between growth and sustainability (Satoshi 2022).

Scope of the Blue Economy:

- Traditional Sectors: Fisheries, aquaculture, and maritime transport.
- Emergent Sectors: Marine biotechnology, offshore renewable energy, and ocean floor exploration.
- Ecology Services: Carbon appropriation, climate regulation, and coastal protection.
- Social Inclusion: Livelihood creation for coastal communities and development equity.

- Conservation Sustainability: Biodiversity conservation and pollution management.
- Policy Integration: Collaborative frameworks for marine resource governance.

2.2 Principles of Sustainable Blue Economy

The central principles on which the sustainable blue economy functions are based on global sustainability goals, such as that by the United Nations Sustainable Development Goal 14 (United Nations 2020):

- **Sustainability**: balancing the pursuit of economic development with the preservation of marine ecosystems (Satoshi 2022).
- **Inclusivity**: balancing resource access and opportunities for coastal communities (UNEP 2021).
- **Resilience**: adapting to environmental change—such as from climate change—in the pursuit of long-term sustainability (OECD 2016).
- **Transparency**: Open and accountable governance of marine resources (Commission 2020).
- **Innovation**: Utilization of advanced technologies and innovative practices for sustainable development (UNEP 2021).

2.3 Emerging Technologies: Blockchain and AI

Blockchain and artificial intelligence are two emerging technologies that are strong enough to deal with the most critical issues in the blue economy. Blockchain is a decentralized, immutable digital ledger that promotes the transparency, traceability, and accountability of various marine sectors, particularly fisheries management and maritime trade. Making transactions secure while recording them and keeping data tamper-proof can help combat IUU fishing and ensure ethical sourcing as well as optimize marine supply chains (Treiblmaier and Beck 2019). AI complements blockchain by providing sophisticated tools of data analysis for predictive decision-making, adding value through real-time monitoring. Marine conservation can rely on AI-based technologies to analyze enormous oceanic data for accurate tracking of marine species as well as levels of pollution and illegal activities (van Putten et al. 2021). AI further encourages predictive analytics concerning vessel routing, wave energy management, and climate models, hence providing better operational efficiency and sustainability. It fosters innovative approaches to the management of marine resources through blockchain and AI integration that allows for data-driven, transparent, and accountable practice in the handling of environmental and socio-economic challenges in the blue economy (Zheng et al. 2021).

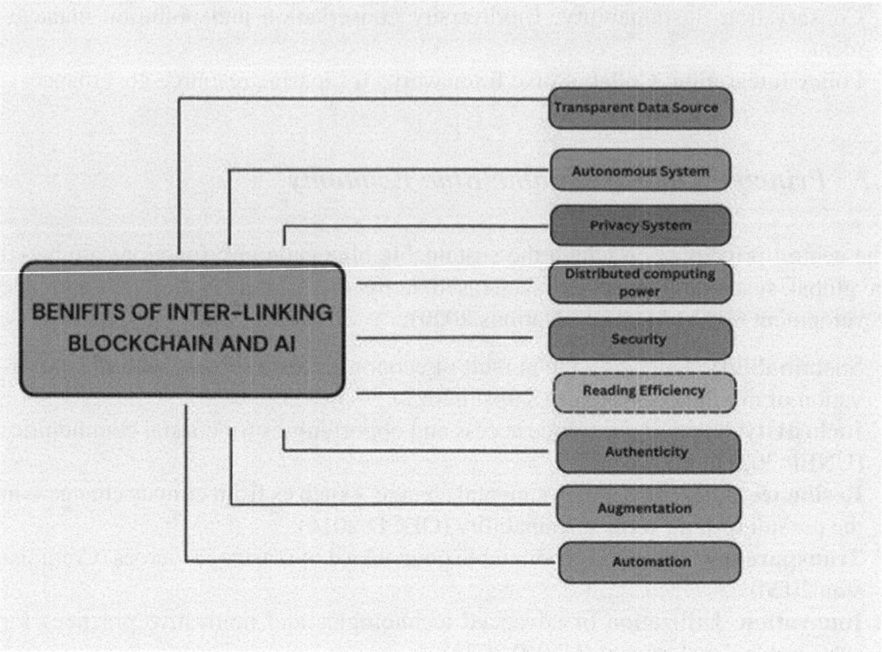

Fig. 1 Benefits of interlinking blockchain and AI. *Source* By authors

Figure 1 illustrates how linking blockchain with AI creates a secure, transparent data framework that enhances predictive analytics and decision-making. This integration drives automated, real-time monitoring and incentivizes sustainable practices across marine resource management.

2.4 Overview of Blockchain Technology

Blockchain refers to a decentralized digital ledger technology that allows the recording of transactions across several nodes in a safe, transparent, and immutable manner. It started as the core of Bitcoin but has developed into diverse uses in various areas, including finance, supply chains, and environmental management (Nakamoto 2008). These advancements have paved the way for the adoption of blockchain technology across multiple industries, including finance, healthcare, cybersecurity, and business, supporting practical applications (Ghosh et al. 2023; Girdhar et al. 2023; Jiang et al. 2023; Tyagi et al. 2023). It is based on blocks of data connected using cryptographic hashes so that any alteration in previous blocks would be noticeable and virtually impossible to achieve, which increases the integrity of data (Swan 2015).

A critical aspect of blockchain technology is the consensus mechanism, which achieves a mutual agreement among the network participants on the validity of transactions. Common consensus protocols are Proof of Work, Proof of Stake, and Practical Byzantine Fault Tolerance (Narayanan et al. 2016). Moreover, the exploration into scalability and interoperability issues highlights the ongoing efforts to enhance DLT functionalities, ensuring their seamless integration across different platforms and industries (Ghosh et al. 2023; Girdhar et al. 2023; Jiang et al. 2023; Tyagi et al. 2023). Blockchain also enables the execution of smart contracts—self-executing agreements coded on the blockchain. These smart contracts automatically activate predefined actions based on conditions being met. In return, this ensures automated compliance and reduces the costs associated with transactions.

3 Blue Economy—Existing Problems

The blue economy is essentially the safe and effective exploitation of ocean resources to contribute to economic development, social well-being, and environmental conservation. The related gamut of challenges impacts marine ecosystems, affecting millions of people involved in varied coastal and ocean-based activities. Long-term sustainable resources of oceans depend on resolving these interrelated challenges.

3.1 Overfishing and Resource Depletion

One of the worst impacts on marine ecosystems is overfishing. In this regard, fish stocks are being fished at a rate that surpasses their reproduction ability. Overfishing disturbs marine food chains, lowers fish populations to unsustainably low levels, and reduces biodiversity. The Food and Agriculture Organization reported that illegal, unreported, and unregulated (IUU) fishing contributes to approximately 26 million tons of catches of fish per year, thereby significantly damaging attempts to regulate fisheries and maintain ecological balance (FAO 2022).

Impacts of overfishing go past the depletion of fish stocks: removing keystone species upsets the fragile balance in predator–prey relationships, which leads to a domino effect on marine food webs. Biodiversity loss eventually leads to ecosystems collapsing, as oceans become challenged to recover their essential functions: carbon sequestration, climate regulation, and coastal protection (Chaum 1983). It also puts coastal communities that rely on fishing for livelihood and food security in a precarious situation. Declines in fish stocks lead to economic hardship among small-scale fishers, and with this comes more poverty and inequality. Competition between industrial fishing operations and small-scale fishers worsens the situation since such operations often overexploit marine resources at an unsustainable rate, leaving little marine resource access for local fishers.

3.2 Environmental and Ecological Pollution

Globally, the biggest concerns are the degrading health and sustainability of ecosystems in the world's oceans through marine pollution. The aggregation of plastic waste and oil spills, accompanied by chemical waste, destroys the beauty of marine organisms. Every year, about 14 million tons of plastic become entrapped within the ocean habitat, which brings about habitat destruction and the obliteration of its biodiversity while undermining food webs (Jambeck et al. 2015).

Plastic pollution is particularly problematic because it lasts longer and spreads out so widely. Many marine species, such as fish, seabirds, and turtles, ingest plastic debris because they mistake it for food, leading to starvation, internal injuries, and death. There are also ghost nets of fishing gear that still capture and kill marine organisms even after being discarded. Besides plastic debris, oil spills and chemical irrigation from farming and industry also poison the critical habitat of the world's oceans—from coral reefs and seagrass beds to mangrove forests. These sites are critical not only for maintaining marine biodiversity but also for contributing toward carbon sequestration that alleviates climate change. The contamination of ocean waters also affects human communities by reducing the availability of clean drinking water and seafood, which poses serious public health risks (Panti et al. 2020).

3.3 Lack of Transparency and Traceability

Transparency and traceability in global seafood supply chains represent a significant obstacle to sustainability in the blue economy. Seafood fraud and mislabeling occur frequently, as studies have estimated that up to 20% of seafood products are mislabeled, thereby causing economic losses and complicating efforts to monitor sustainable practices (Oceana 2020). Since legally caught fish cannot be distinguished from illegally caught ones, and without tracking systems, it creates an opportunity for IUU operators to take undue advantage of the marine resources, which ultimately is a challenge in conservation and compliance with regulation.

A more transparent supply chain would serve the interests of all stakeholders involved, from a consumer seeking ethical seafood to the government enforcing the regulation of fisheries. Blockchain has been proposed as a solution because it provides digital records that cannot be tampered with and would allow for the tracking of seafood from the ocean to the plate, meeting environmental and ethical standards (Bailey et al. 2018).

3.4 Climate Change and Ocean Degradation

Climate change fundamentally impacts ocean health, and its changing trends affect the stability of the marine ecosystems and industries operating within the blue economy. The most critically prominent climate-related challenges include rising ocean temperatures, acidification, and sea-level rise (Hughes et al. 2018). One of the most evident impacts is coral bleaching, where rising sea temperatures cause corals to expel the symbiotic algae that provide them with nutrients and color. Such massive coral bleaching events have been on the rise, resulting in the decline of coral reef ecosystems, which harbor 25% of all marine species and offer critical coastal protection (Sumaila et al. 2020).

3.5 Ineffective Resource Use and Financial Systems

Many of the problems in the blue economy can be traced back to poor resource management and weak financial frameworks that reward undesirable practices. Industrial fishing operations still receive subsidies from governments, further over-exploiting marine stocks rather than encouraging sustainable substitutes (Österblom and Folke 2013).

Inefficient regulatory frameworks and data collection mechanisms hinder ocean governance. Most developing countries cannot monitor fishing, protect marine life, and prevent IUU fishing. In addition, economic barriers hinder the implementation of innovative technology, such as AI-based ocean monitoring and blockchain-based seafood traceability systems (Treiblmaier and Beck 2019).

3.6 Marine Conservation Governance Challenges

Although a strong governance framework is needed to effectively conserve the marine environment, 64% of the world's oceans fall within the high seas, barely regulated, and more vulnerable to exploitation (UNEP 2021). Ambiguous jurisdictional limits make it difficult to enforce the laws on environmental protection, observe fishing quotas, and protect the biodiversity of marine life.

International cooperation is weak, and economic interests sometimes conflict. As there are pro-conservation countries, others continue to explore the oceans to reap short-term economic benefits; this creates 'empty pockets' of regulations in unsustainable practices (Tyagi et al. 2023).

Figure 2 highlights the major challenges in the blue economy, including over-fishing, pollution, and inadequate regulatory frameworks. These obstacles undermine sustainable development and threaten the long-term health of marine ecosystems.

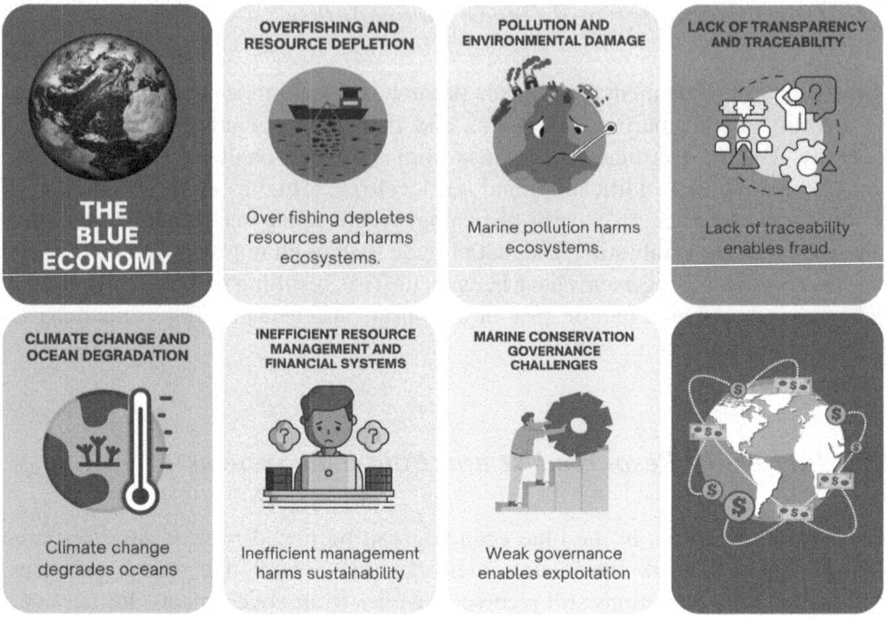

Fig. 2 Challenges faced by the blue economy. *Source* By authors

4 Blockchain in the Blue Economy

4.1 Fundamentals of Blockchain: Decentralization and Immutability

Blockchain is a decentralized digital ledger that stores data securely across several nodes, providing transparency and minimizing data manipulation risks. Its immutability ensures that the data stored cannot be deleted or modified, promoting accountability and trust. These features render blockchain extremely useful for marine resource management, fisheries, and pollution monitoring, increasing operational efficiency through smart contracts.

4.2 Blockchain Ocean Resource Management

Blockchain improves marine resource management through verifiable, unalterable records of data. It makes seafood traceability possible, stopping illegal fishing and mislabeling. Open data sharing between stakeholders such as fisheries and regulatory bodies encourages accountability and sustainable use.

4.3 Fisheries and Aquaculture Transparency

Blockchain provides end-to-end traceability across seafood supply chains through the capture of critical information, including fishing practice, location, and vessel ID. Consumers can check for product authenticity using QR codes, while government authorities use blockchain to conduct checks on compliance, keeping fraud out and encouraging responsible sourcing.

4.4 Reducing Illegal Fishing

Blockchain stores an unalterable history of fishing activity, which makes it possible to monitor vessels, catch quantities, and timestamps in real-time. Transparency enables authorities to enforce fishing regulations, achieve compliance with environmental norms, and deter illegal, unreported, and unregulated (IUU) fishing.

4.5 Blockchain for Marine Pollution Monitoring

By combining blockchain with IoT sensors, real-time pollution data (e.g., chemical levels and water temperature) is safely recorded and examined. Such a system allows for the early identification of pollution incidents, ensuring quick responses and bringing responsible individuals to account for any environmental damage.

Figure 3 illustrates how blockchain enhances the Blue Economy by ensuring transparency, traceability, and secure data management across marine industries. It enables sustainable fisheries, combats illegal fishing, and supports decentralized financial solutions for ocean conservation.

5 Artificial Intelligence in the Blue Economy

5.1 AI and Data-Driven Decision-Making in Marine Conservation

Artificial intelligence allows for data-driven decision-making in marine conservation. It aggregates huge amounts of complex data to identify trends, patterns, or anomalies—thus yielding real-time information on marine ecosystems and empowering conservationists with informed action. For example, AI-integrated satellite imagery can monitor changes in ocean surface conditions, while underwater sensors can track biodiversity and environmental health metrics.

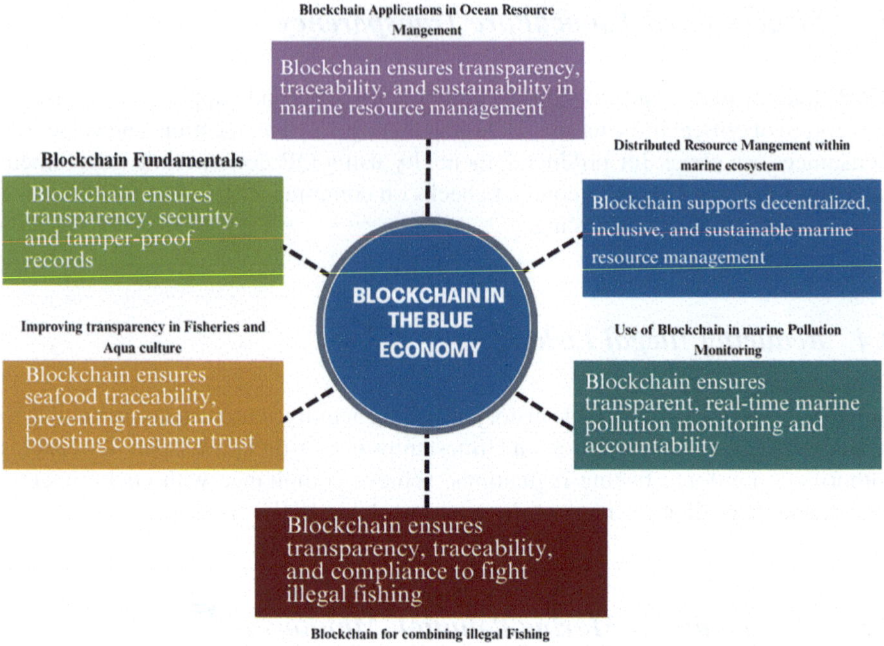

Fig. 3 Blockchain in the blue economy. *Source* By authors

Drones with AI machinery are critical equipment in surveying marine environments and tracking cases of poaching and loss of habitation. AI tools are highly critical in identifying points of priority where marine protected areas should be established by calculating biodiversity metrics, oceanographic conditions, and human activities. The rate at which data is computed through AI guarantees that this is done with precision and timely accuracy (Fig. 4).

5.2 AI in Predictive Analytics for Fisheries and Ecosystem Monitoring

Predictive analytics, powered by AI, transforms the way fisheries and marine ecosystems are monitored and managed. It uses advanced models of AI in the integration of historical catch data, environmental variables, and patterns of species' behaviors to make attempts at trying to predict dynamics in fish populations and migration routes. For instance, deep learning algorithms can try to predict spawning or seasonal migrations based on trends of data that extend over multiple years. It offers predictive capabilities to the advantage of the regulatory agency and the fisheries manager in

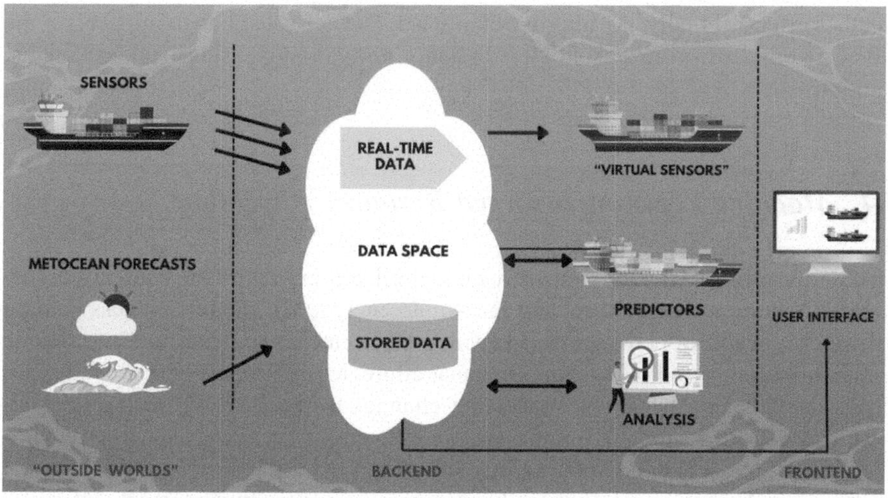

Fig. 4 Marine conservation using AI. *Source* By authors

the setting of realistic allowable catch allocations consistent with normal reproductive cycles, hence providing for maximum opportunities to limit the likelihood of over-exploitation or degradation of threatened fishery resources.

Other applications relate to environmental monitoring through tracking such water parameters as temperature, salinity, and nutrient content. The AI systems would continue to analyze these streams of data in real-time for anomalies that might be indicative of ecological disturbances, such as sudden temperature changes that would be associated with an outbreak of harmful algal blooms or low-oxygen events. This is a much earlier intervention that stakeholders might otherwise take as adjustments in their fishing efforts or through the enforcement of conservation activities that could avoid further devastation. This means AI-driven predictive analytics is the proactive management tool that keeps fisheries and marine ecosystems resilient to environmental variability.

5.3 AI for Climate Modelling and Climate Change Mitigation Strategies

AI is critical for climate change impact analysis through the processing of enormous climate data, such as ocean temperatures, sea levels, and pH changes, to forecast long-term marine environmental change. Predictive models based on AI assist in forecasting habitat shifts, changes in species distribution, and impacts on fisheries. Through the simulation of different climate scenarios, AI assists in adaptation planning, for example, determining vulnerable areas for restoration. Moreover, AI maximizes renewable energy systems such as offshore wind farms by forecasting weather effects. Finally, AI improves knowledge of climate change impacts on the blue economy, allowing stakeholders to adopt mitigation measures and sustainable solutions.

5.4 Machine Learning in Green Resource Allocation

Machine learning (ML) maximizes resource utilization in the blue economy, especially in fisheries, aquaculture, and renewable energy. ML models process data on fish stocks, health of ecosystems, and volumes of catches to determine sustainable fishing quotas to avoid overfishing. In aquaculture, ML tracks water quality, disease trends, and feeding behavior, suggesting changes to maximize growth and minimize waste. Furthermore, ML improves the efficiency of renewable energy by optimizing offshore wind and wave energy operations with the help of environmental information. Through continuous enhancement of resource management, ML guarantees sustainable use of marine resources while balancing economic gains and environmental protection.

5.5 Detection and Evaluation of Marine Pollution Using AI

It poses an extreme level of threat.

1. Explore specific machine learning algorithms that enhance efficiency in solar panel energy capture. 2. Discuss the role of data analysis in optimizing wind turbine performance and reducing maintenance costs. 3. Examine case studies where machine learning successfully identified sources of marine pollution. 4. Highlight innovative technologies that combine renewable energy solutions with pollution monitoring systems. 5. Consider the potential for machine learning to forecast environmental changes and guide future renewable energy initiatives to ocean health, and AI significantly contributes to the identification and measurement of such threats. Machines with AI techniques—including image recognition and pattern analysis, satellites, drones, and other means of remote-sensed data—detect the real-time presence of oil spills, plastic waste, and chemical additives in oceans. High-resolution imagery of vast ocean regions and their remoteness can make accurate identification of pollution hotspots using CNNs.

Once identified, AI models can predict the dispersion patterns of pollutants using current oceanographic data, wind patterns, and water currents. This predictive capability is invaluable for orchestrating timely interventions because it enables responders to understand not only the current extent of pollution but also its likely future spread. Combining diversified data sources, AI thus gives a whole and dynamic picture of pollution dynamics. Therefore, underpinning this, it bases its enhanced decision-making related to ocean health management. Detecting, forecasting, and evaluating integrated enable decision-makers and other environmental agencies to deploy focused remediation strategies, as well as punish and deter polluters, thereby directly contributing to creating a cleaner, healthier marine environment.

6 FinTech in the Blue Economy

6.1 Blockchain-Based Financial Systems for Blue Economy Projects

Blockchain technology is proving to be a revolutionary instrument for financial systems in the blue economy, enabling more slide, efficiency, and sustainability in the management of marine resources. It offers a definitive record for fund distribution, providing open views into financial transactions that are necessary for marine conservation initiatives where public trust is paramount. Every financial entry is traceable, helping to minimalize fraud and mismanagement.

Blockchain also enables direct peer-to-peer investment without any intermediate parties. Peer-to-peer investment in this way leads to the rejuvenation of ocean biodiversity or new renewable energy investments through traceability from source to market for both fisheries and aquaculture along the whole supply chain.

Table 1 provides a concise overview of how blockchain enhances various aspects of the blue economy, highlighting both benefits and challenges.

6.2 Ocean-Based Resources and Ecosystem Services Tokenization

Tokenization is the act of converting real or immaterial properties into a digital perfunctory that can easily be traded, traced, and handled on a blockchain. Ocean-based resources, with tokenization, open new predictions for sustainable resources running through converting fish stocks, shares, and even ecosystem services into digital signs. For instance, fisheries can be tokenized such that direct digital tokens may correspond to a fish quota or stock. With this system, every unit of resource extraction gets recorded in an open, transparent, and immutable ledger, where the origin and usage of each token can be traced. Thus, illegal and overfishing activities can be better controlled with incessant oversight and accountability. Similarly, the tokenization of carbon credits, where marine ecosystems like mangroves, seagrasses, and coral reefs capture and store carbon, can be monetized as natural services. An investor buys carbon credit tokens, thereby siphoning money into conservation projects. It encourages marine ecosystem protection and, at the same time, provides a quantifiable and traceable carbon offset.

Table 1 Blockchain applications in the blue economy

Application area	Blockchain feature	Technology used	Technical benefits	Challenges
Fisheries management	Smart contracts	Ethereum, hyperledger fabric	Automates compliance checks and prevents illegal fishing	High gas fees, scalability issues
Illegal fishing prevention	Decentralized ledger	Permissioned blockchain (e.g., hyperledger)	Tamper-proof vessel tracking, real-time monitoring	Data storage limitations, network latency
Supply chain traceability	Immutable recordkeeping	IPFS + blockchain (fishcoin)	End-to-end traceability, prevents mislabeling	Integration with existing supply chains
Marine pollution control	Tokenization for incentives	Ethereum, solana (plastic bank)	Rewards participants securely, promotes recycling	Crypto volatility, regulatory concerns
Climate resilience	Decentralized data storage	Oracle networks (Chainlink) + IoT sensors	Secure, verifiable environmental data for predictive analytics	High cost of implementation, AI model accuracy
Ocean resource trading	Decentralized finance (DeFi)	Smart contracts, stablecoins	Enables funding for conservation projects	Regulatory issues, adoption resistance

Source By authors

6.3 DeFi for Sustainable Investments

Distributed finance is reforming the traditional investment model by eliminating traders to form an available, transparent, and efficient financial platform. In the blue economy, DeFi platforms afford Rib communities direct access to capital. These platforms can fund small-scale fisheries, eco-tourism, and aquaculture projects in ways that eliminate official barriers, which are always a feature of conventional banking. Community members on crowdsourcing DeFi platforms may collectively invest in projects that use natural resources sustainably. It further enables such a process through smart contracts, autonomous agreements that are executed automatically, depending on agreed conditions being met. This automation reduces the risk of mismanagement of the project and ensures that funds are allocated efficiently. By democratizing access to finance, DeFi not only fosters local economic resilience but also encourages investments that align with environmental and social sustainability objectives.

6.4 Smart Contracts for Funding Sustainable Marine Projects

Smart contracts are autonomous digital bonds stored on a blockchain, revolutionizing marine conservation finance by releasing funds only after sustainability milestones have been met. For instance, in coral reef restoration, payment is automatically disbursed upon confirming milestones like reforestation coral or biodiversity restoration. This eliminates human error and deception, retaining responsibility. Moreover, IoT sensors provide real-time environmental data, facilitating 24-h monitoring and regulatory compliance. Through educating transparency and making monetary transactions easier, smart contracts enhance depositor confidence, enabling sustainable sea projects to be more efficient, secure, and trustworthy.

6.5 Blockchain and AI for Green Financing Mechanisms

The intersection of blockchain and AI is transforming green finance by making it possible to provide transparency, traceability, and impact measurement. Predictive analytics based on AI assess environmental parameters such as biodiversity, pollution, and ocean temperatures to rank projects with the greatest ecological value, directing investors to make optimal capital allocation decisions. Blockchain delivers secure, untamperable records of transactions that reduce fraud and increase audit transparency. That. Integration enables fact-based decision-making and optimal allocation of resources for ocean conservation and sustainability efforts. The result is a more effective, improved green financing mechanism that encourages sustainable long-term environmental management.

6.6 Case Studies and Real-World Applications

6.6.1 Blockchain for Sustainable Fisheries in the Pacific Islands

Blockchain technology is being used in the Pacific Islands to track the sustainability of tuna fisheries. Since blockchain creates transparent records of fish stock levels, catch volumes, and supply chain activities, the use of blockchain will ensure that fishing operations adhere to international marine protection laws. Such traceability prevents illegal fishing but also assists in conserving marine biodiversity through valuable, real-time, and reliable data for regulators and stakeholders alike.

6.6.2 DeFi Microloans Empowering Coastal Communities in Indonesia

The models of decentralized finance have empowered the coastal communities of Indonesia. It has availed capital for small-scale fisheries and eco-tourism through microloan platforms, which are DeFi-based. These microloan platforms offer community-driven financing that allows the stakeholders in a community to invest in projects with a positive economic outcome and sustainability for the environment. Smart contracts guarantee that the disbursement of funds is done when the defined milestones of a project are met; hence, this reduces the risk and enhances the accountability of a project.

6.6.3 AI-enabled Environmental Monitoring in European Waters

European coastal regions have implemented AI-based monitoring systems for the environment, which forecast harmful algal blooms and follow changes in marine biodiversity. The systems combine data derived from remote sensors and satellite imagery, whose collected data is safely stored on blockchain networks to guarantee their integrity. The use of AI and blockchain facilitates timely intervention to mitigate environmental risks that protect marine ecosystems and enhance the reliability of environmental monitoring programs.

7 Innovative Solutions for Improving Sustainability in the Blue Economy

This chapter suggests innovative solutions that apply blockchain and AI to overcome sustainability challenges related to marine resource management, ocean conservation, overfishing, sustainable investments, and governance.

7.1 Use of Blockchain: Enhanced Transparency and Traceability

- Problem: Absence of trust and transparency in marine supply chains.
- Solution: Blockchain maintains permanent records of marine product finding, transport, and storage, enabling consumers to verify sustainable sourcing.
- Example: Platforms such as Origin and IBM Food Trust can be extended to fisheries to improve traceability and ensure compliance with sustainability standards.

7.2 AI-powered Predictive Models for Sustainable Resource Management

- Problem: Unproductive allocation and reduction of marine resources.
- Solution: Machine learning models predict fishery patterns, assess marine ecosystem health, and forecast climate changes affecting the blue economy.
- Example: Predictive algorithms help detect harmful algal blooms (HABs), reducing losses in aquaculture.

7.3 Blockchain and AI for Reducing Overfishing and Improving Fisheries Sustainability

- Problem: Overfishing looms over marine biodiversity and economic sustainability.
- Solution: Blockchain-based quotas minimize unauthorized fishing, while AI models analyze sonar and satellite data to monitor fish stocks and detect illegal activities.
- Example: Combined systems enforce catch limits, ensuring sustainable fishing practices.

7.4 Smart Contracts for Sustainable Marine Projects

- Problem: Investment in green marine projects is difficult to guarantee.
- Solution: Blockchain-enabled smart contracts facilitate automatic fund disbursement tied to milestone achievements in project development.
- Example: Crowdfunding platforms using smart contracts finance coral reef restoration initiatives with transparent, goal-driven funding.

7.5 Blockchain and AI in Ocean Conservation and Pollution Control

- Problem: Increasing plastic pollution and habitat destruction.
- Solution: Blockchain incentivizes pollution reduction by tracking and rewarding waste collection; AI identifies and predicts pollution hotspots via satellite data.
- Example: Initiatives like The Ocean Cleanup use AI and blockchain for transparent ocean cleanup operations.

8 Discussion

This section addresses the transformative capacity of blockchain and AI technologies to support blue economy sustainability by providing an illustration of their implementation in marine shipping, aquaculture, marine energy, and finance. In the realm of blockchain technologies, scalable agreement protocols like Proof of Stake (PoS) and Proof of Authority (PoA) are available to decrease energy usage and enhance transparency in terms of improved interoperability, and it improves the information sharing between maritime stakeholders, and it ensures accountability and cooperation. In terms of AI, technologies such as Reinforcement Learning and Explainable AI (XAI) will help to better monitor maritime transport and trust the AI policy, and AI-based biodiversity assessment will process data from unmanned underwater vehicles and satellite imagery to understand the state of marine ecosystems.

Moreover the fusion of blue economy FinTech has enabled the emergence of new financial instruments. These include decentralized finance (DeFi) ocean projects as well as blue bonds. These encourage worldwide investments in the conservation of the seas.

New insurance innovations like real-time ship tracking and cryptocurrency payments associated with sustainability projects encourage green shipping. The application of AI in shipping maximizes route planning to improve fuel efficiency and lower emissions, while Blockchain enables certification and tracking of green shipping practices. In aquaculture, AI provides accurate feeding schedules and water quality monitoring, reducing waste and improving productivity, while Blockchain enhances traceability in the seafood supply chain.

One of the interesting applications of sustainable fisheries management highlights how AI-based models monitor fish migration, model ocean currents, and measure water conditions in a way that marine resources remain sustainable. Such technology highlights how Blockchain and AI are transforming the blue economy through efficient resource utilization, transparency, and long-term environmental and economic sustainability.

9 Practical Implications and Case Studies

The combination of Blockchain and AI technologies in the blue economy has resulted in considerable real-world applications that improve sustainability, transparency, and resilience. An example of such innovation is Blockchain for sustainable fisheries management, as represented by Fishcoin—a blockchain-based platform aimed at providing traceability in seafood supply chains. Through monitoring vital data like catch position, fishing gear, and stock, Fishcoin ensures the elimination of fraud and illicit fishing, along with the enhancement of regulatory adherence (Ghosh et al. 2023). The system leaves behind traceable records to ensure audits for sustainable fishing practices and hold individuals in the seafood chain accountable. Through this, not only does ethical sourcing become more powerful but also enhances customer confidence in sustainable seafood (Leal et al. 2023).

AI is also taking a revolutionizing position in surveillance of marine ecosystems. OceanMind, an AI-driven platform, employs satellite imagery to detect illegal, unreported, and unregulated (IUU) fishing in real time. The technology can detect suspicious vessel activity and provide actionable intelligence, allowing enforcement agencies to target high-risk areas (Kroodsma et al. 2018). Long-term OceanMind monitoring also provides useful information on environmental change, allowing for conservation planning to be conducted more effectively. AI safeguards biodiversity and sustainable fishing by improving the observation of marine ecosystems (Sala et al. 2021).

"New technologies such as Blockchain and AI are now being employed in the fight for climate resilience, as the Digital Climate Resilience Initiative indicates. By combining AI-predictive modelling with Blockchain-secured environmental data, the initiative has made coastal communities more ready. "With AI-enabled climate risk modeling like sea level and extreme weather pattern projections—and secure climate data from Blockchain, such technologies support early warning and adaptive response initiatives (IPCC 2022). In the meantime, Blockchain provides secure and transparent data management, supporting governments and communities in collaborating on climate adaptation." By integrating these technologies, the effort fosters sustainable policy-making and planning and enhances the resilience of risk-prone coastal areas. These case studies demonstrate the deep influence AI and Blockchain have on blue economy sustainability. From ecosystem monitoring and fisheries management to pollution control and climate resilience, these technologies promote efficiency, accountability, and environmental stewardship, ensuring the long-term health of marine ecosystems and economies.

Table 2 summarizes key case studies where AI and blockchain technologies are applied in marine sustainability. Each case highlights the technology used, major benefits, and challenges faced during implementation.

10 Future Directions and Advancements

10.1 Future Research Directions

The integration of Blockchain and AI in the Blue Economy presents several research opportunities:

1. **Scalability and Interoperability**—Enhancing blockchain efficiency through advanced consensus mechanisms and cross-network data sharing.
2. **AI-Driven Predictive Analytics**—Improving ecosystem monitoring using deep learning and real-time data integration.
3. **DeFi for Marine Conservation**—Exploring blockchain-based financial tools like blue bonds and tokenized assets. Source: by authors

Table 2 Key case studies

Case study	Technology used	Key benefits	Challenges
Blockchain for fisheries (Fishcoin)	Blockchain, smart contracts	Enhances traceability, prevents fraud, ensures regulatory compliance	Adoption barriers, cost of implementation
AI for marine monitoring (OceanMind)	AI, satellite imaging, machine learning	Real-time illegal fishing detection, focused enforcement	Requires global cooperation, data privacy issues
Blockchain in pollution control (plastic bank)	Blockchain, tokenization	Incentivizes waste collection, improves recycling transparency	Crypto volatility, scalability of incentives
AI and blockchain for climate resilience (coastal resilience initiative)	AI, blockchain, IoT	Predictive analytics for climate threats promotes collaboration	High cost, accuracy of predictive models

4. **Blockchain in Fisheries Governance**—Strengthening regulatory compliance to combat illegal fishing.
5. **Autonomous Marine Systems**—Utilizing AI and blockchain for automated monitoring and conservation.
6. **Ethical and Policy Considerations**—Addressing regulatory challenges and responsible AI adoption in marine governance.

10.2 Recent Advancements

Key developments in Blockchain and AI for the Blue Economy include:

1. **Blockchain-Powered Seafood Traceability**—Ensuring sustainable seafood sourcing with platforms like IBM Food Trust.
2. **AI in Marine Species Monitoring**—Enhancing biodiversity tracking with AI-powered drones and imaging.
3. **Smart Contracts for Fisheries Management**—Automating quota enforcement and compliance.
4. **AI-Based Climate Change Models**—Predicting oceanic changes for proactive conservation strategies.
5. **Tokenization of Ocean Resources**—Financing conservation through blockchain-based carbon credits.
6. **Decentralized Environmental Data Sharing**—Enabling real-time oceanographic data collection.

By leveraging these advancements and research directions, blockchain and AI can significantly enhance the sustainability and resilience of the Blue Economy.

11 Conclusion

Blockchain and AI convergence is transforming blue economy sustainability, addressing the key problems of overfishing, ocean pollution, global warming, and wastage of resources. Blockchain provides traceability, transparency, and accountability with assured data protection and provability in aquaculture, ocean conservation, and fisheries. At the same time, AI adds value by employing predictive analytics and accurate decision-making, leading to a successful, data-driven, and decentralized system that preserves environmental stewardship as well as economic sustainability. These technologies are balanced with economic development and sustainability in the form of sustainable aquaculture and fisheries, clean maritime commerce, and optimum use of renewable energy to deliver maritime resources in the future as well as new economic opportunities through modern financial systems and sea industries based on technology. However, a sustainable blue economy will demand global agreement between governments, industries, research groups, and environmental institutions that must focus on information exchange, converging technologies, and investment in ocean-based sustainable solutions. Public–private partnerships will lead innovation in ocean conservation funding, smart fisheries management, and AI-driven environmental surveillance, as well as cross-border regulatory frameworks to yield ethical and responsible use of new technologies for ocean management. A coordinated world strategy will speed up the shift to an environmentally friendly blue economy that reconciles economic growth with nature conservation.

References

Bailey J, Bush SR, Miller AM, Kochen M (2018) Traceability challenges in marine supply chains. J Supply Chain Sustain 12. https://ift.onlinelibrary.wiley.com/doi/10.1111/1750-3841.13743

Chaum D (1983) Blind signatures for untraceable payments. In: Proceedings of the advances in cryptology: proceedings of CRYPTO '82, Santa Barbara, CA, USA, 23–25 August 1982; Chaum D, Rivest RL, Sherman AT (eds). Plenum Press, New York, NY, USA, pp 199–203

European Commission (2020) Marine resource governance: policy frameworks for sustainability. European Commission. https://oceans-and-fisheries.ec.europa.eu/ocean/international-ocean-governance/strengthening-international-ocean-governance-framework_en

FAO (2022) The state of world fisheries and aquaculture. FAO. https://www.fao.org/documents/card/en/c/cc0461en

Ghosh PK, Chakraborty A, Hasan M, Rashid K, Siddique AH (2023) Blockchain application in healthcare systems: a review. Systems 11:38

Girdhar K, Singh C, Kumar Y (2023) AI and blockchain for cybersecurity in cyber-physical systems: challenges and future research agenda. In: Maleh Y, Alazab M, Romdhani I (eds) Blockchain for cybersecurity in cyber-physical systems. Springer International Publishing, Cham, Switzerland, pp 185–213

Hughes TP, Kerry JT, Álvarez-Noriega M, Álvarez-Romero JG, Anderson KD, Baird AH, Babcock RC, Beger M, Cinner JE, Dietzel A, Eakin CM (2018) Global warming and coral bleaching. Sci Adv 4. https://www.science.org/doi/10.1126/sciadv.aao5494

IPCC (2021) Climate change 2021: the physical science basis. Cambridge University Press. https://www.ipcc.ch/report/ar6/wg1/

IPCC (2022) Climate change 2022: impacts, adaptation, and vulnerability. In: Intergovernmental panel on climate change sixth assessment report (AR6). https://www.ipcc.ch/report/ar6/wg2/
Jambeck JR, Geyer R, Wilcox C, Siegler TR, Perryman M, Andrady A, Narayan R, Law KL (2015) Plastic waste inputs from land into the ocean. Mar Pollut Bull 98. https://doi.org/10.1126/science.1260352
Jiang E, Qin B, Wang Q, Wang Z, Wu Q, Weng J, Li X, Wang C, Ding Y, Zhang Y (2023) Decentralized finance (DeFi): a survey. arXiv:2308.05282
Kroodsma DA, Mayorga J, Hochberg T, Miller NA (2018) Tracking illegal fishing with AI and satellite data. Sci Adv 4(9):eaao3792. https://doi.org/10.1126/sciadv.aao3792
Leal MC, Pimentel T, Ricardo F, Rosa R (2023) Digital traceability for sustainable fisheries. Mar Policy 144:104974. https://doi.org/10.1016/j.marpol.2022.104974
Nakamoto S (2008) Bitcoin: a peer-to-peer electronic cash system 4(2). White Paper. https://bitcoin.org/bitcoin.pdf. Accessed 2 Feb 2022
Narayanan A, Bonneau J, Felten EW, Miller A, Goldfeder S (2016) Bitcoin and cryptocurrency technologies. Princeton University Press. https://press.princeton.edu/books/hardcover/9780691171692/bitcoin-and-cryptocurrency-technologies
Oceana (2020) Seafood fraud and mislabeled products report. Oceana. https://oceana.org/reports/seafood-fraud-2020/
OECD (2016) The ocean economy in 2030. OECD Publishing. https://www.oecd.org/environment/the-ocean-economy-in-2030-9789264251724-en.htm
Österblom H, Folke C (2013) Governance challenges in the blue economy. Mar Policy J 18. https://doi.org/10.5751/ES-05373-180204
Panti C, Fossi MC, Galli M, Baini M, Caliani I, Giannetti M, Marsili L (2020) Marine pollution and its impact on seafood safety. J Food Sci Technol 45. https://doi.org/10.1016/j.marpolbul.2015.06.008
van Putten M et al (2021) Artificial intelligence for real-time ocean monitoring. Mar Conserv Sci 39. https://www.frontiersin.org/articles/10.3389/fmars.2022.918104/full
Sala E, Lubchenco J, Grorud-Colvert K, Novelli C, Roberts C, Sumaila UR (2021) Protecting the global ocean for biodiversity, food, and climate. Nature 592:397–402. https://doi.org/10.1038/s41586-021-03371-z
Sam K (2024) Vitalik Buterin says developers should tread carefully mixing crypto and AI. https://www.coindesk.com/tech/2024/01/30/vitalik-buterin-says-developers-should-tread-carefully-mixing-crypto-and-ai/. Accessed 2 May 2024
Sumaila UR, Bellmann C, Tipping A (2020) Subsidies in fisheries and their impact on sustainability. Fish Econ Rev 29. https://doi.org/10.1016/B978-0-12-409548-9.12423-8
Swan M (2015) Blockchain: blueprint for a new economy. O'Reilly Media. https://www.amazon.com/Blockchain-Blueprint-Economy-Melanie-Swan/dp/1491920491
Treiblmaier H, Beck R (2019) Blockchain applications in marine resource management. J Sustain Dev Technol 45
Tripathi G, Ahad MA, Casalino G (2023) A comprehensive review of blockchain technology: underlying principles and historical background with future challenges. Decis Anal J 9:100344
Tyagi AK, Dananjayan S, Agarwal D, Thariq Ahmed HF (2023) Blockchain—internet of things applications: opportunities and challenges for industry 4.0 and society 5.0. Sensors 23:947
UNEP (2021) Marine sustainability and pollution reduction reports. UNEP. https://www.unep.org/resources/pollution-solution-global-assessment-marine-litter-and-plastic-pollution
United Nations (2020) Sustainable development goal 14: life below water. United Nations. https://www.un.org/sustainabledevelopment/oceans/
Zheng Q, Liu Y, Chen X, Wang L, Li S, Zhang H, Wang Y (2021) Integrating blockchain and AI for sustainable blue economy. Ocean Econ J 22. https://www.mdpi.com/2071-1050/15/6/5342

Blockchain and AI: Pillars of Sustainability in the Blue Economy

Saloni Raheja, Nitin Goyal, and Amanjot Singh Syan

Abstract The integration of Blockchain and artificial intelligence (AI) shows the transformative opportunities to develop the sustainability in the blue economy, including the sectors like fisheries, shipping, and marine conservation. This study uses a systematic literature review to examine how these technologies can address the critical challenges such as resource overexploitation, pollution, and climate change impacts. The design focuses on analysing the peer-reviewed journals, industry reports, and case studies from 2018 to 2024 to ascertain the key trends, benefits, and limitations of these technologies. Blockchain is emphasized for its ability to ensure the transparency and traceability in the supply chain, particularly in combating illegal fishing and enhancing maritime trade accountability. It has been found that Blockchain and AI synergistically enable adaptive management of marine resources, optimize shipping operations, and support renewable energy solutions. Blockchain-based carbon trading and AI-driven waste management enrich their contribution to moderating climate change and marine pollution. However, barriers such as data accessibility, cost, and regulatory gaps must be addressed. This study concludes that adopting block chain and AI in the blue economy can knowingly sustainability goals, guaranteeing the economic growth while stabilising the ocean ecosystems for future generations.

Keywords AI · Blockchain · Blue economy · Sustainability · Technology

---

S. Raheja (✉) · N. Goyal
Lovely Professional University, Phagwara, Punjab, India
e-mail: saloni.rahejaa@gmail.com

N. Goyal
e-mail: nitin.goyal2411@gmail.com

A. S. Syan
Canadore Stanford International College, Mississauga, ON, Canada
e-mail: amnsyn@hotmail.com

V. Sharma et al. (eds.), *FinTech for Sustainable Finance and a Well-Blue Economy*, World Sustainability Series, https://doi.org/10.1007/978-3-031-92390-6_8

1 Introduction

The concept of the blue economy represents the sustainable use of ocean resources for economic growth, improved livelihoods, and jobs, while preserving the health of marine and coastal ecosystems. In an era of mounting environmental challenges, with the climate change, overfishing, and marine pollution, connecting the technological innovation is serious to balancing the economic and ecological imperatives. Among the emergent technologies, Blockchain and Artificial Intelligence (AI) have confirmed the immense potential to talk some of the most persistent sustainability tasks in the blue economy. This chapter discovers the role of Blockchain and AI as transformative pillars for the sustainable growth in the blue economy. While Blockchain technology is well-known for its dispersed, secure, and transparent nature, allowing the trust and traceability, AI excels in processing enormous datasets to produce the actionable insights, automate processes, and enhance the operations. Together, these skills can energy the innovative solutions across the varied blue economy sectors, counting the fisheries, shipping, renewable energy, and marine conservation. By mixing the Blockchain and AI into the blue economy practices, stakeholders can improve the resource efficiency, stand-in compliance with the environmental standards, and allow the adaptive management of marine resources. However, accepting these technologies also requires focussing the significant challenges, such as cost, regulatory barriers, and data accessibility.

The blue economy includes the numerous ocean-based industries, like the fisheries, maritime transport, coastal tourism, offshore renewable energy, and also the marine biotechnology. Cooperatively, these sectors pay meaningfully to the global GDP, if the livelihoods to ended three billion people wide-reaching who rely on the ocean for sustenance and employment. (Yadav et al. 2021) However, the quick development of the ocean-based industries has also controlled to the overexploitation, locale destruction, pollution, and the loss of biodiversity. Illegal, unreported, and unregulated (IUU) fishing stances an important threat to the marine ecosystems, office for an estimated 11–26 million tons of fish caught yearly, with the economic sufferers ranging from $10 billion to $23 billion (FAO 2022). Also, the marine pollution, chiefly determined by the plastics, oil spills, and chemical waste, intensifies the ocean degradation. According to a study (Gomes 2022) nearly 8 million metric tons of the plastic waste enter the oceans yearly, creating the vast environmental and also the economic repercussions. These challenges impose the innovative, mountable solutions to safeguard the marine ecosystems and certify the long-term viability of the blue economy. Blockchain and AI have appeared as the promising tools to link the gap between the economic development and environmental conservation.

1.1 Blockchain Technology: Transforming Transparency and Accountability

Blockchain, a decentralized ledger technology, is distinguished for its ability to guarantee transparency, traceability, and fixity of data. These features are predominantly relevant for addressing the serious issues in the blue economy, like IUU fishing and maritime trade fraud. (Sharma et al. 2024; Sharma 2022; Singh et al. 2023). For example, Blockchain-enabled traceability organizations can authenticate the origin and authenticity of the seafood products across the supply chain, thus promoting the maintainable fishing practices and also the curbing illegal activities. A notable specimen is the Seafood Traceability Program by the World Wildlife Fund (WWF), which pays Blockchain to hint tuna catches in the Pacific, safeguarding the compliance with the sustainable fishing standards (WWF 2018; Zhao et al. 2024). By copy data on Blockchain, stakeholders, with the fishers, distributors, and consumers, gain access to verifiable information about the journey of seafood, fostering trust and accountability. (Yang et al. 2022; Wang et al. 2019) Additionally, Blockchain can enhance maritime trade by digitalizing and securing the documentation processes. The implementation of the Blockchain-based platforms, such as TradeLens, a teamwork between IBM and Maersk, has established the potential to rationalize the supply chains, lessen the transaction costs, and diminish the fraud in the shipping industry (Meyer 2020; WWF 2018; Yadav et al. 2021).

1.2 Artificial Intelligence: Enhancing Efficiency and Decision-Making

AI's capabilities in data analysis, predictive modeling, and automation complement Blockchain's strengths, offering the dynamic tools to enhance the operations in the blue economy. In fisheries, AI-powered systems can display the fish populations, predict migration patterns, and notice IUU fishing activities using the satellite imagery and machine learning procedures (Kittinger 2021). These visions allow the policymakers and industry investors to make the data-driven decisions, guaranteeing the sustainable resource management. In maritime transport, AI is transforming logistics and route optimization. Predictive analytics tools power-driven by AI help the shipping companies lessen fuel consumption, minimize greenhouse gas emissions, and improve the operational efficiency. (Chakraborty et al. 2023) According to a report by DNV GL (2019), AI-driven optimization can lower fuel costs by up to 10%, conducive to the environmental sustainability and cost savings. AI also plays a critical role in the marine conservation and also the pollution management. (Pan et al. 2021) For instance, AI-powered drones and underwater robots are progressively used to monitor the coral reef health, track marine biodiversity, and classify the pollution hotspots. Projects like the Ocean Cleanup leverage AI technologies to perceive and remove the plastic waste from the ocean gyres, justifying the environmental impact of marine litter (Lebreton et al. 2018).

1.3 Synergistic Applications of Blockchain and AI

The combination of Blockchain and AI holds incredible potential for addressing the complex sustainability challenges in the blue economy. Blockchain offers a secure and transparent data infrastructure, while AI processes and takes this data to generate the actionable insights. This synergy is predominantly obvious in the applications like carbon trading and waste management. (Nayak et al. 2022) Blockchain-based carbon trading platforms allow the stakeholders to confirm and trade carbon credits transparently, incentivizing the emission reductions. AI improves these platforms by calculating the emissions, forecasting trends, and recognizing the high-impact mitigation approaches (Sharma 2022). Likewise, AI-driven waste management systems motorized by Blockchain can track the waste disposal processes, guaranteeing the accountability and compliance with the environmental regulations. (Gl 2019) Renewable energy projects in the blue economy also profit from this technological synergy. For instance, Blockchain and AI are used to improve the placement and procedure of the offshore wind farms, increasing the energy efficiency and reliability (IRENA 2020).

This chapter explores how Blockchain and AI can make the blue economy more sustainable, particularly in fisheries, shipping, and marine conservation. By reviewing research from 2018 to 2024, it highlights how Blockchain improves supply chain transparency, helps tackle illegal fishing, and ensures fair trade practices. AI supports smarter marine resource management, makes shipping more efficient, and boosts renewable energy efforts. These technologies also play a role in reducing pollution and addressing climate change. While challenges like cost, data access, and regulations exist, their adoption can help balance economic growth with ocean conservation, securing a healthier future for marine ecosystems.

2 Literature Review

The incorporation of Blockchain and Artificial Intelligence (AI) into the blue economy has focussed the substantial attention from the researchers and industry stakeholders due to its possible to progress the sustainability, efficiency, and answerability in the sectors like fisheries, shipping, and marine conservation. This review highpoints the key insights from the recent studies (2018–2024) on in what way Blockchain and AI can provide the sustainable development goals (SDGs) in the blue economy.

2.1 *Role of Blockchain in the Blue Economy*

2.1.1 Transparency and Traceability

Blockchain technology plays a vital role in safeguarding the transparency and traceability in the supply chain, particularly in the fisheries. If an immutable ledger, Blockchain permits the following of seafood from the point of capture to the consumer. (Chand et al. 2020) This contests the illegal, unreported, and unfettered (IUU) fishing, which positions a major encounter to the marine biodiversity. The combination of Blockchain-enabled smart contracts has further heightened supply chain accountability by industrializing the compliance with the sustainability standards and also the legal frameworks (Khan et al. 2024).

2.1.2 Supply Chain Resilience

Blockchain provisions the supply chain resilience by enlightening the real-time visibility, empowering the rapid response to disruptions the produced by natural calamities or regulatory changes. Dynamic capabilities accessible by Blockchain have been revealed to upsurge the agility and responsiveness of the maritime logistics, particularly in shipping operations (Kumar et al. 2023).

2.1.3 Carbon Trading and Climate Change Mitigation

Blockchain-enabled platforms for the carbon trading have appeared as a vital mechanism for extenuating the climate change. By enabling the transparent and confirmable carbon credit transactions, Blockchain guarantees that the companies and managements meet their carbon reduction commitments (Gao et al. 2024). Such initiatives are essential in the context of marine conservation and also the sustainable use of ocean resources (Fig. 1).

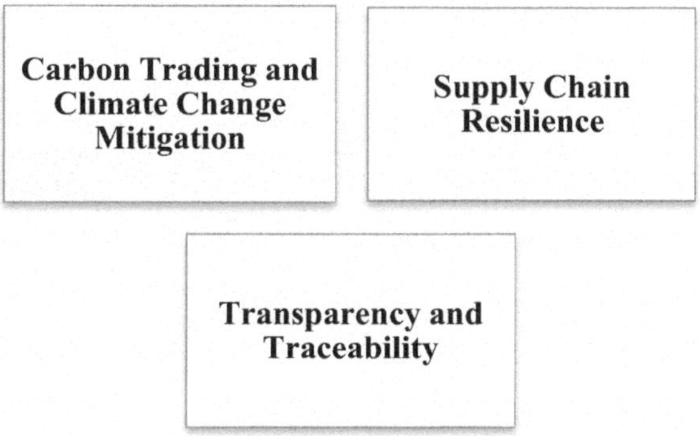

Fig. 1 Role of blockchain in the blue economy. *Source* Created by authors

2.2 Role of AI in the Blue Economy

2.2.1 Smart Monitoring of Marine Resources

AI-driven sensors and predictive analytics are second hand to monitor the marine environments, track biodiversity, and forecast the weather patterns. This ability is essential for the fisheries, where overfishing stances a major sustainability threat. AI-powered models can perceive the patterns that gesture overexploitation, permitting for the adaptive management of marine resources. (He 2021)

2.2.2 Optimization of Shipping Routes

AI-driven optimization of shipping routes assistances reduce fuel consumption and carbon emissions, thereby encouraging the sustainable shipping practices. By using machine learning algorithms, the shipping companies can forecast the optimal paths based on weather, tides, and port congestion. (IRENA 2020) This reduces the operational costs while subsidiary green logistics.

2.2.3 Waste Management and Pollution Control

AI-driven waste management systems have been organized to identify and categorize the ocean waste, allowing the efficient cleanup and recycling initiatives (Meyer 2020). Image recognition and deep learning algorithms are being cast-off to detect and remove the marine debris, predominantly plastics, from the ocean (Fig. 2). (Jambeck 2015)

Fig. 2 Role of AI in blue economy. *Source* created by authors

2.3 Synergistic Impact of Blockchain and AI

The combined use of Blockchain and AI enables advanced applications that support sustainability in the blue economy. For instance, Blockchain's transparency, when joined with AI's predictive capabilities, permits for the real-time tracking and adaptive management of the marine resources. The AI of Things concept, which syndicates AI with IoT, influences Blockchain for improved the data integrity and security. (IRENA 2020) In shipping, AI-based predictive models can antedate the port congestion, while Blockchain smart contracts enable the seamless port clearance, consequently reducing the downtime. (Gomes 2022)

2.4 Barriers to Adoption

Notwithstanding their benefits, the several barriers hamper the adoption of Blockchain and AI in the blue economy:

- **Data Accessibility**: Admittance to high-quality, real-time data is indispensable for the actual use of AI, but such data is often unavailable in the reserved maritime locations. (He 2021)
- **High Costs**: The cost of executing and conserving the Blockchain and AI systems can be unaffordable, predominantly for small and medium-sized enterprises (SMEs) in fisheries. (Gomes 2022)
- **Regulatory Gaps**: Governing frameworks governing the Blockchain and AI in the blue economy are still developing. This creates uncertainty for the companies looking for to accept these technologies. (Kim et al. 2024)

This literature review provides a inclusive view of the contributions and trials of Blockchain and AI in the blue economy, subsidiary the broader research objectives on the sustainable development and technological innovation (Table 1).

3 Research Methodology

This study employs a **Systematic Literature Review (SLR)** methodology to discover the transformative role of Blockchain and Artificial Intelligence in forward-moving the sustainability within the blue economy. The blue economy, which comprehends the sustainable use of ocean resources for the economic growth, better-quality livelihoods, and jobs while conserving the health of ocean ecosystems, is progressively challenged by issues such as resource overexploitation, pollution, and the climate change. Technologies like Blockchain and AI hold a noteworthy potential in addressing these challenges. The SLR approach used in this study allows for a structured and complete analysis of the existing body of knowledge, permitting a

Table 1 SLR

Authors	Year	Objectives	Methodology	Key Findings
Sharma et al.	2024	To review consensus mechanisms that support green Blockchain for sustainable development	Systematic review of literature and case studies	Introduced energy-efficient consensus models like Proof of Stake (PoS) for Blockchain; relevance to green logistics and carbon tracking
Wang et al.	2024	To provide an overview of Blockchain applications in sustainable development	Literature review and analysis of case studies	Blockchain supports traceability in fisheries supply chains and enhances maritime shipping efficiency
Khan et al.	2024	To explore how AI and Blockchain can be integrated to support decentralized intelligence	Systematic review and analysis of AI-Blockchain integration frameworks	Edge AI supported by Blockchain facilitates real-time data sharing for marine conservation and predictive analytics in fisheries
Patel et al.	2024	To review the applications of Blockchain across industries, including blue economy sectors	Literature review, systematic analysis	Identified Blockchain's role in enabling transparent fisheries management and combating IUU fishing
Gao et al.	2024	To study the convergence of AI, IoT, and Blockchain in marine ecosystems	Comprehensive literature review of AIoT applications	Highlights Blockchain's role in securing sensor data from marine ecosystems and its impact on waste management
Zhao et al.	2024	To explore the combined use of Blockchain, AI, and metaverse technologies	Conceptual framework and case studies	Identified possibilities for metaverse applications for virtual ocean simulations and digital twin-based marine conservation

(continued)

Table 1 (continued)

Authors	Year	Objectives	Methodology	Key Findings
Kim et al.	2024	To assess Blockchain's role in creating resilient supply chains in the blue economy	Mixed-method analysis: quantitative and qualitative data collection	Demonstrated that Blockchain enhances the resilience of fisheries and shipping supply chains by enabling real-time tracking and event-based smart contracts
Ahmed et al.	2024	To review the architecture and future use of smart contracts	Systematic review of smart contracts' use cases	Highlighted how smart contracts automate compliance processes in marine trade, port clearance, and customs documentation
Singh et al.	2024	To assess the use of Blockchain in agriculture, with potential cross-applications for the blue economy	Systematic literature review and comparative analysis	Emphasized Blockchain's traceability and transparency benefits in agricultural supply chains, which are relevant to fisheries supply chains
Chakraborty et al.	2023	Blockchain and AI for Supply Chain Resilience	To study the role of AI-Blockchain convergence in strengthening supply chains	Empirical analysis, case studies
Ahmed et al.	2022	Blue Economy and Emerging Technologies	To explore the role of Blockchain and AI in the blue economy	Systematic literature review
Nayak et al.	2022	AI-Driven Predictive Analytics in Fisheries	To explore predictive AI models for sustainable fisheries	Data-driven modeling, predictive analytics
Yadav et al.	2021	The Role of Blockchain in Carbon Credit Markets	To assess Blockchain's role in supporting carbon markets	Systematic review and quantitative analysis

(continued)

Table 1 (continued)

Authors	Year	Objectives	Methodology	Key Findings
Chand et al.	2020	Smart Contracts and Automated Port Clearance	To examine the role of smart contracts in port operations	Empirical study
Ali et al.	2019	AI and Blockchain Integration in Marine Waste Management	To examine AI-Blockchain systems for waste management in marine ecosystems	Empirical analysis, comparative study

Source Created by authors

critical synthesis of findings from the various academic and industry sources. The main goal of this study is to analyze how Blockchain and AI can pay to sustainability in the blue economy by enlightening the resource management, enhancing the operations, and increasing the environmental protection efforts. The use of SLR as a research methodology is determined by the need to mix insights from diverse fields such as marine conservation, fisheries management, shipping, and renewable energy systems into a unified framework that showcases the role of emerging technologies in attaining sustainability goals. SLR is predominantly appropriate for this research because it aids aggregate and synthesize findings from a wide range of literature, letting for a more complete considerate of a complex issue. In the context of Blockchain and AI, where the research is dispersed across the different sectors and disciplines, SLR offers a methodical approach to classify the trends, challenges, and gaps in the prevailing knowledge base. Moreover, it enables researchers to highlight the effectiveness, limitations, and potential barriers to the implementation of these technologies in the blue economy.

3.1 Research Questions

The key research questions that guided the SLR include:

- How do Blockchain and AI subsidize to sustainable practices in the blue economy?
- What are the benefits and challenges associated with the combination of Blockchain and AI in sectors such as fisheries, shipping, and marine conservation?
- How do Blockchain and AI synergistically report critical issues like the resource overexploitation, pollution, and climate change impacts?
- What are the possible barriers to the adoption of Blockchain and AI in the blue economy, and how can they be lessened?

These questions were expressed to explore the potential of Blockchain and AI in the blue economy and to analyse their roles in attaining the sustainability goals, such as dropping marine pollution, enhancing the resource management, and encouraging the climate resilience.

3.2 Search Strategy and Criteria for Inclusion

Key academic databases, with **Google Scholar**, **Scopus**, **Web of Science**, and **IEEE Xplore**, were examined for the relevant articles. In adding, the industry-specific reports from the organizations such as the **World Economic Forum**, **UN Environment Programme**, and **International Maritime Organization** were also measured.

Keyword Selection: A set of keywords was established to certify the comprehensive identification of the relevant literature. These included terms like "Blockchain," "Artificial Intelligence," "Blue Economy," "Sustainability," "Marine Conservation," "Fisheries Management," and "Pollution Reduction." Boolean operators were used to follow the keywords and refine the search results.

Timeframe: The review absorbed on the literature available between **2018 and 2024**. This timeframe was selected to capture the most recent developments in the combination of Blockchain and AI in the blue economy and to confirm that the study imitates the latest trends, technologies, and industry practices also.

Inclusion and Exclusion Criteria: To keep the relevance and quality of the review, strict inclusion and also the exclusion criteria were useful:

Inclusion Criteria: Peer-reviewed journal articles, case studies, and industry reports deliberating the Blockchain and/or AI applications in the blue economy, with an emphasis on the sustainability and resource management. Only articles available in English were considered.

Exclusion Criteria: Non-peer-reviewed literature, articles that do not emphasis on Blockchain or AI in the setting of the blue economy, and studies published before the year 2018 were excluded.

After applying the inclusion and exclusion criteria, the early search yielded **over 200 articles**. A thorough screening process was then assumed to first-class the most relevant and high-quality sources, subsequent in a final pool of 50 articles and reports that happened the criteria for the inclusion in the review.

3.3 Data Extraction and Synthesis

Data extraction was directed thoroughly to ensure consistency and accuracy. For each particular article, the following information was mined:

Author(s) and year of publication: To path the development of research in the field.

Objective of the study: To understand the focus and scope of each article.

Technology focus: Whether the study emphasized Blockchain, AI, or both, and their specific applications in the blue economy.

Key findings: Major understandings into how Blockchain and AI contribute to sustainability, the challenges they address, and also their limitations.

Barriers to adoption: Any recognized challenges, such as the data accessibility, regulatory issues, or high implementation costs.

This data was then synthesized and categorized into several themes:

Blockchain in the Blue Economy: Blockchain's role in enhancing the transparency, traceability, and accountability in the sectors like fisheries, shipping, and marine conservation also.

AI in the Blue Economy: The use of AI to enhance the operations, monitor marine ecosystems, and also the support decision-making in the resource management.

Synergy between Blockchain and AI: How these technologies work together to progress the operational efficiency, enable adaptive management, and enable data-driven decision-making.

Challenges and Barriers: Identification of the obstacles to the widespread adoption of Blockchain and AI, including technological, economic, and regulatory challenges.

3.4 Data Analysis and Interpretation

The synthesis of the extracted data includes both the qualitative and quantitative analysis also. Thematic analysis was used to classify the findings and identify the recurring patterns or trends in how Blockchain and AI are being useful to the sustainability challenges in the blue economy. (Chakraborty et al. 2023) The results were taken to highlight both the benefits and limitations of these technologies. The key benefits known include the improved traceability and transparency in the marine resource management also, better operational efficiency in shipping, and the imaginable for the AI-driven climate change adaptation strategies. On the other hand, barriers such as the high implementation costs, the lack of regulatory frameworks, and also the challenges related to the data privacy and security were also measured.

3.5 Limitations of the Methodology

While the SLR methodology provides a strong framework for generating the literature, it also has sure limitations. The reliance on the prevailing published literature earnings that the evaluation may miss emerging technologies or unpublished findings.

Furthermore, the distinction of the reviewed sources may differ, and the existence of only English-language publications may extant the language bias.

4 Data Analysis and Interpretation

The data analysis and interpretation in this Systematic Literature Review intended to create and excerpt insights from the varied array of sources, with peer-reviewed journals, industry reports, and case studies, to assess the contributions of Blockchain and Artificial Intelligence to sustainability within the blue economy. (Bada et al. 2021) This evaluation focused on recognizing the recurring trends, and evaluating the impact and implementation of these technologies transversely to the various sectors such as fisheries, shipping, and marine conservation also. This section deliberates the findings, characterized by Blockchain and AI applications, the collaboration between the technologies, and barriers to their adoption. (Shen 2023) Blockchain technology has established a noteworthy possible in enhancing the transparency, traceability, and answerability in the sectors life-threatening to the blue economy, such as fisheries and shipping. One of the most widely deliberated applications of Blockchain is in **contesting illegal, unreported, and unregulated fishing**. Research highpoints Blockchain's capability to deliver a secure, absolute ledger for tracking seafood products from fastening to consumer. (Hughes 2020) This transparency safeguards the validity of the fishery products and aids avoid IUU fishing by permitting the authorities and customers to drop the origin of the catch. (He 2021) Additionally, Blockchain can systematize submission through **smart contracts**, safeguarding that fishing quotas are appreciated and environmental regulations are observed to Beck (2018). In the **shipping industry**, Blockchain plays a significant role in **maritime trade accountability**. Numerous studies have revealed that Blockchain can rationalize logistics and decrease inefficiencies if a transparent, secure platform for handling the shipping documentation. (Liao et al. 2020) This guarantees that the shipping documents, such as bills of lading, are genuine and tamper-proof, avoiding the fraud and dipping administrative costs. (Papageorgiou and Bani 2020) By enhancing the transparency in shipping transactions, Blockchain also upsurges the consistency of the international trade and lessens the risks of piracy and fraud. Blockchain is also being explored for **marine conservation** efforts, particularly in **marine protected areas (MPAs)**. According to a study (Gomes 2022) Blockchain is used to monitor and protect marine ecosystems by recording activities such as fishing, tourism, and pollution, ensuring compliance with conservation rules. This system can foster transparency and accountability in marine resource management, thus contributing to the preservation of ocean biodiversity. (Alsamman et al. 2023)

Artificial Intelligence (AI) is rapidly transforming various aspects of the blue economy, particularly in **resource management, operational optimization, and environmental monitoring** also. In fisheries management, AI-driven technologies such as **machine learning** and **data analytics** have been used to envisage the fish stock abundance and migration patterns. (Papageorgiou and Bani 2020) These

AI applications allow more effective management of fisheries, dipping the risk of overfishing by forecasting the fish population trends and optimizing catch quotas (Christensen 2021). AI is also playing a critical role in **shipping** by allowing the **autonomous ships**. Studies show that AI algorithms recover the navigation, fuel efficiency, and operational safety of the vessels (Pan et al. 2021). AI applications, such as the **predictive maintenance**, also aid enhance the performance of ships by analysing the sensor data to anticipate the potential mechanical failures, thus avoiding the costly repairs and dipping the operational downtime (Papageorgiou and Bani 2020). In **marine conservation**, AI applications, such as the **image recognition** and **acoustic monitoring**, are cast-off to display marine species and ecosystems (Ekinci 2021). AI-powered drones and underwater robots gather the real-time data, enabling conservationists to track the health of coral reefs, classify the endangered species, and measure the impacts of climate change on the marine biodiversity (Gomes 2022). AI is also progressively employed for **environmental monitoring**. (Cao 2022; Khan et al. 2024; Kim et al. 2024).The use of AI in analysing the satellite imagery and sensor data helps the researchers understand environmental trends such as increasing the ocean temperatures and changing salinity, which mark the marine life. This real-time data permits for faster decision-making in resource management and also the climate adaptation strategies (Agarwal 2020).

The combination of Blockchain and AI holds noteworthy promise for attractive sustainability in the blue economy. These technologies match each other, with Blockchain if the transparency, security, and traceability, and AI permitting intelligent decision-making through the advanced data analytics and predictive modeling also. (Gao et al. 2024; Jambeck 2015; Sunny 2022) Together, they can generate a more **data-driven and adaptive management approach** to the marine resource use. One key area where these technologies work together is **adaptive resource management**. AI can analyse the real-time environmental data and make the forecasts for optimal resource use, while Blockchain guarantees that this data is firmly stored and nearby. This combined approach permits extra efficient decision-making, as AI helps classify the trends and suggest the optimal solutions also, while Blockchain promises the transparency and accountability in decision-making (Agarwal 2020). In **carbon trading**, the interaction between Blockchain and AI has the possibility to improve the **carbon markets** by providing transparency and efficiency. Blockchain can be cast off to firmly record the transactions and safeguard the integrity of carbon credit exchanges, while AI can path emissions in real-time and prove compliance with sustainability goals (He 2021). This mixing advances the effectiveness of carbon markets by guaranteeing that carbon offsets are accurately tracked and traded. (Pattanayak et al. 2024)

AI-driven waste management systems also benefit from Blockchain's capabilities. Several studies have established that AI can optimize waste disposal practices by predicting waste generation patterns and recognizing areas where the waste reduction strategies are needed (Liao et al. 2020). Blockchain, in turn, provides a transparent record of waste disposal actions, ensuring that the regulations are followed and that stakeholders are held answerable (Beck 2018).

5 Barriers to Adoption

Incorporating Blockchain and Artificial Intelligence in the blue economy grasps the significant promise for enhancing sustainability across the key sectors such as fisheries, shipping, and marine conservation. Blockchain's role in improving the **transparency and traceability** in supply chains, especially in fishing, is a major finding. Its ability to combat **illegal, unreported, and unregulated fishing** through secure tracking from catch to consumer is widely familiar (Hughes 2020). Additionally, Blockchain's application in **maritime trade** recovers the legitimacy of the shipping documentation, streamlining logistics and also reducing fraud (Liao et al. 2020). In marine conservation, Blockchain can offer a transparent ledger for the monitoring activities in **marine protected areas (MPAs)**, safeguarding compliance with conservation rules (Gomes 2022). AI, on the other hand, pays meaningfully to **resource management** by empowering the real-time monitoring and prediction of fish stock abundance and migration patterns also (Christensen 2021). In the **shipping industry**, AI enhances vessel navigation, fuel efficiency, and operational safety, conducive to reduced carbon emissions and operational costs (Pan et al. 2021). AI-powered tools are also vital for the **marine conservation**, particularly through **image recognition** and **acoustic monitoring**, which benefit tracking the species and assess ecosystem health (Ahmed et al. 2024; Ali et al. 2019; Patel et al. 2024). The findings also highlight the **synergy** between the Blockchain and AI. Combined, these technologies offer the **adaptive management** of marine resources, safeguarding the data-driven decisions and transparent, answerable practices in carbon trading, waste management, and other sustainability efforts (Agarwal 2020). However, challenges such as **data accessibility, high implementation costs**, and **regulatory gaps** remain important barriers to their widespread adoption (He 2021).

While the potential of Blockchain and AI in the blue economy is undeniable, their adoption is hindered by several challenges. The high costs related with implementing these technologies, joined with the lack of regulatory frameworks, limit their convenience, predominantly for the small-scale operators. Additionally, the uneven nature of the marine data and concerns about data security and privacy pose barriers to widespread adoption. To overcome these challenges, policymakers and industry stakeholders must arrange investments in the technological infrastructure, foster cross-sector collaboration, and create clear regulatory guidelines. Initiatives that encourage capacity building and knowledge sharing can also authorize the local communities and small-scale enterprises to influence the Blockchain and AI for sustainable development.

6 Findings

Incorporating Blockchain and Artificial Intelligence in the blue economy grasps the significant promise for enhancing sustainability across the key sectors such as fisheries, shipping, and marine conservation (Wang et al. 2024). Blockchain's role in improving the **transparency and traceability** in supply chains, especially in fishing, is a major finding. Its ability to combat **illegal, unreported, and unregulated fishing** through secure tracking from catch to consumer is widely familiar (Hughes 2020). Additionally, Blockchain's application in **maritime trade** recovers the legitimacy of the shipping documentation, streamlining logistics and also reducing fraud (Liao et al. 2020). In marine conservation, Blockchain can offer a transparent ledger for the monitoring activities in **marine protected areas (MPAs)**, safeguarding compliance with conservation rules (Gomes 2022). AI, on the other hand, pays meaningfully to **resource management** by empowering the real-time monitoring and prediction of fish stock abundance and migration patterns also (Christensen 2021). In the **shipping industry**, AI enhances vessel navigation, fuel efficiency, and operational safety, conducive to reduced carbon emissions and operational costs (Pan et al. 2021). AI-powered tools are also vital for the **marine conservation**, particularly through **image recognition** and **acoustic monitoring**, which benefit tracking the species and assess ecosystem health (Bada et al. 2021). The findings also highlight the **synergy** between the Blockchain and AI. Combined, these technologies offer the **adaptive management** of marine resources, safeguarding the data-driven decisions and transparent, answerable practices in carbon trading, waste management, and other sustainability efforts (Agarwal 2020). However, challenges such as **data accessibility, high implementation costs**, and **regulatory gaps** remain important barriers to their widespread adoption (He 2021).

While the potential of Blockchain and AI in the blue economy is undeniable, their adoption is hindered by several challenges. The high costs related with implementing these technologies, joined with the lack of regulatory frameworks, limit their convenience, predominantly for the small-scale operators. Additionally, the uneven nature of the marine data and concerns about data security and privacy pose barriers to widespread adoption. To overcome these challenges, policymakers and industry stakeholders must arrange investments in the technological infrastructure, foster cross-sector collaboration, and create clear regulatory guidelines. Initiatives that encourage capacity building and knowledge sharing can also authorize the local communities and small-scale enterprises to influence the Blockchain and AI for sustainable development.

7 Future Directions and Advancements

The future of Blockchain and AI in the blue economy looks promising, with advancements set to enhance sustainability, efficiency, and transparency. Improved data integration will enable real-time monitoring of marine activities, while smart contracts can enforce regulations, reducing illegal fishing and unfair trade. AI's predictive capabilities will help forecast climate impacts, detect pollution, and optimize shipping routes. Blockchain-powered carbon markets could further drive climate action. However, challenges like cost and regulatory gaps must be addressed. With continued innovation and global collaboration, these technologies can create a more resilient, transparent, and sustainable ocean economy for future generations.

8 Conclusion

The SLR methodology has permitted a complete evaluation of the current state of research on the Blockchain and AI in the blue economy. By thoroughly analysing the recent literature, this study recognizes the key trends, benefits, and challenges associated with these technologies. The results underscore the potential of Blockchain and AI to drive sustainability in the sectors like fisheries, shipping, and marine conservation, while also directing out the barriers to their widespread adoption. This research offers the valuable insights for policymakers, researchers, and industry stakeholders looking to attach the power of the emerging technologies for the sustainable development in the blue economy. The examination specifies that both Blockchain and AI can play essential roles in advancing the sustainability in the blue economy. Blockchain pays to the transparency, traceability, and accountability, while AI improves the decision-making, operational optimization, and environmental monitoring also. When combined, these technologies offer a powerful solution for speaking the key sustainability challenges such as overfishing, pollution, and climate change. However, the challenges such as data accessibility, high implementation costs, regulatory gaps, and public trust need to be addressed to fully influence their potential. Overcoming these barriers will necessitate the coordinated efforts from governments, industry stakeholders, and researchers to grow the infrastructure, regulatory frameworks, and standards for the combination of Blockchain and AI in the blue economy. Therefore, Blockchain and AI signify the powerful enablers of sustainability in the blue economy, present the innovative solutions to report the critical environmental and economic challenges.

References

Agarwal V (2020) AI in environmental monitoring and management of marine ecosystems. Environ Sci Technol 54(10):5123–5130

Ahmed N, Hassan R, Javed S (2022) Blue economy and emerging technologies. ResearchGate
Ahmed N, Roy P, Das S (2024) Blockchain-enabled smart contracts: architecture, applications, and future trends. IEEE Xplore
Ali M, Rehman M, Zafar F (2019) AI and Blockchain integration in marine waste management. ScienceDirect
Alsamman M, Fazea Y, Mohammed F (2023) Towards a Blockchain-based agricultural ecosystem: a systematic review. Int Conf Innov Intell Inform Comput Technol 3:381–388. https://doi.org/10.1109/3ICT60104.2023.10391332
Bada AO, Damianou A, Angelopoulos CM, Katos V (2021) Towards a green Blockchain: a review of consensus mechanisms and their energy consumption. Int Conf Distrib Comput Sens Syst 17:503–511. https://doi.org/10.1109/DCOSS52077.2021.00083
Beck T (2018) Blockchain technology for fisheries management: enhancing traceability and transparency. Mar Policy 90:25–32
Cao L (2022) Decentralized AI: edge intelligence and smart Blockchain, metaverse, Web3, and DeSci. IEEE Intell Syst 37(3):6–19. https://doi.org/10.1109/MIS.2022.3181504
Chakraborty S, Das T, Ghosh A (2023) Blockchain and AI for supply chain resilience. Springer
Chand P, Kaur A, Gill S (2020) Smart contracts and automated port clearance. IEEE Xplore
Christensen V (2021) Predicting fish populations using AI models: A new approach to sustainable fisheries management. Fish Res 235:88–97
DNV GL (2019) AI in shipping: the future of smart shipping. Report by Det Norske Veritas
Ekinci A (2021) AI-based monitoring systems for marine conservation: real-time applications and future potential. J Ocean Technol 16(2):102–115
FAO (2022) The state of world fisheries and aquaculture 2022: towards blue transformation. Food Agric Organ Rome
Gao R, Chen X, Liu Z (2024) Blockchains for artificial intelligence of things (AIoT): a comprehensive survey. Springer
Gomes R (2022) Blockchain applications in marine conservation: case studies from marine protected areas. Conserv Technol 8(4):199–210
He Y (2021) Blockchain and AI integration for sustainable fisheries management and carbon trading. Sustainability 13(7):2024
Hughes T (2020) Blockchain in combating IUU fishing: transparency and traceability in the supply chain. Ocean Coast Manage 196:105257
IRENA (2020) Innovation outlook: offshore wind. Int Renew Energy Agency
Jambeck JR (2015) Plastic waste inputs from land into the ocean. Science 347(6223):768–771
Khan M, Ali T, Lee J (2024) Decentralized AI: edge intelligence and smart Blockchain. IEEE Xplore
Kim H, Park S, Choi J (2024) Blockchain technology and its relationship with supply chain resilience. Springer
Kittinger JN (2021) Technology and innovation in monitoring illegal fishing. Front Mar Sci
Kumar KPV, Lakshmi B, Sai Kumar S, Muralidhar V, Raghavendra Sai N, Nagamalleswara V (2023) Blockchain technology: a comprehensive review and future directions. Int Conf Augment Intell Sustain Syst 2:1362–1368. https://doi.org/10.1109/ICAISS58487.2023.10250707
Lebreton L et al (2018) Evidence that the Great Pacific Garbage Patch is rapidly accumulating plastic. Sci Rep 8(1):4666
Liao H et al (2020) Optimizing maritime logistics with Blockchain and AI: applications and challenges. Int J Ship Transp Logist 12(5):457–473
Meyer D (2020) Maersk's TradeLens Blockchain is shipping industry's Internet of Things. Forbes
Nayak R, Reddy D, Mishra B (2022) AI-driven predictive analytics in fisheries. ScienceDirect
Pan S et al (2021) Artificial intelligence in autonomous maritime shipping: trends and opportunities. Ocean Eng 238:108001
Papageorgiou A, Bani A (2020) The role of Blockchain in enhancing maritime trade accountability. Transp Res E Logist Transp Rev 142:102013
Patel S, Kumar D, Shah N (2024) A systematic review of Blockchain applications. IEEE Xplore

Pattanayak S, Arputham RM, Goswami M, Rana NP (2024) Blockchain technology and its relationship with supply chain resilience: a dynamic capability perspective. IEEE Trans Eng Manage 71:10398–10412. https://doi.org/10.1109/TEM.2023.3235771
Sharma R (2022) Blockchain and AI in carbon trading: enhancing transparency and sustainability. J Clean Prod
Sharma P, Singh R, Kumar V (2024) Towards a green Blockchain: a review of consensus mechanisms. IEEE Xplore
Shen M (2023) Blockchains for artificial intelligence of things: a comprehensive survey. IEEE Internet Things J 10(16):14483–14506. https://doi.org/10.1109/JIOT.2023.3268705
Singh A, Verma R, Gupta S (2023) Towards a Blockchain-based agricultural ecosystem: a systematic review. ResearchGate
Sunny FA (2022) A systematic review of Blockchain applications. IEEE Access 10:59155–59177. https://doi.org/10.1109/ACCESS.2022.3179690
Wang S, Ouyang L, Yuan Y, Ni X, Han X, Wang FY (2019) Blockchain-enabled smart contracts: architecture, applications, and future trends. IEEE Trans Syst Man Cybern Syst 49(11):2266–2277. https://doi.org/10.1109/TSMC.2019.2895123
Wang J, Li Y, Zhang H (2024) Blockchain technology: a comprehensive review and future directions. IEEE Xplore
WWF (2018) Blockchain in tuna supply chain. Retrieved from [WWF website]
Yadav S, Gupta R, Sinha K (2021) The role of Blockchain in carbon credit markets. ResearchGate
Yang Q, Zhao Y, Huang H, Xiong Z, Kang J, Zheng Z (2022) Fusing Blockchain and AI with Metaverse: a survey. IEEE Open J Comput Soc 3:122–136. https://doi.org/10.1109/OJCS.2022.3188249
Zhao Y, Wong K, Tan L (2024) Fusing Blockchain and AI with Metaverse. Springer

Blockchain and FinTech Synergies: Driving Sustainability in the Blue Economy

Samoon Khan, Rupinder Katoch, and Diya Mahendru

Abstract The blue economy is a concept that promotes the sustainability and sustainable use of ocean resources for economic growth, improved livelihoods, jobs while preserving ocean ecosystem health, is a crucial step and critical area of global sustainability efforts. This study explored synergies between blockchain technology and artificial intelligent (AI) in improving sustainability within the blue economy. The Blockchain technology with its inherent qualities of transparency, immutability, and decentralised nature, has potential to address major issues of traceability and accountability in marine industries. Artificial Intelligence (AI) on the other hand enhances data analysis and predictive models, decision making processes that can optimise resources management and mitigate environmental degradation. This Combination of these technologies could enable real-time monitoring of oceanic conditions improve the traceability of sustainable of seafood and sea-products, reduce illegal fishing and support carbon footprint reduction in maritime transportation. Few key case studies and real world applications of blockchain and AI in blue economy has been examined, including the use of blockchain for certifying sustainable seafood (e.g., "Fishcoin" initiative), AI applications in predicating fish populations and addressing marine pollution. Analysis of blockchain and AI in the blue economy is not only important for advancing sustainability but also for promoting a more equitable distribution of ocean resources, empowering stakeh9lders and protecting biodiversity for future generations. By providing a detailed analysis of these technologies roles in marine sustainability, this chapter aims to contribute to ongoing discourse on integrating digital innovation into environmental governance and policies.

Keywords Blue economy · Blockchain technology · Artificial intelligence (AI) · Sustainability · Marine conservation

S. Khan (✉) · R. Katoch
Mittal School of Business, Lovely Professional University, Phagwara, Punjab, India
e-mail: thesamoonkhan@gmail.com

D. Mahendru
Department of Political Science, School of Social Sciences and Languages, Lovely Professional University, Phagwara, Punjab, India

V. Sharma et al. (eds.), *FinTech for Sustainable Finance and a Well-Blue Economy*, World Sustainability Series, https://doi.org/10.1007/978-3-031-92390-6_9

1 Introduction

The Blue Economy concept has surfaced as a crucial pillar of sustainable development, balancing the competing realities of sustainable economic development, environmental protection and community well-being. Oceans cover 71% of the Earth's surface, are essential to the ecological balance of the planet and provide sustenance to billions of individuals (Gaill et al. 2016). Blue Economy contributes an estimated $1.5 trillion to the global economy each year, a number that is expected to double by 2030 (OECD 2016). But that promise is increasingly endangered by threats such as overfishing, marine pollution and climate change. As a result, the sustainability of ocean resources is now a top agenda for the government, industry, and international organizations, while emerging digital technologies like blockchain and financial technology (FinTech) are increasingly being seen as transformative tools.

The Blue Economy includes various sectors such as fisheries, aquaculture, coastal and maritime tourism, maritime transport and renewable ocean energy. These sectors not only offer economic opportunities, but they are also vital to tackling one of the world's greatest challenges of food security, climate change, and loss of biodiversity. Yet, unsustainable practices, like overfishing and habitat destruction, destroy the ecological balance and marine pollution, which is only aggravated by the dumping of 11 million metric tonnes of plastic waste in the ocean every year (Landrigan et al. 2020; https://www.unep.org/interactives/pollution-to-solution/). (Furthermore, the impacts of climate change, including rising sea levels, ocean acidification, and coral bleaching, jeopardize the livelihoods of coastal communities. To attain sustainability in this regard, endorsing latest technologies in the Blue Economy is the way forward.

1.1 The Role of the Blue economy in Combating Sustainability Challenges

With its attributes of transparency, immutability, and decentralization, blockchain technology is helping to answer the challenges of the Blue Economy. Blockchain technology also offers secure and tamper-proof records for stakeholders to trace the source and path of seafood products to counter illegal, unreported, and unregulated (IUU) fishing and pollute oceans and seas (Pratiwi et al. 2024). For example, Fishcoin aims to incentivize sustainable fishing practices through the provision of rewards to fishers (https://www.seafoodsource.com/news/environment-sustainability/fishcoin-incentivizing-blockchain-based-traceability-in-developing-nations). Green energy trading platforms, based on blockchain technology, promote investment in renewable energy and minimize carbon emissions. Likewise, FinTech innovation is improving financial inclusiveness and efficiency in the Blue Economy. Mobile banking and decentralized finance (DeFi) platforms enable coastal communities to access credit, insurance, and green financing opportunities, empowering steam and small-scale fishers and entrepreneurs (Basly 2024) The synergy between blockchain and FinTech promotes sustainable development and economic growth. While the Blue Economy is a highly promising area, it is confronted with several challenges to achieve sustainability. Overfishing is still a major problem, with the Food and Agriculture Organization (FAO) stating that 347% of the global fish stocks are overfished (FAO 2024). Marine pollution not only harms marine biodiversity, plastic waste also disrupts ecosystems. Climate change aggravates these figures: increasing sea temperatures, as well as extreme weather, put coastal infrastructure and ecosystems at risk (Das and Swain 2024). Also, economic inequalities in coastal and island communities complicate their responses to environmental change. Women represent almost half of the global fisheries workforce (Chambon et al. 2023), yet they experience constraints to equal participation and access to resources (Chambon et al. 2023; UN Women et al. 2020). Solutions to these challenges need to be integrated, inasmuch as technology has come of age, so have laws and policies after all.

1.2 Introduction to Distributed Ledger Technology and Fintech as Disruptive Forces

Blockchain and FinTech play a crucial role in the Blue Economy beyond economic efficiency; these tools encourage environmental stewardship and social equity. Blockchain technology improves transparency and accountability across industries and allows stakeholders to track the sustainability of supply chains and guarantee adherence to environmental rules (Difrancesco et al. 2022). Decentralization ensures that data is secure and accessible, enabling communities to participate in resource management; for instance with blockchain. This is where FinTech complements these capabilities with respect to financial tools that provide equitable access to resources and investments. Green financing platforms link investors with projects that foster marine conservation, renewable energy, and sustainable aquaculture (Shiiba et al. 2021). Additionally, AI-powered predictive analytics are improving resource allocation, enabling stakeholders to make data-backed decisions tailored to mitigate environmental destruction.

1.3 International attempts at Economic Sustainability and Ocean Conservation

Around the world, governments, international organizations, and private-sector actors are harnessing blockchain and FinTech to advance sustainability in the Blue Economy. The United Nations Sustainable Development Goals (SDGs) and currently, SDG 14: Life Below Water places a strong emphasis on these aspects of conservation and sustainable use of ocean resources (https://www.un.org/sustainabledevelopment/oceans/). Climate mitigation across all landscapes and ecosystems, including terrestrial, freshwater and oceans (Difrancesco et al. 2022), is intrinsic to the Paris Agreement (Shiiba et al. 2021) and the Convention on Biological Diversity (CBD) (https://www.un.org/sustainabledevelopment/oceans/), which seeks to protect at least 30% of the oceans by 2030 (Protected Planet Report 2024). On a regional level, for example, the European Union has its Blue Growth Strategy, which is centered on driving innovation in marine industries and investing in renewable ocean energy. In a similar vein, the Pacific Island nations spearhead ocean conservation by championing sustainable fisheries and by climate adaptation.

Private sector members and non-governmental organizations (NGOs) also play a crucial role in advancing the Blue Economy. Blockchain and cryptocurrencies Companies such as IBM and Maersk use blockchain to make the supply chain more transparent and to make maritime transport more efficient (Maersk 2024). Environmental NGOs like Oceana and the World Wildlife Fund (WWF) do research and advocacy work focused on fighting overfishing and protecting habitats in the oceans. Both of which are not just necessary for scaling innovative solutions, but also crucial to realizing the vision The Blue Economy. This has been called the Blue Economy, which is the one development work that brings together economic development and environmental conservation. Stakeholders will be able to tackle important issues, including resource management, pollution management, and financial inclusion, via the integration of blockchain and FinTech (Liu et al. 2021). Such technologies contribute to increasing the efficiency and sustainability of ocean-based industries, while also promoting social equity and environmental stewardship. With momentum building worldwide towards the conservation of our oceans and sustainable economic development, the Blue Economy provides the framework for how we can use innovation to protect our oceans and their ecosystems for our future generations.

2 The Blue Economy: An Emerging Paradigm for Sustainable Development

2.1 Definition and Scope

The Blue Economy is a game changing aim to embrace the wealth of Ocean resources whilst ensuring their sustainability for the long term. Coined in the context of aligning with global efforts related to sustainable development, this phrase captures any activity that makes use of ocean resources to deliver economic profit, enhance livelihoods, and maintain ecosystem health (Elegbede et al. 2022). Blue Economy has to be more than just traditional maritime industries, but engraining principles of environmental stewardship and social equity into the economic growth strategy. The Blue Economy is the sustainable use of ocean resources that contributes to economic growth, job creation and improved livelihoods while preserving the health of ocean ecosystem (World Bank 2017). This importance is further reflected in the Sustainable Development Goals (SDGs), particularly SDG 14, which highlights the need to conserve and sustainably use the oceans, seas, and marine resources. OECD reports claim that ocean-based industries generate around $1.5 trillion to the global economy per annum, a number projected to exceed $3 trillion by 2030 (OECD 2016).

Fisheries and Aquaculture: The fisheries sector underpins global food security with fisheries providing nearly 3.3 billion people with at least 20% of the animals used for protein intake (World Bank 2022). Innovations continue to reduce the environmental impact and improve the production efficiency in the already-exploding field of sustainable aquaculture.

Maritime Transport: Responsible for 80% of global trade (by volume), maritime transport is a cornerstone of international trade (World Bank Group 2024). As lifelines of global economies, ports and shipping lanes need efficiency and reduced carbon emissions.

Coastal and Marine Tourism: Coastal tourism is a huge revenue source for many countries, in particular for Small Island Developing States (SIDS). The long-term growth of this sector depends on marine biodiversity and untouched environments, which require conservation efforts.

Use of Renewable Ocean Energy: Wind, wave, and tidal energy is being developed in ocean where they are considered as useful sustainable alternative to fossil fuels (Osman et al. 2022). Not only do these technologies decrease greenhouse gas emissions, but they also offer economic opportunities for coastal communities.

Marine Biotechnology: Exploiting marine organisms for pharmaceuticals, cosmetics, and biofuels is a cutting-edge industry in the Blue Economy that could provide solutions to global health and energy needs.

There is a range of activities with a wide spectrum of global ecosystems and markets. But its achievement hinges upon a mechanism for balancing economic aspirations with the imperative to conserve marine biodiversity and further social equity.

2.2 Challenges in the Blue Economy

However, the Blue Economy faces many challenges that endanger its sustainability. All these issues are induced by human activities, environmental pressures and socio-economic inequalities and highlight the importance of integrated and site-based management approaches.

Overfishing Among the most critical issues of the Blue Economy are overfishing. According to the Food and Agriculture Organization (FAO), 34% of global fish stocks are overfished, and 60% are fully exploited (Gatta 2022). Overfishing threatens marine ecosystem balance, diminishes biodiversity, and jeopardizes the livelihoods of millions who rely on fisheries. Small-scale fishers—who comprise 50% of global fish catch (Basurto et al. 2025)—are especially impacted, with dwindling fish stock leading these fishers to spend additional time and resources for lower yield.

Marine Pollution: Marine pollution is a widespread problem, and pollution by plastic waste is one of the most aspected causes of marine pollution. Eleven million metric tons of plastic enter the ocean every year and that number is projected to triple by 2040 if current trends do not change according to the United Nations Environment Programme (UNEP) (United Nations Environment Programme 2022). Pollution levels are further exacerbated by oil spills, chemical runoff and untreated sewage, which harm both marine life and human health (Landrigan et al. 2020). Microplastics especially have drawn into the food chain, threatening biodiversity and public safety.

Illegal, Unreported and Unregulated (IUU) Fishing: IUU fishing leads to unsustainable practices in fisheries management, resulting in losses of approximately $23 billion per year to the world's economy (Popescu 2022). This activity drains fish stocks, risks biodiversity and erodes the livelihoods of law-abiding fishers. This challenge is compounded by weak enforcement mechanisms and insufficient supply chain transparency, highlighting the need for innovative and effective solutions to improve traceability and accountability for brands.

Climate change: Climate change has wide-ranging effects on the Blue Economy. Around 93% of excess heat from greenhouse gases goes into oceans, causing increased sea temperatures, ocean acidification, and coral bleaching (IUCN 2017). These shifts break down marine ecosystems, deplete fish stocks and jeopardize the chances of coastal communities. Extreme weather conditions, e.g., hurricanes and typhoons, can also endanger maritime infrastructure and human life.

Economic Inequalities: Coastal and island communities often face socio-economic disparities due to limited access to financial resources, infrastructure, and technology. Women, who make up almost half of the fisheries workforce, face substantial barriers to equal participation and access to resources (https://iucn.org/content/one-third-reef-building-corals-face-extinction; UN Women et al. 2020b). Doing so is critical for supporting inclusive and equitable growth in the Blue Economy.

Loss of Biodiversity: Marine biodiversity is also under great stress: a third of reef-forming corals and more than a third of marine mammals are threatened with extinction. (IUCN) Habitat destruction, invasive species, and pollution contribute to biodiversity loss, which erodes the resilience of marine ecosystems.

These challenges demonstrate the necessity to develop the new and innovative approaches to the development of sustainable, equitable ocean resources. These are formidable obstacles to efficiency, and have to date proved limiting to the full realization of the full potential in the blue economy, which in turn cannot be achieved without good governance, technological integration and collaboration among multiple stakeholders.

2.3 How Technology Can Help in Overcoming Challenges

Blue Economy and Finding Technological Solutions Technology can be a game changer to the Blue Economy challenges; in fact, technological solutions are needed to donectivity and scale where necessary. Technological innovations range from digital platforms to decentralized systems, bringing huge transformation to the management and conservation of ocean resources.

Digital Innovation as an Enabler of Sustainable Practices: Technologies such as blockchain, AI, and IoT are instrumental in promoting sustainable practices in the Blurred economy. Blockchain, for example, enables transparency and traceability throughout seafood supply chains, verifying adherence to sustainability standards. By utilizing AI in this space, stakeholders can use these insights to anticipate future environmental hazards and optimize resource management. Internet of Things IoT devices, such as smart sensors, allow scientists to monitor oceanic conditions in real time, garnering the information needed for adaptive management.

Decentralized and Transparent Systems: The decentralized nature of blockchain technology is immensely revolutionary, as it removes the reliance on centralized authorities and injects trust into the relationships between stakeholders. Blockchain provides an unchangeable record of transactions that ensures data is accurate, accessible, and tamper-proof (Nwariaku et al. 2024). The ability to do that is vital for cracking down on illegal fishers, managing pollutants and verifying the sustainability of seafood.

Predictive Analytics and Decision Making: Through utilizing AI powered predictive analytics, AI will completely develop and enhance the decision-making process by allowing a wider scope of patterns and trends in marine ecosystems to be analyzed at a much faster rate. For instance, AI can help predict fish population dynamics, evaluate the effects of climate change on biodiversity, and can optimize shipping routes to minimize fuel consumption (Nahar 2024). Stakeholders can utilize these insights to make informed decisions that consider both economic growth and environmental preservation.

Financial Inclusion and Green finance: FinTech platforms are transforming how financial services are delivered in the Blue Economy, enabling coastal communities to access credit, insurance, and green finance opportunities (Loukoianova 2024). DeFi allows micro fishers and small-scale entrepreneurs to access the global economy, and crowdfunding platforms link investors with sustainable projects.

Renewable Ocean Energy Technologies: Development of renewable ocean energy technologies, such as offshore wind farms and wave energy converters, that lessen dependence on fossil fuels and help combat climate change. Such technologies would promote carbon-neutrality while generating economic opportunities for coastal communities.

Connectivity and Collaboration on an Global Scale: Digital platforms encourage interaction between actors, allowing governments, sectors and NGOs to exchange data and best practices (Alsharabi et al. 2024). Such platforms can facilitate multi-stakeholder governance in which diverse perspectives and interests are integrated into decision-making processes.

Although technology has great potential, its use is subject to several obstacles such as expensive costs of implementation, low digital literacy rates and regulatory constraints. The Blue Economy can utilize technology at a global scale for a sustainable and resilient development, by driving innovation, investing in capacity-building and by promoting cross-sectoral partnerships.

3 An Exploration of Blockchain Technology in the Blue Economy

Blockchain technology, recognized for its revolutionary effects in various industries, has become an essential facilitator for the Blue Economy to become sustainable. Its decentralization, immutability, and transparency make it a potential solution to complex issues such as traceability, illegal practices, and resource management inefficiencies in the marine space (Bajare et al. 2024). With the Blue Economy entering an era of unprecedented growth, the potential of blockchain to drive transparency, collaboration, and sustainable practices across ocean-based industries cannot be overstated. Here we provide an overview of the features and applications of blockchain technology in the Blue Economy, along with detailed descriptions of the principles of data storage and security, the current state of research on maritime blockchains, and the benefits and challenges of implementing blockchain technology.

3.1 Blockchain: Disruptive Technology Built for Transparency and Trust

Phrase: Blockchain is a distributed ledger that stores transactions in a network of computers. Its key characteristics transparency, immutability, and decentralization offer a solution to eradicate inefficiencies while also ensuring accountability among various stakeholders. These characteristics have utterly transformed conventional systems, removing intermediaries, automating processes via smart contracts, and offering tamper-proof records open to participants (Bajare et al. 2024). Transparency allows for all interested parties to refer to a single source of truth—this is particularly valuable in industries like fisheries where it is often challenging to verify where products come from, and whether they are sourced sustainably and legally. Immutability means that when data is placed on blockchain, it cannot be changed or deleted, making it extremely compelling in terms of fraud prevention and accountability. Decentralization is the removal of any central authority, leading to local stakeholders' involvement, lower operational costs, and promoting inclusion (Boumaiza and Maher 2024). The core characteristics of blockchain are particularly relevant to the Blue Economy, wherein a multitude of constituents, from artisanal fishermen and environmentalists to international corporations and states, must work together to sustainably govern ocean resources.

3.2 Disruptive Roles of Blockchain in the Blue Economy

3.2.1 Improving Traceability in Seafood Supply Chains

Traceability issues in seafood supply chains have long plagued consumers with mislabelling, illegal harvesting and fraud damaging trust and undermining sustainability. Amounting to 20% of the global seafood is considered mislabelled, which may pose a risk for overfishing and illegal, unreported and unregulated (IUU) fishing, according to the Food and Agriculture Organization (FAO) (https://www.fao.org/iuu-fishing/fight-iuu-fishing/en/). Blockchain proves to be a great solution by ensuring end-to-end transparency and accountability. Then there are initiatives like Fishcoin, where they reward fishers for entering data about their catches on a blockchain platform. Individually verified by parties throughout the supply chain, each entry details the origins, handling, and sustainability of seafood products for consumers. In a similar way, the Seafood Alliance Blockchain Initiative makes sure that certified sustainably sourced seafood is in accordance with methods that meet strict environmental and social standards (Alsharabi et al. 2024). Through a verified data chain on blockchain, stakeholders can thus eliminate fraud, encourage compliance, and foster consumer trust in seafood market.

By implementing it, the seafood supply chain can be protected. Such systems also empower small-scale fishers by utilizing the information in a way that enables them to prove that their catches are legal and sustainable, thus opening the door to premium markets.

3.3 Fighting Illegal Unreported and Unregulated (IUU) Fishing

Financial losses due to IUU fishing are estimated at $23 billion annually worldwide (Widjaja et al. 2020), while it is responsible for approximately 26 million tons of fish logged from the ocean each year, making it a primary risk factor for both development and biodiversity. This practice disrupts attempts to fish sustainably and threatens the livelihoods of those who fish legally. Blockchain technology and Satellite tracking coupled with Internet of Things (IoT) devices are powerful allies in the fight against IUU fishing. Blockchain-connected real-time vessel tracking systems can provide visibility into fishing activities and allow authorities to reinforce restrictions by ensuring that the vessels are operating in authorized areas and are following regulations. Blockchain-enabled Electronic Catch Documentation Schemes, for example, offer a tamper-proof ledger of information on catch data from harvest to sale. It promotes transparency, allowing regulators to track down illegal activities, trace seafood back to the boats that caught it, and enforce penalties for violations.

Research from The Pew Charitable Trusts shows how the introduction of blockchain and digital monitoring technologies can improve the efficiency of fisheries management systems. Particularly, such tools have an immense role in deterring illegal practices and fostering international collaboration for conservation of common marine resources.

3.4 Incentives for Marine Pollution Management Via Tokenization

Plastic waste marine pollution is now one of the biggest environmental challenges to ocean health. According to the United Nations Environment Programme (UNEP), 11 million metric tons of plastic are dumped into the ocean each year, severely disturbing marine ecosystems and threatening biodiversity (Williams and Rangel-Buitrago 2022). What is ideal about blockchain is that its tokenized incentives and rewards can give digital value to actions that actively contribute to saving the environment. Reinforcing the point, programs like Plastic Bank offer an incentive to communities and individuals to collect ocean-bound plastic waste. Confusion about who the tokens were for but when I accepted the tokens however, immediately turned it into a shower of tokens that could be redeemed for goods and services so there's an economic incentive to reduce the pollution. More recently, blockchain technology is being applied to tracking how waste materials are collected, recycled and reused. Already, these initiatives have shown tangible impacts. In areas severely impacted, up to 50% of plastic waste leakage into oceans has been curtailed through blockchain-based waste management systems (Gong et al. 2022). Enhancing local waste management strategies with blockchain systems allows stakeholders to leverage these initiatives to expand circular economies that work towards both environmental and economic objectives.

3.5 Sustainable Maritime Transportation Promote

Maritime transport is an integral part of the Blue Economy, representing more than 80% of total global trade by volume (UNCTAD 2023). However, it also plays a major role in greenhouse gas emissions as well as marine pollution. Logistics optimization and maritime regulatory compliance monitoring through blockchain technology increases reliability, efficiency, and sustainability in maritime transport while simultaneously significantly improving global shipping sustainability. Shipping companies can use the blockchain systems to monitor fuel across vessels, getting real-time data on their emissions and their carbon footprint. These data are crucial to comply with international environmental regulations, e.g., that imposed by the International Maritime Organization (IMO) (Farah et al. 2024). Moreover, these blockchain based supply chain solutions help eliminate inefficiencies, enhance coordination among stakeholders, and promote trust within global trade networks.

Blockchain in maritime transport reduces operating costs and accelerates the transition to low-carbon shipping. Combining blockchain with disruptive technologies such as IoT and AI offers the maritime industry a resilient solution to environmental challenges while also remaining competitive in an environmentally sustainable trade communications strategy.

The Benefits of blockchain for the Blue Economy

The unique challenges of the Blue Economy lend themselves well to the various benefits offered by blockchain technology:

Secure and Auditable: Blockchain provides a secure and auditable ledger of transactions that can be explored by all parties involved, which improves accountability. This transparency is critical in industries like fisheries, which suffer from rampant mismanagement and fraud.

Enhanced Operational Efficiency: Smart contracts and automated processes reduce the need for administrative tasks, saving both time and resources. Licensing, certifications, payments, and other requirements can be handled smoothly with less room for delays and human error, for instance.

Decentralized Data Sharing and Collaboration: Blockchain is an enabler for sharing data across stakeholders, such as governments, industries, NGOs, etc. A collaborative approach is essential for addressing complex problems such as marine conservation.

Sustainability Incentives: API integrations with tokenization to offer sustainability incentives and rewards systems built on the use of an ESG token to drive innovation that makes it easier for customers to align with ESG goals over the longer term.

Challenges in Blockchain adoption in the Blue Economy

Like all technologies with potential, blockchain adoption faces great obstacles:

Difficulty to Implement: Implementation of blockchain systems is an expensive exercise due to the need for infrastructure and training. These costs may hinder adoption by previously unserved low income small-scale fishers and coastal communities that may not be able to afford the benefits of blockchain.

Low Digital Literacy: Actors, especially in underdeveloped regions, lack the skills and knowledge to use blockchain platforms adequately. Closing this knowledge gap through educational initiatives and capacity building is crucial for equitable adoption.

Regulatory Uncertainties: Blockchain operates across borders, it often conflicts with local regulations. Without standardized legal frameworks, businesses struggle to scale and investments are discouraged.

Limited size space: Some blockchain systems encounter scalability challenges, especially as transaction volumes rise, leading to processing delays. Moreover, issues of sustainability due to energy-intensive systems like some types of blockchain systems ifmanifest against the motto of the Blue Economy.

Resistance to Change: Transitioning from traditional systems to blockchain requires significant shifts in mindset and operations. Unless there is widespread buy-in, chances are success will be elusive—first because stakeholders may oppose adoption because of ignorance or scepticism, and second, because a bear market could sour sentiment towards blockchain technology overall. Blockchain technology has the potential to help step up sustainability in the Blue Economy. It addresses all the main challenges of traceability, pollution management, and maritime transport, also it fosters trust, collaboration, and innovation (Guan et al. 2024). But unlocking this potential means overcoming obstacles like high costs, regulatory uncertainties, and fear of change. To safeguard our oceans and harness the potential of the Blue Economy, we need to invest in digital literacy, develop supportive policies, and promote multi-stakeholder partnerships and initiatives that incentivise blockchain to enable a more sustainable and inclusive Blue Economy. In an era where the call for safeguarding our oceans is louder than ever, blockchain technology emerges as a promising tool, providing frameworks that can reconcile economic growth with ecological responsibility.

4 Marine Sustainability: The application of Artificial Intelligence (AI)

Artificial intelligence (AI) stands as a groundbreaking tech in various domains, such as health care, finance, and transportation. AI has a unique potential in the area of marine sustainability solutions to address increasingly complex challenges, optimize usage and management of vital resources and strengthen the long-term health of marine ecosystem (Bibri et al. 2023). In recent years, the Blue Economy—the economic model through which sustainable economic growth and the ocean can be married—has begun to incorporate artificial intelligence (AI) tools to enhance decision-making, optimize operations, and strengthen conservation. This section dives into the potential and role of AI in marine sustainability, including applications, contribution to policy and governance, and limitations.

4.1 *Articles on Artificial Intelligence (AI)*

4.1.1 Defining AI and What It is Capable of

An artificial intelligence is the capability of a machine to imitate intelligent human behavior, meaning that it can learn, reason, solve problems and make decisions. AI refers to any technologies, including machine learning (ML), neural networks, natural language processing (NLP), or computer vision, designed to stimulate the ability to perceive, comprehend, learn, and decide on actions like humans in the context of marine sustainability (Soori et al. 2023). They leverage new technologies to analyze massive datasets, uncover trends, and identify actionable insights, all of which are critical for owning and managing complex marine reserves.

4.1.2 Synergies with Other Digital Technologies

AI becomes strongest when paired with other emerging technologies including blockchain, the Internet of Things (IoT) and remote sensing. For example, sensors empower IoT with the capability to gather real-time data on oceanic conditions; AI algorithms then analyze the sensors' data, enabling decision-making. Likewise, AI-based systems can analyse blockchain-validated data to verify traceability and compliance in seafood supply chains (Gouiza et al. 2024). The wide applicability of AI renders it an indispensable tool to tackle multi-dimensional problems in the Blue Economy. It allows stakeholders to identify risks, optimize operations, and develop focused conservation strategies due to its capacity to process and analyze large datasets.

4.2 *The Quest for Blue Gold: Applications of AI in the Blue Economy*

4.2.1 Modeling Fish Population Dynamics and Ecosystem Response

The world's fish stocks and marine ecosystems face a significant threat from overfishing and habitat destruction. This includes AI-powered predictive models that use data on population dynamics of individual fish species to inform sustainable fisheries management.

Stock assessment models: estimates of so-called grey literature can be found in the literature and it can be time-consuming and resource intensive In contrast, AI models describe environmental variables, fishing behavior, and historical fish stock dynamics to predict fish stock trends (Kühn et al. 2024). For example, NOAA Fisheries uses AI to assess stocks of species, such as Atlantic cod and Pacific sardine, and establish the catch limits designed to meet conservation objectives.

Habitat mapping and species distribution AI systems, combined with satellite imagery and, in some cases, data collected from underwater sensors are used to map out critical habitats and tracking species distribution; One example of an application of AI can be found in Global Fishing Watch, a tool that tracks fishery activities in real-time (Kühn et al. 2024) and provides data for enforcement both by governments and NGOs of regulations to protect sensitive ecosystems.

Climate Change: Climate change has an immediate and significant impact on marine ecosystems by changing water temperature, salinity, and nutrient availability. AI models mimic these alterations to forecast their effect on biodiversity, facilitating proactive precautions.

4.2.2 Marine Pollution and Its Sources Assessment

Marine pollution is a grave problem across the world, notably plastic waste. AI presents novel means to detect pollution hotspots, trace sources, and build successful remediation strategies.

Plastic Detection and Classification: AI-based computer vision systems analyze satellite imagery and drone footage to detect and classify plastic waste in oceans and coastal regions. For example, AI can optimize waste collection efforts in the ocean, as is currently done by projects like The Ocean Cleanup, which specifically targets areas with the highest density of plastic debris (de Vries 2022).

Chemical Contaminant Monitoring: AI algorithms analyze data captured from IoT sensors integrated into marine habitats to identify and measure chemical contaminants, including oil spills and heavy metals (Olawade et al. 2024). This information quickens the respond Time Customer to pollution events, reduce ecological damage.

Predictive Pollution Models: AI-driven models can predict the dispersion of pollution using ocean currents, weather conditions, and human activities. These predictions inform cleanup efforts and policies to prevent future contamination.

4.2.3 Improving Navigation and Lowering Fuel Consumption in Marine Transport

For a landlocked economy, sea transportation is crucial for the global flow of goods and also a significant contributory factor to greenhouse gas emissions. AI is revolutionizing this industry through enhanced navigation, streamlined logistics, and minimized ecological footprints.

Route Optimization: AI systems analyse variables such as weather conditions, ocean currents, and vessel performance to identify the most efficient shipping routes. This cuts down on the need for fuel and shrinks the industry's carbon footprint. Example:

Maersk leverages AI-based onboard tools to reduce fuel costs by about 5% per shipping voyage (Durlik et al. 2024a).

Smart Port Management and Automation: AI helps to automate the operations of a port area, helping reduce waiting time and enhancing cargo handling efficiency. Smart ports, such as Rotterdam and Singapore, utilize AI to improve operational sustainability (Durlik et al. 2024b).

Autonomus shipping: ***AI*** is the backbone of autonomous vessels development, as AI allows technology that reduces human vagaries, and thus makes shipping safer. And these vessels can also optimize fuel use, aiding environmental agenda.

4.3 AI in Policy and Governance

4.3.1 Make Decisions Based on Evidence

AI improves the ability of policymakers and conservationists to make decisions based on data. AI delivers actionable insights through the analysis of both historical and real-time data, shaping regulatory frameworks, resource allocation, and conservation strategies. This resource is aimed at helping the already-produced objectives discover ways to identify or monitor listening areas in the news with respect to topics including, but not limited to, Marine Protected Areas (MPAs). AI can be used to identify ecologically important for areas and assess the effectiveness of MPAs in protecting biodiversity. AI-assisted tools evaluate fluctuations in fish populations, coral conditions, and water quality throughout MPAs, informing policy modifications to optimize conservation effectiveness (Rees et al. 2020). From satellite and radar data, AI systems detect illegal fishing in waters, detecting violations of international agreements. Governments use these insights to better allocate enforcement resources.

4.3.2 Improving collaboration with Stakeholders

AI brings together various stakeholders such as governments, industries, and NGOs. Advancements in AI facilitate creation of digital infrastructure that allows for sharing of data, common understanding and collective action.

Global Fishing Watch: AI-based platform that enables stakeholders to monitor fishing activities globally for accountability and cooperation in fisheries (Wing and Woodward 2024).

AI facilitates the data gathering and analysis required for Stakeholder reporting related to the Sustainable Development Goals (SDGs), specifically SDG 14 (Life Below Water) (Nahar 2024).

4.3.3 Policy Compliance and Enforcement

Compliance Monitoring AI automates regulatory compliance by monitoring and reporting on the process itself, greatly reducing the burden on regulatory agencies. AI systems examine vessel behaviour to identify improper activity (e.g., fishing in the closed areas or "catch quotas") (Wing and Woodward 2024). This data underlies enforcement actions to maintain policy compliance.

4.4 AI Marine Applications and Their Limitations

4.4.1 Data Availability and Quality

Being useful AI to a very large extent relies on data and the quality of data. In many parts of the world, marine ecosystem data is sparse or outdated, limiting the potential for AI to deliver accurate insights. The inability to register all data limits such aims and is a barrier in implementing a plan, thus strengthening the need to expand data collection networks and implement similar methodologies.

4.4.2 The Need for Ethical Considerations and Potential for Misuse

The application of AI in marine sustainability has ethical implications, including the potential for use in harmful ways. For instance, surveillance technologies can violate privacy rights and predictive models can be employed to pillage rather than conserve marine resources. The ethical use of this technology must be established through safeguards and guidelines.

4.4.3 Technological and Financial Challenges

Implementing AI systems can incur ɟrнvest of time and money, ₰nd in pitfalls of losing data and resources. Resource and technical constraints with coastal/developing nations is also one of the stumbling blocks for AI (artificial intelligence) assisted solutions. There is a need for international support and capacity-building initiatives to close this gap. AI is transforming marine sustainability by providing cutting-edge solutions to key Blue Economy challenges. Its transformative potential is evidenced by examples in fisheries management, pollution control, and maritime transport (Nguyen et al. 2024). In addition, AI in policy and governance ensures that conservation efforts are informed, effective and inclusive. However, unlocking the true capabilities of AI comes with its challenges such as data, ethics, and accessibility issues. Through collaboration, capacity-building investments, and robust regulatory frameworks, stakeholders can leverage AI to protect marine ecosystems and ensure sustainable development for future generations.

Environmental Impact

1. ***Improved Traceability and Resource Management***

Hence, the traceability of resources from origin to end consumer improves through the adoption of blockchain technology. In the Blue Economy, this entails tracing the supply chain of seafood, tracking where and how materials for marine buildings are obtained, and even regulating the waste emitted from ships. The World Wildlife Fund (WWF) conducted a pilot project in Fiji, New Zealand, and Vanuatu, using blockchain to trace tuna from ocean to plate (https://www.wwfpacific.org/our_work/sustainable_blue_economy/). This helped guarantee that the fish were caught legally, from sustainable sources, and with a stop to illegal and unreported fishing, a big contributor to overfishing.

2. ***Energy Consumption Concerns***

Although blockchain can help establish sustainability by allowing for better traceability and compliance, it also has a substantial energy footprint, especially systems with a Proof of Work (PoW) consensus mechanism. PoW is used by the Bitcoin network, which is estimated to require some 91 terawatt-hours of electricity a year, this is greater than the total annual energy consumption of Finland (Treiblmaier 2023).

3. ***Mitigation Through Technology***

Many newer blockchain systems shifting to more energy-efficient consensus mechanisms like Proof of Stake (PoS) or hybrid models can significantly address the energy consumption of the blockchain systems. Ethereum's transition from PoW to PoS is expected to reduce its energy use by more than 99% (Vries 2022). Documentation handling is reduced as much as 60% based on a single-blockchain for a given shipping route, which can indirectly decrease port stay times and fuel costs (Difrancesco et al. 2022).

Social Impact

1. ***Economic Inclusion***

FinTech, especially via mobile banking and decentralized financial solutions, can provide unprecedented access to financial means for migrant and underprivileged coastal communities. These tools can connect small-scale fishers and aquaculture operators to credit, insurance and markets. In Indonesia, a mobile payment system has allowed fishers to sell their catch directly to consumers, bypassing intermediaries and increasing their profits by up to 20% (Zelasney et al. 2020). Fishcoin has been adopted in several areas, helping local fishermen increase revenue by as much as 30% through improved access to markets and better pricing.

2. ***An Economy that Creates Jobs and Transforms Our Workforce***

Within the Blue Economy, Blockchain and FinTech do not merely displace traditional sectors (fishers, traders, shippers) but most importantly create new jobs in the realms of data, technology and maintenance. For example, in the Philippines, new blockchain training programs are teaching fishers how to leverage technology in tracking their catches and finding new markets. Research has shown that under 20% of small-scale fishers in sub-Saharan Africa have access to the technology and connectivity needed to take advantage of blockchain-based solutions.

3. ***Risk of Contributing to Worsening Inequalities***

Source: These technologies have many advantages, but if they are not implemented thoughtfully, they may add to existing inequalities. Making sure the benefits of technology are accessible and equitable is paramount. A survey of small island developing states found that just 35% of fishers had access to the technology required to plug into a blockchain-enabled supply chain system (Chunduri et al. 2024).

Ethical Issues

1. ***Data Privacy and Security***

By collecting and storing large volumes of data, as is the case with blockchain and FinTech technologies, major concerns arise with regards to secure information management and privacy. Although blockchain provides better security, the irreversible characteristic of this technology makes it crucial for individuals to behave responsibly regarding data handling. When they performed a pilot on seafood traceability using a blockchain, participants worried that sensitive information like the whereabouts of valuable fishing grounds could be made public, resulting in overfishing. In a survey conducted by Deloitte (2019), more than 40% of blockchain-proficient executives claim that protection risks are the main roadblock to keep blockchain technologies more widespread.

2. ***Consent and Data Ownership***

This is how blockchain constitutes a biointerned community and therefore is increasingly artified and it also makes sure that all participants in blockchain systems within vulnerable communities know how their data is going to be used and remain in control of it. When deploying a blockchain solution for microloans in coastal East Africa, we had to explain and obtain informed consent from participants who had little prior exposure to this technology. (World Bank Group 2018) The World Bank reports that there are approximately 1.7 billion adults worldwide that remain unbanked and nearly half of these adults are located in areas where the Blue Economy could be enhanced.

3. ***Complying with Regulations and adjustments***

Blockchain and FinTech work in fast-moving legal environments. They will also need to comply with both local and international regulations, to ensure their ethical and sustainable integration into the terms of the Blue Economy. The General Data Protection Regulation (GDPR) has a significant impact on how blockchains, especially public ones, are systematized involving personal data of EU citizens, even if the blockchain itself isn't operated within the EU (Finck 2019). The Blue Economy has an effective tool for sustainability in the integration of blockchain and FinTech, (Bajare et al. 2024) which can be an efficient technology for supporting the Blue Economy as it offers several advantages for environmental, social, and economic performance. Yet, in order to fully achieve these benefits, we must thoroughly address the environmental, social, and ethical challenges as well (United Nations 2022). This needs an multi stakeholder approach: representatives from government, industry, academia, and communities, working together to use these technologies responsibly and inclusively, towards a sustainable and equitable Blue Economy.

Key Insights on Synergies Between Blockchain and FinTech Advancing Sustainability in the Blue Economy

Some of the major findings from convergences of blockchain technologies, FinTech, and sustainability efforts related to the blue economy. These findings clearly demonstrate the potential of these technologies to contribute to the double bottom line of increasing economic growth with increasing environmental sustainability.

1. ***Bridging the Blue Financing Gap***

Financing Gap: $2 trillion financing gap for SMEs in developing areas needed to fuel the growth of the blue economy This is a key gap for assistance to local actors, crucial for sustainable development (Pacific. 2022).

SME Blue Impact Asia (Blended Finance Solutions) SME Blue Impact Asia suggests creating blended finance platforms that could help match international capital with local SMEs. This method is designed to improve funding access while supporting sustainable strategies in areas such as marine ecosystems and resilient infrastructure.

2. ***The role of FinTech for Sustainability***

Enhancing Financial Inclusion: How are FinTech's positively impacting the finance industry? This step is essential to ensuring equitable, sustainable economic development, and for the realization of the Sustainable Development Goals (SDG).

Enabling Greener Investments: The financial technologies promote the transparency and efficiency of the transactions that can translate higher investments in green products. Particularly through digital platforms, FinTech can provide insight into environmental impact and create incentives for responsible financial behaviours.

3. ***Synergies Between Blockchain and Fintech***

Expected Results from Blockchain and FinTech: The use of blockchain with FinTech can provide new financial solutions for financing sustainable projects. This can improve processes, reduce costs and build trust among blue economy stakeholders.

Data traceability and accountability: Blockchain technology offers a distributed ledger that enhances transparency in financial transactions associated with environmental sustainability. The reform is necessary to attract investors that are interested in environmentally, socially and governance (ESG) criteria.

4. ***Collaborative Ecosystems***

The development of innovative ecosystems that cultivate communication and collaboration between startup initiatives, traditional financial institutions and regulatory bodies contributes to providing financial services. The end result is a new line of products that better fits the nuances of sustainable finance as it's being defined through our collaborative approach.

Supportive regulation—Supportive regulation is key to fostering a culture of innovation in the FinTech space, especially with regards to sustainability efforts. Example: Regulatory sandboxes, which allow experimentation with new technologies, can ensure that venture capital investment does not come at the expense of consumer protection and data privacy.

5 Future Research Directions

Emerging Research Domains: FinTech is still a nascent area of research and we need more research in the intersection of FinTech and sustainability—climate solutions in particular. Thus, future studies should extend to how emerging technologies can be used to further close the gap between financial practices and sustainability.

Impact Assessment: Further research is needed in the assessment of long-term impacts of blockchain and FinTech integration on sustainability in the blue economy. Assessing these impacts will help in fine-tuning strategies to leverage technology for sustainable development.

Such amalgamation of blockchain ideology with FinTech acts as an enabler, driving sustainability in the blue economy, filling the financing gap, enhanced cleaning of investment, improved financial inclusion, collaborative ecosystem, creating significant opportunities. These results highlight the need for progress in this field to attain sustainable economics.

6 Conclusion

The Study emphasized that blockchain and artificial intelligence (AI) have the potential to be game changers in solving some of the greatest challenges of the Blue Economy. Using blockchain technology allows tracking seafood from its source to the dinner table, which is essential for fighting illegal fishing and contributing to sustainable ocean management. It enables the ability to use data to inform decisions, make better use of resources and predictive analysis to help understand marine environments, and protect them. The case studies, such as Fishcoin initiative and Maersk blockchain implementation, give practical examples of the way with these technologies are ensuring operational efficiencies and better compliance with sustainability standards. It is a transformational change towards more transparency, efficiency, and sustainability in the marine and maritime economies. This integration enables a level of monitoring and compliance that was impossible in the past, and that is deeply critical in an industry whose activities tend to be spread over broad and often isolated tracts of land. The unique contribution of your research is to the extent that these technologies are harnessed for the service of economic activities and their able to reaffirm the aforementioned goals, while in the process promoting ecological conservation and equitable development for all as advocated in the UN global goals for sustainability which recognizes the Sustainable Development Goals (SDGs) focusing on SDG 14: Life Below Water. Our findings highlight the necessity for empowering policy frameworks that may promote and facilitate the resilient adoption of blockchain and AI for the Blue Economy. Meanwhile, governments and international bodies are encouraged to establish standards and regulations that balance innovation with preventing these technologies from deepening environmental destruction or widening social inequalities. Broadly, these encompass policies around data governance, initiatives for cybersecurity and ensuring that the benefits of technology are distributed equitably.

You have written a study acknowledging both the potential of blockchain and AI but also the challenge of implementation, such as the high cost of technology deployment but also technological literacy, and potentially the environmental impact—given the energy consumption associated with these technologies. To tackle the many problems these create, stakeholders must work together, developing infrastructure, education, and renewable energy solutions that can reduce the ecological impacts of digital technologies. Our article is not a promenade of technology utopianism, and our discussion provokes further research not just on the level of improving the cyber-physical systems but also on the socio-economic impacts of blockchain technology and AI, especially in coastal underserved and vulnerable communities. In addition, the implications for future research areas are evident as future studies may focus on how such technologies can be scaled sustainably and inclusively or whether, and how, these could and should connect with other emerging technologies, e.g., the Internet of Things (IoT) and big data analytics. According to the study, the Blue Economy must not only undergo technological upgrades with the inclusion of the blockchain and AI; they are indispensable tools needed to ensure the future sustainability of ocean resources. These technologies are supporting new methods of managing and conserving marine resources, by promoting greater transparency, accountability, and efficiency. But their successful inclusion requires careful orchestration of policy, inclusive engagement of stakeholders, and iterative adjustment to technological change and feedback from the environment.

References

Alsharabi N, Ktari J, Frikha T, Alayba A, Alzahrani AJ, Jadi A, Hamam H (2024) Using blockchain and AI technologies for sustainable, biodiverse, and transparent fisheries of the future. J Cloud Comput Adv Syst Appl 13. https://doi.org/10.1186/s13677-024-00696-8

Bajare D, Zsembinszki G, Yiatros S, Kaewunruen S, Cidik MS, Schiller G, Zhang N, Rizzo A, Tambovceva T, Hendawy M, Cavdar AD, Borg PR (2024) Stakeholders' role, inter-Relationships, and obstacles in the implementation of circular economy. In: Springer tracts in civil engineering, pp 629–646

Basly S (2024) Introduction: Blockchain, decentralized finance, and entrepreneurship. In: Springer eBooks, pp 1–9

Basurto X, Gutierrez NL, Franz N, Del Mar Mancha-Cisneros M, Gorelli G, Aguión A, Funge-Smith S, Harper S, Mills DJ, Nico G, Tilley A, Vannuccini S, Virdin J, Westlund L, Allison EH, Anderson CM, Baio A, Cinner J, Fabinyi M, Hicks CC, Kolding J, Melnychuk MC, Ovando D, Parma AM, Robinson JPW, Thilsted SH (2025) Illuminating the multidimensional contributions of small-scale fisheries. Nature. https://doi.org/10.1038/s41586-024-08448-z

Bibri SE, Krogstie J, Kaboli A, Alahi A (2023) Smarter eco-cities and their leading-edge artificial intelligence of things solutions for environmental sustainability: a comprehensive systematic review. Environ Sci Ecotechnol 19:100330. https://doi.org/10.1016/j.ese.2023.100330

Boumaiza A, Maher K (2024) Leveraging blockchain technology to enhance transparency and efficiency in carbon trading markets. Int J Electr Power Energy Syst 162:110225. https://doi.org/10.1016/j.ijepes.2024.110225

Chambon M, Miñarro S, Fernandez SA, Porcher V, Reyes-Garcia V, Drouet HT, Ziveri P (2023) A synthesis of women's participation in small-scale fisheries management: why women's voices matter. Rev Fish Biol Fish 34:43–63. https://doi.org/10.1007/s11160-023-09806-2

Chunduri V, Raparthi M, Yellu RR, Keshta I, Byeon H, Soni M, Bhatt MW (2024) Blockchain-based secure optimized traceable scheme for smart and sustainable food supply chain. Discov Sustain 5. https://doi.org/10.1007/s43621-024-00287-2

Das A, Swain PK (2024) Navigating the sea level rise: exploring the interplay of climate change, sea level rise, and coastal communities in India. Environ Monit Assess 196. https://doi.org/10.1007/s10661-024-13191-z

De Vries A (2022) Cryptocurrencies on the road to sustainability: ethereum paving the way for Bitcoin. Patterns 4:100633. https://doi.org/10.1016/j.patter.2022.100633

De Vries R. Using AI to monitor plastic density in the ocean. https://theoceancleanup.com/updates/using-artificial-intelligence-to-monitor-plastic-density-in-the-ocean/

Deloitte: Deloitte's 2020 global blockchain survey. https://www2.deloitte.com/jp/en/pages/financial-services/articles/bk/2020-global-blockchain-survey.html

Difrancesco RM, Meena P, Kumar G (2022) How blockchain technology improves sustainable supply chain processes: a practical guide. Oper Manag Res 16:620–641. https://doi.org/10.1007/s12063-022-00343-y

Durlik I, Miller T, Kostecka E, Tuński T (2024a) Artificial intelligence in maritime transportation: a comprehensive review of safety and risk management applications. Appl Sci 14:8420. https://doi.org/10.3390/app14188420

Durlik I, Miller T, Kostecka E, Łobodzińska A, Kostecki T (2024b) Harnessing AI for sustainable shipping and green ports: challenges and opportunities. Appl Sci 14:5994. https://doi.org/10.3390/app14145994

Elegbede I, Kaullysing D, Aderonke L-A, Saba AO, Ositimehin KM, Akinwunmi MF, Jinad M, Kies F, Pedro S, Adewumi I (2022) Ocean sustainability. In: Springer eBooks, pp 1–7

Farah MB, Ahmed Y, Mahmoud H, Shah SA, Al-Kadri MO, Taramonli S, Bellekens X, Abozariba R, Idrissi M, Aneiba A (2024) A survey on blockchain technology in the maritime industry: challenges and future perspectives. Futur Gener Comput Syst 157:618–637. https://doi.org/10.1016/j.future.2024.03.046

Financing the Blue Economy: investments in sustainable blue small–medium enterprises and projects in Asia and the Pacific (2022)
Finck M (2019) Panel for the future of science and technology, European Parliamentary Research Service, Scientific Foresight Unit (STOA): Blockchain and the General Data Protection Regulation. EPRS | European Parliamentary Research Service
Gaill F, Causse C, Copin C, Escoffier LF, Grillet C, Ras L, Daniel R, Voyé J, Schumm R (2016) Ocean and climate
Gatta PP (2022) The state of world fisheries and aquaculture 2022. Food and Agriculture Organisation (FAO)
Gong Y, Wang Y, Frei R, Wang B, Zhao C (2022) Blockchain application in circular marine plastic debris management. Ind Mark Manage 102:164–176. https://doi.org/10.1016/j.indmarman.2022.01.010
Gouiza N, Jebari H, Reklaoui K, Little Lion Scientific (2024) Integration of IoT-enabled technologies and artificial intelligence in diverse domains: recent advancements and future trends. J Theor Appl Inf Technol 102
Guan P, Wood LC, Wang JX, Duong LNK (2024) Blockchain adoption in the port industry: a systematic literature review. Cogent Bus Manage 11. https://doi.org/10.1080/23311975.2024.2431650
IUCN (International Union for Conservation of Nature): Ocean warming. IUCN (International Union for Conservation of Nature), Gland, Switzerland (2017)
IUCN: One third of Reef-Building corals face extinction. https://iucn.org/content/one-third-reef-building-corals-face-extinction
IUU day | Illegal, Unreported and Unregulated (IUU) fishing | Food and Agriculture Organization of the United Nations. https://www.fao.org/iuu-fishing/fight-iuu-fishing/en/
Kühn B, Cayetano A, Fincham JI, Moustahfid H, Sokolova M, Trifonova N, Watson JT, Fernandes-Salvador JA, Uusitalo L (2024) Machine learning applications for fisheries—at scales from genomics to ecosystems. Rev Fish Sci Aquacult 1–24. https://doi.org/10.1080/23308249.2024.2423189
Landrigan PJ, Stegeman JJ, Fleming LE, Allemand D, Anderson DM, Backer LC, Brucker-Davis F, Chevalier N, Corra L, Czerucka D, Bottein M-YD, Demeneix B, Depledge M, Deheyn DD, Dorman CJ, Fénichel P, Fisher S, Gaill F, Galgani F, Gaze WH, Giuliano L, Grandjean P, Hahn ME, Hamdoun A, Hess P, Judson B, Laborde A, McGlade J, Mu J, Mustapha A, Neira M, Noble RT, Pedrotti ML, Reddy C, Rocklöv J, Scharler UM, Shanmugam H, Taghian G, Van De Water JAJM, Vezzulli L, Weihe P, Zeka A, Raps H, Rampal P (2020) Human health and ocean pollution. Ann Glob Health 86:151. https://doi.org/10.5334/aogh.2831
Liu H, Yao P, Latif S, Aslam S, Iqbal N (2021) Impact of green financing, FinTech, and financial inclusion on energy efficiency. Environ Sci Pollut Res 29:18955–18966. https://doi.org/10.1007/s11356-021-16949-x
Loukoianova E (2024) Fintech applications for boosting climate finance. IMF Staff Climate Notes. 2024, 1. https://doi.org/10.5089/9798400293740.066
Maersk: How blockchain technology is beefing up supply chain visibility (2024). https://www.maersk.com/news/articles/2021/07/27/how-blockchain-technology-is-beefing-up
Martin: Goal 14: conserve and sustainably use the oceans, seas and marine resources. https://www.un.org/sustainabledevelopment/oceans/
Nahar S (2024) Modeling the effects of artificial intelligence (AI)-based innovation on sustainable development goals (SDGs): applying a system dynamics perspective in a cross-country setting. Technol Forecast Soc Chang 201:123203. https://doi.org/10.1016/j.techfore.2023.123203
Nguyen HP, Nguyen CTU, Tran TM, Dang QH, Pham NDK (2024) Artificial intelligence and machine learning for green shipping: navigating towards sustainable maritime practices. JOIV Int J Inform Vis 8:1. https://doi.org/10.62527/joiv.8.1.2581
Nwariaku NH, Fadojutimi NB, Lawson NLGL, Agbelusi NJ, Adigun NOA, Udom NJA, Olajide NTD (2024) Blockchain technology as an enabler of transparency and efficiency in sustainable supply chains. Int J Sci Res Arch 12:1779–1789. https://doi.org/10.30574/ijsra.2024.12.2.1454

OECD (2016) The ocean economy in 2030. OECD Publishing, Paris, France
Olawade DB, Wada OZ, Ige AO, Egbewole BI, Olojo A, Oladapo BI (2024) Artificial intelligence in environmental monitoring: advancements, challenges, and future directions. Hyg Environ Health Adv 100114. https://doi.org/10.1016/j.heha.2024.100114
Osman AI, Chen L, Yang M, Msigwa G, Farghali M, Fawzy S, Rooney DW, Yap P-S (2022) Cost, environmental impact, and resilience of renewable energy under a changing climate: a review. Environ Chem Lett 21:741–764. https://doi.org/10.1007/s10311-022-01532-8
Popescu I (2022) European parliamentary research service: illegal, unreported and unregulated (IUU) fishing. EPRS | European Parliamentary Research Service
Pratiwi RAI, Fani LA, Kusasi F (2024) Blockchain technology in fisheries industry: a systematic literature review. BIO Web Conf 134:05004. https://doi.org/10.1051/bioconf/202413405004
Protected Planet Report 2024. UNEP-WCMC and IUCN: Cambridge, United Kingdom; Gland, Switzerland (2024)
Rees SE, Sheehan EV, Stewart BD, Clark R, Appleby T, Attrill MJ, Jones PJS, Johnson D, Bradshaw N, Pittman S, Oates J, Solandt J-L (2020) Emerging themes to support ambitious UK marine biodiversity conservation. Mar Policy 117:103864. https://doi.org/10.1016/j.marpol.2020.103864
Seafood Expo Asia: Fishcoin incentivizing blockchain-based traceability in developing nations. https://www.seafoodsource.com/news/environment-sustainability/fishcoin-incentivizing-blockchain-based-traceability-in-developing-nations
Shiiba N, Wu HH, Huang MC, Tanaka H (2021) How blue financing can sustain ocean conservation and development: a proposed conceptual framework for blue financing mechanism. Mar Policy 139:104575. https://doi.org/10.1016/j.marpol.2021.104575
Soori M, Arezoo B, Dastres R (2023) Artificial intelligence, machine learning and deep learning in advanced robotics, a review. Cogn Robot 3:54–70. https://doi.org/10.1016/j.cogr.2023.04.001
The State of World Fisheries and Aquaculture 2024. FAO, Rome, Italy (2024)
Treiblmaier H (2023) A comprehensive research framework for Bitcoin's energy use: fundamentals, economic rationale, and a pinch of thermodynamics. Blockchain Res Appl 4:100149. https://doi.org/10.1016/j.bcra.2023.100149
UN Trade and Development (UNCTAD): review of maritime transport 2023. UN Trade and Development (UNCTAD) (2023)
UNEA: Pollution to solution: accessing marine litter and plastic pollution. https://www.unep.org/interactives/pollution-to-solution/
United Nations Environment Programme (2022) From pollution to solution: a global assessment of marine litter and plastic pollution. United Nations Environment Programme (UNEP), Nairobi
United Nations: Blue economy: oceans as the next great economic frontier. https://unric.org/en/blue-economy-oceans-as-the-next-great-economic-frontier/
Widjaja S, Long T, Wirajuda H, Van As H, Bergh PE, Brett A, Copeland D, Fernandez M, Gusman A, Juwana S, Ruchimat T, Trent S, Wilcox C (2020) Illegal, unreported and unregulated fishing and associated drivers. World Resources Institute
Williams AT, Rangel-Buitrago N (2022) The past, present, and future of plastic pollution. Mar Pollut Bull 176:113429. https://doi.org/10.1016/j.marpolbul.2022.113429
Wing K, Woodward B (2024) Advancing artificial intelligence in fisheries requires novel cross-sector collaborations. ICES J Mar Sci 81:1912–1919. https://doi.org/10.1093/icesjms/fsae118
UN Women, Biswas N, Rajagopalan R (2020) Women's economic empowerment in fisheries in the blue economy of the Indian ocean rim. Economic Empowerment Section of UN Women
World Bank: Oceans, fisheries, and the coastal economies. World Bank and (FAO 2022) (2024)
World Bank: What is the blue economy?. https://www.worldbank.org/en/news/infographic/2017/06/06/blue-economy
World Bank Group (2018) Financial inclusion on the rise, but gaps remain, global Findex database shows. https://www.worldbank.org/en/news/press-release/2018/04/19/financial-inclusion-on-the-rise-but-gaps-remain-global-findex-database-shows
World Bank Group: Sustainable development in shipping and ports. World Bank (2024)

WWF: Sustainable blue economy. https://www.wwfpacific.org/our_work/sustainable_blue_economy/

Zelasney J, Ford A, Westlund L, Ward A, Riego PO (2020) Securing sustainable small-scale fisheries showcasing applied practices in value chains, post-harvest operations and trade. In: FAO fisheries and aquaculture technical paper, p 652. https://doi.org/10.4060/ca8402en

Blue Economy and Big Data Analytics

Rajkumar Bhola, Shalini Aggarwal, Prerna Dhawan, and Gitana Mečėjienė

Abstract Big data enhances multiple facets of the blue economy by facilitating coordination among various entities in their utilization of natural resources. It can be a vital tool, as it has the potential to create sustainable and practical strategies. The paper uses secondary data to understand the subject under study. Data has been collected from the company's website, newspapers, journals, magazines, and other websites. The present study tries to understand the type of Big Data analytics in companies; to study the companies using big data analytics; to understand the applications of big data into different sectors; to study the key benefits of big data analytics in the blue economy; to study the Data Sources for Blue Economy Analytics. The results show that use of the big data analytics in the blue economy in Sustainable Fisheries Management, Enhanced Shipping Routes, Offshore Energy Development, Marine Pollution Monitoring, Climate Change Impact Assessment, Biodiversity Conservation, Improved Coastal Management.

Keywords Big data analytics · Blue economy · Sustainable development

R. Bhola · S. Aggarwal (✉) · P. Dhawan
Management Department, Chandigarh University, Mohali, Punjab, India
e-mail: shaliniaggar@gmail.com

R. Bhola
e-mail: rajphdscholar2907@gmail.com

P. Dhawan
e-mail: erprerna_0407@rediffmail.com

G. Mečėjienė
Vilniaus Kolegija/Higher Education Institution, Faculty of Economics, Banking Department, Studentų Str. 39A, 08105 Vilniaus, Lithuania

V. Sharma et al. (eds.), *FinTech for Sustainable Finance and a Well-Blue Economy*, World Sustainability Series, https://doi.org/10.1007/978-3-031-92390-6_10

1 Introduction

The ocean plays a crucial role in global trade. They are important for global transportation, food production, tourism, aquaculture, and the preservation of biodiversity. They provide vital resources for both marine and human existences. Furthermore, the ocean economy generates approximately $1.5 trillion in value each year, highlighting its importance to society. Any changes in it can influence industries across the world. Achieving a balance between sustainability and profitability is vital in maintaining the stability of global economies.

The exploitation of ocean resources poses risks to habitats that are critical for ongoing development. Pollution remains a major issue due to its widespread presence. For example, in 2021, the ocean was burdened with 17 million metric tons of plastic, and projections that suggest this could increase two to threefold by 2040. Without proactive measures, this situation endangers industries such as aquaculture, tourism, and fishing. As a result, it becomes important to preserve the ocean. The blue economy aims to conserve and utilize the economic and environmental resources of the ocean. The significance of the blue economy is rooted in ocean ecosystems and industries. The blue economy aims not only for economic benefits but also strives to ensure a sustainable and stable environment for society's advantage. This concept has emerged partly due to the fact that over 80% of international trade is conducted via the oceans. Since the blue economy focuses on the stewardship and advancement of such a vast resource, harnessing big data could be essential.

A blue economy strategy focuses on gathering information to enhance, develop, or oversee essential elements of the ocean's economy and environment. This process is extensive and intricate due to the numerous factors involved. In the end, it conveys the significance of the ocean and coastal areas by transforming it into a practical plan. Various industries stand to gain from better management or growth, particularly as they can utilize contemporary technology to maximize profitability and sustainability. The objective of a blue economy is to safeguard and preserve these resources, thus a strategy should encompass developmental and growth initiatives that adhere to sustainable practices.

Big data highlights the significance of the blue economy strategy by providing a broader perspective. For instance, gathering data on water temperatures can lead to models with practical uses in the fishing sector. A vast array of data can offer crucial insights for various industries, enabling them to align, analyze, and adjust their engagement with ocean resources. While the process of collecting information is crucial, the key elements of big data and the blue economy lie in storage, processing, and distribution. Ensuring that all data is accessible and transparent is a vital component of an effective strategy, as it supports the global ocean economy.

Moving ahead to Big data analytics, it encompasses large, diverse collection of structured, semi structured and unstructured data that continuously grow over the period (Chong et al. 2023). Big data analytics refers to an advanced form of analytics that employs key data analysis methodologies including statistical algoriths, what if analyses, and predictive model, powered by analytics system. It is to process and analyse large volume of complex data, extract meaningful insights and develop solution. Organization can employ technologies like business intelligence (BI) tools to integrate structured and unstructured data collated from various sources (Anubhav et al. 2024). Users can interact with these tools by inputting queries to gain insights into performance and operations of the organization. Big data is not recent, with rudimentary forms of analytics, such as manually analyzing numerical data in spreadsheets, dating back to the 1950s. However, organizations are now recognizing its value in capturing and analyzing real-time data streams to derive actionable insights, especially when complimented with other emerging technologies such as artificial intelligence (Anwar et al. 2024). Big data analytics involves the systematic processing and analyzing of massive, complex **dataset** to glean valuable insights, identify patterns, correlations, and trends within raw data, which, in turn, enable informed decision-making (Akbari et al. 2024). By leveraging data generated from sources, such as social media, smart devices, IoT sensors, and financial transactions, organizations gain actionable intelligence using advanced analytical methods (Akhtar et al. 2024a).

The concept of big data gained prominence in the early 2000s as advancements in technology allowed organizations to manage large volumes of unstructured data. Technologies, such as Spark, Hadoop, and NoSQL databases help to address the challenges of storing and processing these vast datasets. Modern big data analytics relies on tools that handle data characterized by its volume (enormous quantities generated continuously from diverse sources), velocity (the speed of real-time or near-real-time data generation and processing), and variety (heterogeneous data types, including structured, unstructured, and semi-structured formats) (Cui et al. 2024).

The growth of digital technologies, such as IoT, AI, and mobile connectivity has further fueled growth in data, necessitating innovative tools that can swiftly collect, process and analyze data to maximize its value (Conti et al. 2024). Big data is now integral to machine learning, predictive modeling, and advanced analytics, enabling businesses to address challenges and supporting decision-making. The "3 Vs of big data," first identified by Gartner in 2001, remain central to understanding its scope and applications. The first V refers to volume, the most common characteristic of big data, collated from a variety of sources and devices on a continuous basis. The second is Velocity, referring to the speed at which data is generated, whether real-time or near real-time. This necessitates that data must also be accessed, processed, and analyzed at the same rate to have any meaningful impact. The third V refers to variety, implying that data is heterogeneous, or in other words, accessed through several different sources. Also, this data can be structured, unstructured, or semi-structured. More traditional structured data (such as data in spreadsheets or relational databases) is now supplemented by unstructured text, audio, images, video files, or semi-structured formats, such as sensor data that can't be organized in a fixed data schema.

In addition to these original "3 Vs" of big data, there are three more dimensions: Veracity: Big data can be noisy, messy, and error-prone, making it challenging to ensure its quality and accuracy. High veracity signifies reliable and trustworthy data, which is essential for meaningful analysis. Variability: Data meanings can shift overtime due to changes in context, collection methods, or interpretation, which can lead to inconsistencies. Value: The true worth of big data lies in extracting actionable insights that drive informed decision-making. Without determining its business value, big data remains underutilized.

With big data analytics, organizations can reap the following benefits:

a. Enhanced customer experiences: Combining structured and unstructured data helps tailor personalized experiences to meet customer expectations.
b. Improved risk management: Advanced analytics help assess and monitor risks, leading to stronger control and mitigation strategies.
c. Operational efficiency: Faster data processing reveals areas to cut costs, save time, and improve overall efficiency.
d. Big data analytics empowers organizations to harness their extensive data repositories effectively, unlocking transformative insights and driving competitive advantage (Liu et al. 2024a).
e. Improved decision-making: Data-driven insights reveal patterns and trends, driving better operational and strategic choices; Real-time intelligence: Organizations gain the ability to respond instantly to market changes, identify opportunities, and make quick, informed decisions (Makhloufi 2024).
f. Increased agility and innovation: Real-time data analysis ensures quick adaptation, and also supports the development and launch of new products and features. Continuous intelligence: Integrating real-time data streams with advanced analytics ensures ongoing discovery of growth opportunities.

1.1 Research Objective

a. To understand the type of Big Data analytics in companies
b. to study the companies using big data analytics
c. to understand the applications of big data into different sectors
d. to study the key benefits of big data analytics in the blue economy
e. to study the Data Sources for Blue Economy Analytics.

1.2 Research Methodology

The paper uses secondary data to understand the subject under study. Data has been collected from the company's website, newspapers, journals, magazines, and other websites.

1.3 Types of Big Data Analytics in Companies

Big data analytics is classified into four key types, with each type playing a crucial role in business decision-making. The first is descriptive analytics, which summarizes historical data to generate reports and visualizations. These help organizations identify trends and key performance areas, and focus on critical insights. For example, during the pandemic, a pharmaceutical company analyzed office and lab utilization, and identified unused spaces and departments, which helped it save millions (Kurrahman et al. 2024).

The second type, diagnostic analytics mines and analyzes data to examine why specific issues or events occurred. The findings enable organizations to uncover root causes and improve decision-making. For instance, an online retailer discovered through diagnostic analytics that a malfunctioning payment page was causing declining sales despite active shopping carts (Liu et al. 2024b).

The third, predictive analytics, utilizes historical and real-time data along with machine learning, AI, and data mining to forecast future outcomes. This type of analytics provides accurate predictions, helping businesses optimize operations such as scheduling and inventory management. A common application is in manufacturing, where companies can predict equipment failures using algorithms based on historical data, allowing for timely maintenance.

Finally, prescriptive analytics is used to recommend actionable solutions to optimize outcomes. Leveraging machine learning, AI, and statistical modeling, it aids in task prioritization, risk management, and maximizing financial gains. For example, companies in the energy sector analyze factors influencing oil and gas prices to mitigate risks and improve profitability. Together, these analytics types empower organizations to make informed, data-driven decisions and enhance their operational efficiency and strategic planning. For example, utility companies, gas producers, and pipeline owners in the energy sector can identify factors that affect the price of oil and gas to hedge risks.

1.4 Companies Using Big Data Analytics

Big data analytics is central to the operations and strategies of leading global companies like Apple, Amazon, Google, Spotify, and Starbucks, driving their innovation and customer-centric approaches. Apple, for instance, uses advanced big data technologies to guide its strategic decisions. By analyzing how customers interact with its apps and devices, the company refines product designs and aligns new offerings with customer preferences.

Amazon, as a leading e-commerce platform, meticulously tracks and analyzes every detail of customer transactions to understand purchasing behavior. It leverages this data in social media advertising algorithms to recommend products, enhance customer relationships, and improve overall customer experience and services (Zhao et al. 2024).

Google exemplifies the power of big data by analyzing parameters, such as location, search history, and trends. Using sophisticated algorithms, it delivers ranked search results tailored to user intent, emphasizing relevance and authority.

Spotify utilizes big data to revolutionize the music-listening experience. As a fully data-driven company, Spotify gathers insights on streaming locations, song playtimes, devices, and user patterns. This information not only enhances user experiences but also empowers artists through "Spotify for Artists," enabling better promotion and engagement strategies.

Starbucks, a globally recognized coffee brand, relies on big data to manage vast customer interactions. By collecting data from its mobile app and online orders, the company provides personalized recommendations, enhancing customer satisfaction and loyalty. Across these companies, big data analytics drives decision-making, innovation, and tailored customer experiences, setting benchmarks for leveraging data effectively in diverse industries (Akhtar et al. 2024b).

1.5 Applications of Big Data into Different Sectors

The possibilities for utilizing big data are vast across various sectors. Some of the sectors that uses big data analytics are as:

Healthcare: It aids in precise diagnoses and disease prediction, elevating patient care.

Retail: Amazon's use of Big Data Analytics offers personalized product recommendations based on your shopping history, creating a more tailored and enjoyable shopping experience.

Finance: Credit card companies such as Visa rely on Big Data Analytics to swiftly identify and prevent fraudulent transactions, ensuring the safety of your financial assets.

Transportation: Companies like Uber use Big Data Analytics to optimize drivers' routes and predict demand, reducing wait times and improving overall transportation experiences.

Agriculture: Farmers make informed decisions, boosting crop yields while conserving resources.

Manufacturing: Companies like General Electric (GE) use Big Data Analytics to predict machinery maintenance needs, reducing downtime and enhancing operational efficiency.

Blue Economy

The concept of big data is linked to the blue economy as it helps identify trends and leads to the development of beneficial models and strategies. The World Ocean Database, recognized as the oldest and most extensive global repository of publicly accessible oceanic information, encompasses over two decades of research and 20,000 datasets (Alles and Gray 2024). It offers essential insights for a variety of industries. Research related to water can harness big data to perform risk evaluations in water systems, enhance wastewater management, and anticipate pollution sources. Such initiatives can lead to healthier oceans and broader opportunities.

Analyze weather trends to enhance the efficiency of renewable energy sources. Both government agencies and energy firms can leverage these data sets to improve the effectiveness of hydropower projects. Data serves merely as a resource that anyone can apply for purposes pertinent to their industry.

Infrastructure and exploration sectors stand to gain significantly.

1.6 Key Benefits of Big Data Analytics in the Blue Economy

Big data enhances multiple facets of the blue economy by facilitating coordination among various entities in their utilization of natural resources. It can be a vital tool, as it has the potential to create sustainable and practical strategies. Big data analytics is essential to the blue economy, offering crucial insights into oceanic conditions, marine ecosystems, and human activities at sea, which supports improved decision-making for sustainable resource management, streamlined operations in sectors such as fisheries, shipping, and offshore energy, and ultimately contributes to the health and economic sustainability of the ocean environment.

- Sustainable Fisheries Management: Examining data related to fish populations, migration behaviors, and environmental influences enables tailored fishing methods, helping to avoid overfishing and support long-term sustainability.
- Enhanced Shipping Routes: Leveraging up-to-date information on weather patterns, ocean currents, and vessel movement fosters more effective shipping routes, leading to decreased fuel usage and lower emissions.
- Offshore Energy Development: Analyzing large datasets aids in determining the best sites for offshore wind farms and other renewable energy projects, enhancing energy output while reducing environmental consequences.
- Marine Pollution Monitoring: Monitoring ocean currents and evaluating information from sensors facilitates the effective identification of pollution hotspots and the development of mitigation measures.
- Climate Change Impact Assessment: Studying extensive data on ocean temperature, salinity, and changes in sea level improves the ability to predict the impacts of climate change and inform adaptation strategies.

- Biodiversity Conservation: Through data analysis, pinpointing vital marine habitats and species distributions aids in the planning and management of marine protected areas.
- Improved Coastal Management: Applying information about coastal erosion, storm surges, and rising sea levels promotes more effective planning for coastal development and disaster readiness.

1.7 Data Sources for Blue Economy Analytics:

- Satellite Imagery: Offers large-scale data regarding ocean conditions, sea surface temperatures, and chlorophyll concentration.
- Oceanographic Buoys: Gather information on water temperature, salinity, currents, and wave patterns.
- Underwater Sensors: Observe marine ecosystems and species distribution at greater depths.
- Vessel Tracking Systems: Deliver data on ship movements and potential conflicts.

2 Challenges in Implementing Big Data Analytics

Big data offers transformative potential for organizations but also presents significant challenges that must be addressed to maximize its benefits. One critical challenge is making big data accessible. As data volumes grow, organizations need to simplify data usage for stakeholders with varying levels of technical expertise (Malik et al. 2024). Another issue is maintaining data quality, as companies spend considerable time resolving duplicates, errors, and inconsistencies. Moreover, ensuring data security becomes increasingly crucial, requiring robust compliance measures and processes to protect against privacy risks.

Organizations also face difficulties in selecting the right tools and platforms. With new technologies emerging constantly, they must choose flexible solutions that align with existing systems and can adapt to future changes. Compounding this is the lack of skilled professionals, such as data scientists and engineers, which remains a significant barrier to fully leveraging big data. The rapid growth of data further complicates matters, as it demands scalable infrastructure for processing, storage, and security.

Other challenges include data quality issues, where messy or inaccurate raw data can lead to flawed insights and hinder decision-making. Compliance violations pose another risk, as big data often involves sensitive information that must adhere to stringent regulatory requirements like data localization and residency laws (Manroop et al. 2024). Lastly, integration complexity arises from siloed data across various systems and applications, making it essential yet challenging to unify data sources for effective analytics and business use.

Addressing these challenges requires organizations to invest in skilled personnel, robust infrastructure, and advanced technologies while maintaining a focus on security, accessibility and data governance.

Challenges in Implementing Big Data Analytics in the Blue Economy:

- Integrating the Data Integrating data from various sources with different formats and levels of quality can be demanding.
- Access and Sharing of Data ensuring unrestricted access to data is essential\ for fostering effective research and collaboration.
- Advanced Analytical Tools Creating advanced algorithms and models to derive valuable insights from intricate ocean data. Big data analytics has significant potential to revolutionize the blue economy by supporting informed decision-making for sustainable ocean management and economic growth.

2.1 The Future of Big Data Analytics in Blue Economy

The field of big data analytics is rapidly evolving, driven by the expanding scale and complexity of data, coupled with technological advancements. This evolution presents both significant challenges and opportunities. Big data and advanced analytics are central to this transformation, as the sheer volume, velocity, and variety of data demand cutting-edge tools and techniques that can process large datasets in real time (Feng et al. 2022). Technologies such as machine learning and artificial intelligence help to reveal patterns and correlations beyond the reach of traditional methods. However, their application raises ethical concerns, particularly in areas, such as algorithmic decision-making and predictive policing, necessitating careful scrutiny and responsible implementation.

Ethical considerations have become increasingly critical as the power and scope of data analysis grow. Issues, such as data privacy, and the responsible use of AI underscore the algorithmic bias, and the need for ethical practices in data collection, analysis, and application. (Sun et al. 2024) Transparency in data usage, addressing biases, and ensuring equitable benefits from data-driven insights are essential for creating trust and accountability in the field.

The complexity of modern challenges highlights the importance of interdisciplinary collaboration in data analysis. By bringing together expertise from industry-specific domains, data science, and ethics, organizations can develop holistic solutions that are technically robust and aligned with societal goals. For instance, tackling public health issues may involve collaboration among healthcare practitioners, data scientists, and policymakers, ensuring that data analysis serves the broader interests of communities (Fogarty and Campbell 2024).

Finally, the rapid pace of innovation underscores the need for continuous learning and adaptation among data professionals. Staying updated with emerging tools, methodologies, and ethical standards is crucial for leveraging data effectively. Organizations dealing in blue economy can support this by investing in employee training, thus, creating a culture of curiosity, and providing access to advanced technologies. This commitment to ongoing development equips data analysts to navigate future challenges and maximize the potential of data analysis for blue economy (Strauss et al. 2024).

3 Conclusion

The blue economy pertains to the responsible use of ocean resources for economic growth, improving living standards, and creating jobs, all while safeguarding marine ecosystems. It encompasses all economic activities associated with oceans, seas, and coastlines, including sectors like fisheries, shipping, marine renewable energy, and tourism. This concept has gained global traction, with nations like India prioritizing the blue economy as a foundation for future development (Gabrielli et al. 2024).

Current evidence strongly suggests that the future economic and ecological success of our planet relies on a thriving ocean. Concurrently, scientific findings indicate that we have pushed our planet, particularly the ocean, to a critical point where its capacity to deliver sustainable resources and benefits for humanity is at risk. Fortunately, the past year has witnessed unprecedented commitments from governments, businesses, the United Nations, and various stakeholders to tackle these challenges. The UN has designated 2021–2030 as the decade for Ocean science aimed at sustainable development to ensure we acquire the knowledge necessary to rejuvenate and restore the ocean's economic productivity (Jahromi and Ghazinoory 2024). Fourteen nations have collaborated to discover new approaches for a sustainable blue economy. The transformative agenda emphasizes the importance of utilizing ocean science, technology, and data as a main priority, advocating for efforts that facilitate transparent and open sharing and accessibility of ocean information.

Use of Big Data Analytics in blue economy is a transformative force that is reshaping the landscape towards a more intelligent future. Its impact spans across diverse domains, from enhancing healthcare and customizing shopping experiences to fortifying financial security and accurately predicting demand (Samadhiya et al. 2024). Nonetheless, challenges such as the management of vast datasets and the protection of privacy loom large. In a world inundated with data, Big Data Analytics serves as a beacon of guidance, enabling us to make astute decisions, providing tailored interactions, and unearthing invaluable insights (Sariyer et al. 2024). It stands as a robust and dependable instrument that holds the promise of a brighter and more effective future for all. Despite hurdles like data integrity, privacy issues, and the scarcity of skilled professionals, the real-world applications of data analysis in diverse sectors underscore its transformative capacity (Sha et al. 2024). As we navigate the path ahead in data analysis, addressing ethical quandaries, fostering interdisciplinary partnerships, and embracing a culture of continuous learning will be imperative to harnessing the full potential of this potent discipline. Ultimately, data analysis transcends mere numerical figures; it is about metamorphosing data into actionable insights that sculpt a more intelligent and well-informed future (Shaik et al. 2024).

Big data analytics holds significant importance as it enables blue economy to harness their data for the purpose of identifying opportunities for enhancement and optimization. Across various sectors, heightened efficiency results in more sophisticated operations, increased profitability, and contented clientele. The utilization of big data analytics aids blue economy in cost reduction and the development of superior, customer-centric products and services. Data analytics furnishes valuable insights that enhance the functionality of our society (Kim 2024). Within the realm of healthcare, big data analytics not only monitors and scrutinizes individual records but also plays a pivotal role in evaluating outcomes on a global scale. blue economies can employ big data analytics systems and software to make informed decisions based on data. These advantages encompass improved marketing strategies, novel revenue prospects, personalized customer experiences, and enhanced operational efficiency (Kumar and Mittal 2024). Blue economy leveraging big data alongside advanced analytics derive value across multiple dimensions. It helps in **Reducing cost.** Big data technologies like cloud-based analytics can significantly reduce costs when it comes to storing large amounts of data (for example, a data lake) (Mehmood et al. 2024). Real-time data collection can also immediately help you analyze data from various sources and quickly make decisions based on your findings. People in industries with a lot of equipment, such as manufacturing, can use big data to quickly determine when they need to repair machines; **Making faster, better decisions** (Riaz and Ali 2024). The speed of in-memory analytics—combined with the ability to analyse new sources of data, such as streaming data from IoT—helps blue economy analyse information immediately and make fast, informed decisions (Markussen and Teleki 2025).

References

Akbari M, Kok SK, Hopkins J, Frederico GF, Nguyen H, Alonso AD (2024) The changing landscape of digital transformation in supply chains: impacts of industry 4.0 in Vietnam. Int J Logistics Manage 35(4):1040–1072

Akhtar P, Ghouri AM, Ashraf A, Lim JJ, Khan NR, Ma S (2024a) Smart product platforming powered by AI and generative AI: personalization for the circular economy. Int J Prod Econ 273:109283

Akhtar P, Moazzam M, Ashraf A, Khan MN (2024b) The interdisciplinary curriculum alignment to enhance graduates' employability and universities' sustainability. Int J Manage Educ 22(3):101037

Alles M, Gray GL (2024) The marketing on Big 4 websites of big data analytics in the external audit: evidence and consequences. Int J Account Inf Syst 54:100697

Anubhav K, Dwivedi AK, Aashish K (2024) Entrepreneurship education in higher education (2002–2022): a technology-empowered systematic literature review. Int J Manage Educ 22(3):100993

Anwar MA, Zong Z, Mendiratta A, Yaqub MZ (2024) Antecedents of big data analytics adoption and its impact on decision quality and environmental performance of SMEs in recycling sector. Technol Forecast Soc Chang 205:123468

Chong CL, Abdul Rasid SZ, Khalid H, Ramayah T (2023) Big data analytics capability for competitive advantage and firm performance in Malaysian manufacturing firms. Int J Prod Perform Manage

Conti R, de Matos MG, Valentini G (2024) Big data analytics, firm size, and performance. Strategy Sci 9(2):135–151

Cui X, Zhu Z, Liu L, Zhou Q, Liu Q (2024) Anomaly detection in consumer review analytics for idea generation in product innovation: comparing machine learning and deep learning techniques. Technovation 134:103028

Feng T, Yang S, Sheng H (2022) Supply chain integration and novelty-centered business model design: an organizational learning perspective. Eur Manage J

Fogarty TJ, Campbell C (2024) The big data crossroads: accounting education and the challenge of 21st century technology. J Account Educ 68:100914

Gabrielli G, Magri C, Medioli A, Marchini PL (2024) The power of big data affordances to reshape anti-fraud strategies. Technol Forecast Soc Chang 205:123507

Jahromi GS, Ghazinoory S (2024) From superman to Garfield: four scenarios regarding the effects of industry 4.0 on animation companies. Technol Soc 77:102511

Kim ES (2024) Can data science achieve the ideal of evidence-based decision-making in environmental regulation?. Technol Soc 102615

Kumar S, Mittal S (2024) Employee learning and skilling in AI embedded organizations–predictors and outcomes. Dev Learn Organ Int J 38(4):11–15

Kurrahman T, Tsai FM, Sethanan K, Lim MK, Tseng ML (2024) Data-driven natural capital accounting model in Indonesia: impacts of environmentally related economic activities on ecological processes and services. J Clean Prod 143213

Liu H, Xu X, Cheng TCE, Yu Y (2024a) Building resilience or maintaining robustness: insights from relational view and information processing perspective. Transp Res Part E Logist Transp Rev 188:103609

Liu X, Lin J, Jiang X, Chang T, Lin H (2024b) EWOM information richness and online user review behavior: evidence from tripadvisor. J Theor Appl Electron Commer Res 19(2):880–898

Makhloufi L (2024) Predicting the impact of big data analytics capability and green absorptive capacity on green entrepreneurship orientation and eco-innovation. J Enterp Commun People Places Glob Econ 18(4):746–770

Malik M, Andargoli A, Clavijo RC, Mikalef P (2024) A relational view of how social capital contributes to effective digital transformation outcomes. J Strateg Inf Syst 33(2):101837

Manroop L, Malik A, Milner M (2024) The ethical implications of big data in human resource management. Hum Resour Manag Rev 34(2):101012

Markussen BT, Teleki K (2021) How data is transforming the way we care for the ocean. Accessed 20 Jan 2025 from https://www.weforum.org/stories/2021/04/how-a-digitized-version-of-the-ocean-can-help-the-planet-gtgs/

Mehmood K, Kautish P, Mangla SK, Ali A, Kazancoglu Y (2024) Navigating a net-zero economy future: antecedents and consequences of net-zero economy-based green innovation. Bus Strategy Environ

Riaz A, Ali FH (2024) Responsible innovation and sustainable competitive advantage: is big data the missing link? Bus Process Manag J 30(4):1213–1235

Samadhiya A, Agrawal R, Kumar A, Luthra S (2024) Bridging realities into organizations through innovation and productivity: exploring the intersection of artificial intelligence, internet of things, and big data analytics in the metaverse environment using a multi-method approach. Decis Support Syst 114290

Sariyer G, Mangla SK, Chowdhury S, Sozen ME, Kazancoglu Y (2024) Predictive and prescriptive analytics for ESG performance evaluation: a case of Fortune 500 companies. J Bus Res 181:114742

Sha K, Taeihagh A, De Jong M (2024) Governing disruptive technologies for inclusive development in cities: a systematic literature review. Technol Forecast Soc Chang 203:123382

Shaik AS, Nazrul A, Alshibani SM, Agarwal V, Papa A (2024) Environmental and economical sustainability and stakeholder satisfaction in SMEs. Critical technological success factors of big data analytics. Technol Forecast Soc Change 204:123397

Strauss LM, Klein AZ, Scornavacca E (2024) Adopting emerging information technology: a new affordances process framework. Int J Inf Manage 76:102772

Sun P, Yuan C, Li X, Di J (2024) Big data analytics, firm risk and corporate policies: evidence from China. Res Int Bus Financ 70:102371

Zhao G, Xie X, Wang Y, Liu S, Jones P, Lopez C (2024) Barrier analysis to improve big data analytics capability of the maritime industry: a mixed-method approach. Technol Forecast Soc Chang 203:123345

Greening the Future: Leveraging Fintech for Sustainable Blue Economy Investments

Ram Singh, Fazla Rabby, Ashutosh Rath Sharma, Jyotirmoy Banerjee, and Vivek Mishra

Abstract This study examines the ways in which fintech solutions such as blockchain, AI-powered risk assessment, and digital payment platforms can spur long-term investments in blue economy industries like coastal tourism, marine energy, and fisheries. The integration of fintech tools in promoting financial inclusion and risk reduction in green investments is investigated using a mixed-methods approach that combines quantitative data analysis and literature reviews. This chapter offers policymakers, investors, and entrepreneurs' practical insights by highlighting successful implementations and tackling issues including technology and regulatory constraints. Fintech's revolutionary role in bringing financial systems into line with the Sustainable Development Goals (SDGs) of the UN, especially SDG 14 (Life Below Water), is highlighted in the conclusion.

Keywords Blue economy · Green investment · Fintech · Sustainable finance · Marine ecosystem conservation

R. Singh (✉)
Symbiosis University of Applied Sciences, Indore, India
e-mail: ramsinghcommerce@gmail.com

F. Rabby
Stanford Institute of Management and Technology, Sydney, Australia

A. R. Sharma
Birla Global University, Bhubaneswar, Odisha, India

J. Banerjee
Amity Law School, Amity University, Bengaluru, India

V. Mishra
Department of Pure and Applied Mathematics, Alliance University, Bengaluru, India

V. Sharma et al. (eds.), *FinTech for Sustainable Finance and a Well-Blue Economy*, World Sustainability Series, https://doi.org/10.1007/978-3-031-92390-6_11

1 Introduction

As awareness grows of the importance of oceans and coastal areas for sustainable development, the concept of the "blue economy" has gained significant momentum. The "blue economy" is the sustainable utilization of ocean resources for economic growth, job creation, and livelihoods while at the same time preserving the health of marine ecosystems. It spans a wide range of industries, including fisheries, aquaculture, marine biotechnology, renewable energy, and coastal tourism. According to the Organization for Economic Co-operation and Development (OECD), the ocean economy could generate over USD 3 trillion in value a year by 2030 if well-managed sustainably (OECD 2020a). To achieve this vision, we must address key challenges such as overfishing, climate change, ecosystem degradation, and lack of adequate funding sources. There have been a series of challenges that have made financing the blue economy difficult in the past. Traditional financial systems often ignore the special risks of ocean-based investments, such as environmental unpredictability, lengthy payback times, and regulatory complexity. Additionally, the marine environment is poorly understood, usually leading to the perception of marine conservation and restoration projects as low-return, high-risk investments, which weakens opportunities to engage the private sector (World Bank 2017).

1.1 Green Investments in the Blue Economy

Green investments (financial allocations designed to help ecologically sustainable projects) may be one solution to the finance challenges hindering the blue economy. These investments align financial flows with sustainability goals, enabling the development of climate resilience initiatives, marine conservation areas, and renewable energy infrastructure in coastal areas. Investments in offshore wind energy, for example, have caught on as a feasible means of tapping the vast energy potential of seas and mitigating greenhouse gas emissions. Likewise, sustainable forms of aquaculture that are financed with green money are helping to decrease pressure on the beleaguered stocks of wild fish while delivering nutritional security to growing populations (FAO 2021). Green investments in the blue economy show potential but are constrained by various challenges, such as a lack of transparency, high transaction costs, and limited access to international finance markets for local communities and small and medium-sized enterprises (SMEs). That is where fintech, or financial technology, can be a disruptive power. India has a huge 7,500 km coastline, which makes it one of the key actors in the blue economy. India's socioeconomic development will be intimately tied with the industries that comprise its blue economy: fisheries, aquaculture, maritime logistics, renewable energy, and coastal tourism. Such industries, however, face increasingly serious environmental challenges, including habitat destruction, overfishing, pollution of the marine environment, and intensifying impacts of climate change (Ministry of Earth Sciences 2021). India's blue economy is poised to be the next great growth opportunity; this potential, however, is still underfunded against its sustainability needs. For instance, the Indian fisheries, a USD 14 billion-a-year industry, face challenges like hyper-exploitation of resources and shrinking biodiversity. Just like this, to tap the untapped potential of renewable offshore wind and tidal energy in the coastal lines of India, colossal funding is a must for research, infrastructure, and scale (Central Marine Fisheries Research Institute (CMFRI) 2022).

1.2 Innovative Mechanisms

Blue Bonds and Blended Finance are highly innovative mechanisms for financing sustainable marine development. Just as green bonds are financial tools for funding environmental projects, blue bonds are dedicated financial instruments used to fund those that assist marine ecosystems. In (2018), Seychelles was the first to issue sovereign Blue Bonds (BBSB), borrowing money for marine conservation and sustainable fishing. India might take a similar approach to fund its huge potential in the Blue Economy. Projects would include such things as coastal infrastructure, renewable energy, and conservation of biodiversity. Blended finance, combining public and private resources to de-risk investments, has proved effective at attracting private capital into fast-growing industries. Publicly private partnerships in the aquaculture and marine waste management sectors, for instance, fully exploited blended finance's potential to reach sustainability objectives (OECD 2020a).

1.3 Fintech and Digital Innovations in Blue Economy Financing

Financing Financial technology (fintech) is radically transforming financing for the blue economy, increasing accessibility, transparency, and efficiency. Blockchain technology promotes accountability in the disbursement of funds, thus making it more suitable for monitoring investments made on marine conservation and pollution control projects. Crowdfunding platforms allow for communities to garner funds for localized projects, whether that is beach clean-ups or replanting mangroves, ensuring community engagement and ownership. Also, risk analysis tools that use AI help figure out whether blue economy projects are financially and environmentally viable, which makes it easier to make decisions based on data. For India, these new ideas can be used to fill in gaps in financial inclusion with fintech solutions, and small-scale stakeholders in coastal communities can get help (World Bank 2020a). Fintech is the intersection of finance and technology, including a wide spectrum of digital innovations from blockchain and artificial intelligence (AI) to machine learning and lots of mobile banking.

2 Review of Literature

Financing the blue economy starts with a paradigm shift in how humanity views its interaction with its oceans and marine resources. The vast, wild nature of oceans should make us rethink strategies toward resource extraction, according to Smith and other early 2000s scholars. This aspiration was brought to fruition at the United Nations Conference on Sustainable Development (Rio + 20) in 2012, with the blue economy emerging as the international discourse's centrepiece. The Blue Economy focuses on using ocean-based resources while balancing sustainability rather than extracting marine resources. Smith outlines this new economic model in which the focus is on balancing growth with ecological preservation. When the World Bank published its comprehensive report on the potential of the blue economy, the rest of the world is starting to recognize oceans' significant contributions to the economy (World Bank 2017). As noted by OECD, a main concern is the gap between the necessary financing for sustainable marine projects (Organisation for Economic Co-operation and Development (OECD) 2020b). Using blended finance mechanisms to de-risk private investments is one of the most promising solutions. That was echoed by Blended Finance and Coastal Ventures who spoke on how blended finance can catalyze sustainable projects at scale, whether they be offshore wind farms or sustainable aquaculture ventures (Finance and Ventures 2019). UNEP emphasized that overfishing of marine resources poses far-reaching implications, threatening our food security while threatening the livelihoods of millions, especially in coastal communities (UNEP 2017). To turn around fish stock depletion, scholars noted that it requires substantial investment in sustainable fisheries management (Allison et al. 2009). According to Kusworo et al., sustainable aquaculture has its own problems, such as impact on the environment and accessibility for local communities (Kusworo et al. 2020). IRENA acknowledged that offshore wind has the potential to transform the energy mix while providing new economic opportunities in coastal regions (IRENA 2021). With its extensive coastline, India has made significant progress towards offshore wind energy, which has the potential to significantly transform the energy landscape. According to the Ministry of New and Renewable Energy, offshore wind presents an opportunity that could see 50 GW deployed to contribute to the country's renewable energy goals and job opportunities in manufacturing, construction, and maintenance (Ministry of New and Renewable Energy 2020a). India also unveiled its plan to develop 30 GW of offshore wind capacity by 2030 in 2020, and with that, it has taken a big leap towards financing the blue economy in its green infrastructure. Baker and Hughes underlined that offshore wind projects involve high upfront capital costs, but in return, this investment pays off in the economic, environmental, and sustainable long term (Hughes 2019). To address this challenge, an innovative funding solution to generate capital for marine conservation projects, called blue bonds, has recently hit the financial stage. In 2018, Seychelles became the first nation to issue a sovereign blue bond, raising US$ 15 million to support sustainable fisheries and marine conservation (Seychelles Government 2018). That success has prompted other countries to explore similar types of securities. OECD has found that investors are especially keen on blue bonds for the specific allocation of funding to projects with definitive environmental impact (OECD 2020a).

Countries like India are exploring the concept of blue bonds for marine conservation and coastal development. Baker & Hughes also mentioned the usage of blue bonds, including for coral reef restoration, pollution control, or sustainable fishing practices (Baker & Hughes 2019). As highlighted by the World Bank, platforms facilitate raising small amounts of capital from a large number of people to serve as a market for capital to fund local initiatives, such as mangrove restoration and plastic waste removal delivery projects (World Bank 2020b). The Indian government is now worried about cargo movement through the water with a brand-new program called the "Sagarmala Programme," which focuses on port connectivity, coastal shipping, and sustainable maritime infrastructure (Ministry of Shipping 2021). However, the financing of these large-scale infrastructure projects needs not only public funds but also private investors to be involved. But as Salagrama pointed out, the key challenge is to make sure that investments are benefiting local, sustainable, small-scale fishers and coastal communities, which are often sidelined in large infrastructure projects (Salagrama 2006). Tisdell examined the potential for ecotourism, when sustainably managed, to be a significant source of coastal economic development (Tisdell 2013). The lack of adequate regulatory policies and frameworks relates to the misallocation of resources and inefficient resource use and funds (Bavinck et al. 2013), as stated by Bavinck et al. Kusworo et al. also say that policies should take into account the needs of local communities by giving them access to the money that comes in from blue economy activities like fishing and tourism (Kusworo et al. 2020).

2.1 Research Gap

There is a lack of comprehensive frameworks that integrate sustainable financing mechanisms with local and global environmental goals, something that has been identified as a major research gulf in financing the blue economy. Although some research has made the case for the blue economy as an emerging area of economic opportunity, the nuances of scaling up financing for marine conservation, sustainable fisheries, and marine renewable energy at scale are still relatively unexplored. Limited research has been done on ways to harmonize public and private funding, particularly in developing countries like India that usually lack the infrastructure and capacity to house blue economy initiatives on a large scale. Bridging these gaps could offer actionable insights for policymakers, investors, and conservationists working to strike a balance between economic development and ecological sustainability.

3 Objectives and Methodology

The idea of this paper is to investigate how financial technologies (fintech), like blockchain, AI-based risk analysis, and digital payment gateways, can help blue economy sectors like fishing, marine energy, and coastal tourism get more money. This chapter uses a mix of research methods, including literature reviews and a discussion of quantitative data, to look at how fintech tools can be used to help more people get access to money and how risk identification techniques can be used in green investments.

4 Fintech Solutions in Sustainable Investments in the Blue Economy

The sectors that fall under the blue economy are fisheries, marine energy, coastal tourism, and more. Each faces their own finance-related challenges. As the world strives to find harmony between pollution-heavy sectors and economic growth, fintech has emerged as a catalyst for fostering sustainable investments within these industries. Here we explore how fintech can help strengthen three sectors of the blue economy: fisheries, marine energy, and coastal tourism.

4.1 Fisheries: Blockchain and Green Bonds for Sustainable Investments

Fisheries are an important sector in the blue economy for their contribution to global food security, livelihoods, and economic development. If fintech solutions that are being researched in the field of blockchain and green bonds (investments in conservation projects) help to improve traceability for sustainable fisheries management, achieve greater transparency, and leverage more investments for conservation initiatives.

4.1.1 Blockchain for Traceability and Transparency

Blockchain technology has transformed several sectors, fisheries included, by recording transactions through a transparent and non-alterable ledger. To that end, blockchain technology is already being used to trace the journey of seafood products from ocean to plate, preventing unsustainable seafood from reaching the shelves (Sharma et al. 2024). The World Wildlife Fund (WWF), for instance, has collaborated with blockchain startups to monitor seafood worldwide supply chains to enable consumers to verify whether the seafood they buy is sustainably obtained

(WWF 2020). By utilizing blockchain fisheries, we can reduce illegal, unreported, and unregulated (IUU) fishing, a significant issue that results in a $36 billion annual loss for the global economy (FAO 2021).

4.1.2 Green Bonds for Fisheries Conservation

Green bonds are another fintech solution gaining traction in the fisheries sector. These bonds allow governments, corporations, and institutions to raise capital specifically for environmental projects, such as sustainable fisheries management or marine ecosystem restoration. For instance, the Seychelles issued a sovereign blue bond in 2018, raising $15 million to support sustainable fisheries and marine conservation initiatives (Seychelles Government 2018). Similarly, the World Bank's "Blue Bonds" initiative has helped raise funds for sustainable ocean-based projects globally (World Bank 2017).

4.2 Marine Energy: Fintech Solutions for Offshore Wind and Ocean Energy Projects

Marine renewable energy, including offshore wind and tidal energy, is gaining momentum as a sustainable alternative to fossil fuels. However, the high upfront costs of offshore wind infrastructure and the technical risks involved in harnessing ocean energy present significant barriers to investment. Fintech solutions such as blockchain-based funding platforms, digital securities, and green bonds can help mitigate these challenges by improving access to capital and enhancing project transparency.

4.2.1 Blockchain for Project Management and Risk Mitigation

Offshore wind energy projects require substantial investment, often exceeding $1 billion for large-scale developments. Blockchain technology can play a critical role in reducing project costs and improving investor confidence. By creating transparent records of investments, expenditures, and timelines, blockchain helps to ensure that funds are used efficiently (Hughes 2019).

4.2.2 Green Bonds for Offshore Wind Projects

Green bonds have become an essential tool in financing marine energy projects, particularly offshore wind farms. For example, the European Union's green bond market has seen significant growth, with funds allocated to offshore wind projects across the continent. The EU raised €4 billion through green bonds in 2020 alone, with a portion directed toward offshore wind developments (European Investment Bank 2020). Similarly, India's Ministry of New and Renewable Energy has identified offshore wind as a critical component of its renewable energy strategy, with plans to generate 30 GW of offshore wind energy by 2030 (Ministry of New and Renewable Energy 2020a). The EU raised €4 billion in green bonds in 2020, part of which supported offshore wind projects (European Investment Bank 2020). India plans to generate 30 GW of offshore wind energy by 2030, with significant private sector investment expected (Ministry of New and Renewable Energy 2020b).

4.2.3 Green Bonds for Offshore Wind Projects in India

Offshore wind energy is a crucial component of India's renewable energy strategy, especially considering the country's vast coastline and high wind potential. However, financing large-scale offshore wind projects involves significant capital, long gestation periods, and high risks. To address these challenges, green bonds have emerged as a promising financial tool to attract investment for offshore wind energy development. Green bonds offer a structured way to raise capital while ensuring that the funds are exclusively used for environmentally sustainable projects.

4.2.4 Green Bonds in Financing Offshore Wind Energy

India's government has set an ambitious target to generate 30 GW of offshore wind energy by 2030 as part of its broader goal to achieve 500 GW of non-fossil fuel energy by 2030 (Ministry of New and Renewable Energy 2020b). To meet these objectives, financing offshore wind projects will require substantial investment, with estimates suggesting that offshore wind projects in India could require up to $10 billion over the next decade (Finance 2021). Green bonds, which have become a popular mechanism for raising capital for renewable energy projects globally, can play a critical role in financing offshore wind in India.

4.2.5 Green Bonds Issuances in India

The concept of green bonds in India gained traction in 2017, when the first green bond issued by a renewable energy company, ReNew Power, raised ₹2,000 crore to fund its renewable energy projects (Indian Ministry of Finance 2017). Since then, the green bond market in India has expanded, with several issuances targeting renewable energy sectors like solar, wind, and even offshore wind. For offshore wind projects, the potential for green bonds is significant. For example, in 2021, India's National Investment and Infrastructure Fund (NIIF) announced plans to issue green bonds to fund its offshore wind projects (NIIF Green Bond Issuance for Offshore Wind Projects 2021). The NIIF raised ₹3,000 crore (approximately $400 million) through green bonds to support the offshore wind energy infrastructure along India's coasts, particularly in states like Gujarat and Tamil Nadu, which have high offshore wind potential (NIIF 2021).

4.2.6 International Green Bond Market and Offshore Wind

The global green bond ecosystem supports India's green bond market, especially in the context of offshore wind energy. International green bonds are also usually more liquid and more accessible to a wider range of investors. Another example is the Asian Development Bank's decision to issue a $500 million green bond in 2020 through the International Finance Corporation (IFC), a member of the World Bank Group, with the aim of raising funds for either onshore or offshore wind projects in emerging markets such as India (International Finance Corporation (IFC) 2020a).

4.3 Coastal Tourism: Crowdfunding and Digital Platforms for Sustainable Development

Coastal tourism, an important segment of the blue economy, generates billions every year and provides livelihoods for millions of people worldwide. But uncontrolled tourism can cause environmental degradation in sensitive coastal ecosystems. Innovative fintech tools such as crowdfunding platforms, digital payment solutions, and blockchain help countries to capitalize on tourism in a sustainable way to better match economic development goals with environmental protection goals.

4.3.1 Crowdfunding for Eco-Tourism Initiatives

Crowdfunding has emerged as a financial tool that has recently been used to finance ecotourism projects, permitting them to promote sustainable travel while protecting natural diversity and cultural heritage. Through sites like Kickstarter, GoFundMe, or Indiegogo, landscape projects, including ecotourism, can access a global market of individuals who value and are willing to invest in environmental conservation and sustainable tourism (Global Impact Investing Network (GIIN) 2021). In the Philippines and the Maldives, crowdfunding has also been used to put money towards coral reef restoration projects, where those who are willing to go back (or visit) the region for eco-friendly travel experiences donate a portion of their funds to restoring the natural habitat. This type of model not only democratises access to funding but also encourages community engagement and awareness about sustainable tourism (GoFundMe 2020).

4.3.2 Blockchain for Eco-Tourism Transparency

Blockchain technology can also be used in coastal tourism to strengthen trust and transparency. Through a decentralised ledger, blockchains can track the use of money in ecotourism operations and show that the collected funds have been used for sustainable practices, such as waste disposal, energy efficiency, and conservation measures. Some eco-resorts have already adopted blockchain in order to ensure that the money they collect from tourists is going to conservation projects such as marine biodiversity protection and coastal habitat restoration. GoFundMe has successfully raised over $1.3 million for reef restoration programs in the Philippines (Tisdell 2013). The Maldives itself is a proponent of blockchain, as it is currently holding more than $50 million worth of summary ecotourism investments for marine conservation projects based on its new blockchain technology that keeps an eye on where this money goes, thus ensuring transparency of funds (GoFundMe 2020). Fintech-based solutions are increasingly essential for financing sustainable investments in the blue economy (Hughes et al. 2019).

4.4 AI-Driven Risk Assessment in Sustainable Investments in the Blue Economy

Artificial Intelligence (AI) is rapidly changing the landscape of sustainable investments in the blue economy, particularly about the assessment and mitigation of risks from marine-related projects. AI-powered risk assessment solutions, on the other hand, make it possible to conduct more accurate, data-driven evaluations that allow investors to make better decisions and minimize their risk of losses (Sharma et al. 2024).

4.4.1 Enhancing Risk Prediction with AI

AI algorithms can be trained on varied large datasets that can derive accuracy in risk prediction from satellite images, sensor readings, oceanography studies, and market patterns. AI technology can also start to pick up trends and anomalies that would be overlooked by conventional methods altogether. Investors can therefore understand their long-term investment potential much better. AI, for example, has been used in the fishery sector in estimating stocks of fish and sustainability trends. This keeps investors away from exhaustion risks while promoting a sustainable approach to fish fattening.

4.4.2 AI in Credit Risk Modeling for Sustainable Financing

In addition, machine learning-based credit risk models are improving the ability to evaluate the financial feasibility of blue economy investments. For green bonds or any such investment vehicle, AI helps assess the financial health of a company working in marine energy or eco-tourism by analyzing a large number of variables, including environmental compliance, past project performance, market conditions, and more. Through automated credit risk analysis, AI allows investors to pick out high-risk investments more accurately and rebalance their portfolios. Further, AI's ability to monitor regulatory and policy shifts across multiple jurisdictions might allow investors to identify potential breakpoints in the blue economy. According to a recent report by PwC, the financial services industry could potentially save up to $1 trillion annually by implementing AI-driven risk management tools, which can significantly enhance risk identification and decrease operational losses (PwC 2021). The importance of AI-based risk assessment solutions for sustainable and resilient investments in the Blue Economy AI drives investors to understand the marine industries better by enhancing the accuracy of risk predictions and providing actionable insights. AI in risk assessment instills confidence and enables companies to move toward data-driven financing for blue economy projects (Accenture 2021).

4.5 Digital Payment Platforms for Sustainable Investments in the Blue Economy

Digital payment platforms have played a crucial role in enabling sustainable investments in the blue economy, ensuring seamless, secure, and efficient financial transactions. With the growing blue economy, the needs for flexible, scalable, and transparent payment solutions are also on the rise. Many digital payment systems, like mobile wallets and blockchain-based transactions, can handle a lot of small investments for very little money. They can also make transactions clear to make the investment process easier, which is very important for keeping marine and coastal economies going.

4.5.1 Enabling Accessibility for Micro-investments

One of the primary barriers to investing in the blue economy is the significant financial investment required for many initiatives, particularly in sectors such as offshore wind vitality or large-scale marine preservation. Digital fee platforms smash down this limitation by enabling tiny stake chances. Platforms, for instance, M-Pesa in Kenya and Crowdcube in Britain, have demonstrated the ability to pool miniature amounts of cash from a huge variety of investors, rendering it simpler for people to invest in blue economic projects without requiring substantial monetary assets. This democratization of investment makes it possible for the investment of sustainable plans like coral reef reclamation, sustainable fisheries, and eco-pleasant travel enterprises, giving access to investors from each nearby and worldwide market. Moreover, crowdfunding platforms have long allowed smaller projects with tighter budgets to get financing, including local conservation groups working to preserve mangrove forests or support sustainable aquaculture.

4.5.2 Facilitating Cross-Border Transactions

The blue economy is, by definition, transnational, as projects usually go across borders, be it marine conservation, an international cruise ship company operating across regions, or a worldwide wind farm developer. Digital wallets allow Party A to send Party B a digital payment, which is then paid in the local currency without the headache of converting to foreign currency, transaction fees, and complicated international banking regulations. Bitcoin, Ethereum, and other cryptocurrency-based platforms are emerging as global transaction alternatives.

4.5.3 Blockchain for Transparent and Secure Transactions

Blockchain technology, one of the main components of digital payment platforms, is increasingly being employed in sustainable investments in the blue economy because the integrity of the technology allows for transparency and security, making it ideal for use in sustainable investments. Blockchain provides a mechanism that guarantees that every transaction connected to each Blue Economy investment, from financing renewable energy projects to donations for marine conservation, is accurately captured through decentralised, tamper-proof ledgers. Such transparency builds confidence among those investing in the blue economy sectors, often skeptical about their financial integrity, critical to attracting funding. A recent report by the International Finance Corporation (IFC) found that in emerging markets such as Kenya, India, and Indonesia, successive mobile money and digital payment platforms have raised over $10 billion in green investments in renewable energy, agriculture, and ecotourism projects (International Finance Corporation (IFC) 2020a). Blockchain technology is reported to reduce transaction costs by about 30% within the renewable energy sector, facilitating quicker and cheaper investments to flow into blue economy projects and such digital payment platforms can transform the recovery process by enabling sustainable investments in the blue economy (PwC 2021). These platforms are contributing to the expansion of marine ecosystems by allowing micro-investments, enabling cross-border transactions, and committing to transparency through blockchain technology.

4.6 Integration of Fintech Tools in Green Investments

Fintech tools have been a significant catalyst in integrating green investments, enabling financial inclusion, and managing environmental, social, and governance (ESG) risks. With the global transition to a more sustainable future, there is a growing need for capital to finance green projects such as renewable energy, green infrastructure, and sustainable agriculture. Leading fintech tools like digital payment methods, blockchain, crowdfunding, and AI analytics are bridging this gap, enabling green finance opportunities with more inclusivity, transparency, and resilience.

4.6.1 Fostering Financial Inclusion Through Digital Platforms

The emergence of digital finance platforms has been a game changer for individual and small business access to capital as they democratised access to funds, especially in developing countries where the traditional banking system has a limited footprint. Mobile money systems, such as M-Pesa in Kenya and Paytm in India, have unlocked access for millions more to the financial ecosystem, including sustainable project investing. Digital platforms provide examples of how small-scale farmers can access micro-loans to implement sustainable agricultural practices, such as investing in renewable energy or water conservation technologies. When these fintech solutions are used in green investments, they greatly increase financial inclusion by making it possible for people who don't have much money to make investments that are good for the environment and get good returns.

4.6.2 AI and Big Data for Risk Mitigation in Green Investments

Green investments are particularly exposed to environmental, social, and government risks, such as climate change, government regulations, and market changes, which represent a major challenge. To achieve high carbon reduction standards, it is vital to identify, evaluate, and mitigate risks using cutting-edge technologies; hence, fintech tools, especially AI and big data analytics, are being increasingly incorporated into green investment strategies. These AI algorithms use data on a much bigger scale satellite data, weather data, financial data, and so on making them more than capable of predicting the future possible risks in green investments.

4.6.3 Blockchain for Transparency and Accountability

Similar to mobile payments, blockchain technology is a vital fintech tool used in risk mitigation in green investments. This reduces the chances of fraud, mismanagement, or corruption, as the funds can only be spent in alignment with the original spend allocation through a decentralised and transparent ledger system of the blockchain. These mechanisms ensure that funders will only release funds once certain criteria around sustainability are met, thus promoting accountability amongst the project developers and countering the risk of funds being misused. In addition, the transparency offered by blockchain allows investors to trace the development of their investments in real time, which creates more trust and encourages greater participation in green investments.

4.6.4 Crowdfunding for Risk Diversification

Crowdfunding platforms also contribute to risk mitigation by allowing investors to diversify their portfolios. By pooling investments from multiple individuals, crowdfunding spreads the risk across various projects, reducing the financial exposure of any single investor. This collective funding model is particularly useful in green investments, which often involve long-term projects with uncertain outcomes. Crowdfunding platforms such as Green Unite and Ecosia support green projects by allowing investors to contribute to multiple initiatives at once, helping to mitigate individual risk. According to a report by PwC, the integration of AI and blockchain in green investments could reduce transaction costs by 30–40%, enhancing the efficiency of investment management and improving risk mitigation (PwC 2021). In addition, a study by the International Finance Corporation found that fintech solutions, such as mobile money platforms and AI-driven analytics, could increase the flow of capital to sustainable projects by 50% over the next decade, significantly boosting financial inclusion in the green economy (IFC Green Bonds for Offshore Wind Projects in Emerging Markets 2020b). Moreover, a report by the Global Impact Investing Network (GIIN) noted that 67% of impact investors are now using AI tools to assess environmental risks and monitor the impact of their investments (Global Impact Investing Network (GIIN) 2021). This reflects a growing trend of integrating fintech solutions to enhance decision-making and improve the risk-adjusted returns of green investments. The integration of fintech tools into green investments is a game-changer for both financial inclusion and risk mitigation.

5 Findings and Suggestions

5.1 Findings

5.1.1 Increased Financial Inclusion Through Digital Payment Platforms

Fintech has greatly improved financial inclusion, particularly in emerging markets with unbanked populations, with tools such as mobile money systems and digital wallets. Services like M-Pesa (Kenya), Paytm (India), and mobile banking solutions across Southeast Asia have enabled people and businesses to make financial transactions, including green project investments (International Finance Corporation (IFC) 2020a).

5.1.2 Risk Mitigation Through AI and Big Data

AI and big data analytics have been deployed under green investment strategies to identify and mitigate risks more effectively. As AI can take huge amounts of data into consideration, it can give a more accurate prediction of risk in projects like renewable energy or sustainable agriculture. Models driven by AI also deal with determining the most efficient location for sustainable energy projects that can be beneficial to them and lower risk for investors (PwC 2021).

5.1.3 Blockchain for Transparency and Accountability

Blockchain technology can offer a high level of transparency in green investments, enabling a clear view of how funds are allocated and where they are used. This largely mitigates the risks of financial fraud and thereby enhances investor confidence in the integrity of green projects (Global Impact Investing Network (GIIN) 2021).

5.1.4 Crowdfunding Platforms for Risk Diversification

Crowdfunding platforms, such as Kickstarter and Green Unite, have become relevant financing instruments for green initiatives, particularly for high-capital or long-term horizon projects. Crowdfunding also allows thousands of people to invest small amounts into a number of different projects, which helps spread out the risk (Global Impact Investing Network (GIIN) 2021).

5.1.5 Scalability and Long-Term Viability of Green Projects

Fintech tools are helping not only to improve access to capital but also to ensure the long-term viability and scalability of green projects. Fintech solutions further promote a more resilient, greener, and profitable economy by providing the means for secure and efficient ways to invest in and monitor projects (Accenture 2021).

5.2 Suggestions

5.2.1 Expand Mobile Payment Platforms to Rural Areas

Establish partnerships between governments and fintech companies to expand the mobile payment systems in the unserved sectors. Such a scheme would allow rural populations, who are often involved in environmentally friendly farming or renewable energy initiatives, to receive financing.

5.2.2 Enhance AI Algorithms for Local Climate Risk Assessment

AI-driven analytics are helpful but still require specific local AI models that can take into account environments and market data. Policymakers and financial institutions should make investments in AI tools that look at regional factors such as local climate risks, market dynamics, and environmental policies (Accenture 2021).

5.2.3 Promote Blockchain Adoption for Green Project Transparency

Blockchain technology is increasingly being used in green investments to improve transparency and accountability. New companies adopt blockchain systems for tracking the progress of green projects, governments and international financial institutions may also, in return, provide them with incentives or funding (PwC 2021).

5.2.4 Scale Crowdfunding for Green Investments Globally

The crowdfunding model presents an opportunity to increase sustainability efforts on a global scale. Existing platforms have demonstrated potential in cutting costs and incorporating impact tracking, yet there is still room for further efforts to encourage small investors globally to support green initiatives. Transaction fees remain high in some regions, and investor protections require strengthening, particularly for those in developing markets.

5.2.5 Encourage Collaboration Between Financial Institutions and Fintech Providers

A key determinant of fintech tools' accessibility and effectiveness will depend on building cross-sector partnerships; those traditional financial institutions with their expertise in handling market risk scenarios might well collaborate with innovative technological providers. These partnerships can also help fill the knowledge gap in financial literacy, especially in emerging markets where an understanding of sustainable finance may be absent (Global Impact Investing Network (GIIN) 2021).

6 Conclusion

Fintech technology could be newly implemented into sustainable investments within the blue economy and completely change how we make investments, monitor them, and how they scale. Fintech provides these tools via digital payment platforms, AI-driven analytics, blockchain technology, and crowdfunding to help address financial inclusion, risk mitigation, and sustainability of green projects in the long run. Therefore, these tools provide underserved communities, particularly those in remote or rural areas, with access to financing for eco-friendly initiatives, thereby filling a significant gap in traditional financing systems. Investors are now able to evaluate and forecast environmental and market risks more effectively with the use of AI and big data analytics, resulting in more rational decisions. Through the transparency features of blockchain, accountability in respect of green investments is facilitated, thereby limiting the chances of fraud and mismanagement of funds. Crowdfunding, by gathering small amounts of money from a broad base, opens access to green investments to many people, letting them participate in sustainable development projects while reducing their risk. Innovative financial technologies can bridge the blue economy with the capital it needs to scale sustainably and project viability, transparency, and inclusiveness. For the blue economy to reach its full potential, fintech itself will be an essential element of the smart and sustainable future we can create together. With such innovations and opportunities, the relatively new field of fintech can effectively contribute to positive environmental change in the blue economy, which, in turn, underscores new opportunities for revenue generation, investment diversification, and risk management that can be harnessed for positive change in our oceans and waterways, benefitting generations to come.

7 Directions for Further Research

There are several areas in which further research should be conducted about the respective integration of fintech tools in sustainable investments within the blue economy. There is also a need for detailed skills studies to focus on the impact of AI and big data analytics for prediction and mitigation of risks related to each marine and coastal industry sector in the maritime domain, as climate change poses unique challenges in each of these sectors. Second, there is scope for research that addresses the creation of more localized fintech solutions that are better suited to the environmental, economic, and cultural context in which they operate, especially in emerging markets where investment opportunities for green action are plentiful but undercapitalized.

References

Accenture (2021) AI in renewable energy and environmental sustainability. Accenture. https://www.accenture.com/us-en/insights/ai-renewable-energy
Allison EH et al (2009) The economics of coastal and marine resources in developing countries. Mar Policy 33(4):571–579. https://doi.org/10.1016/j.marpol.2008.09.012
Baker & Hughes (2019) Financing offshore wind: harnessing the blue economy. Baker & Hughes. https://www.bakerhughes.com/l
Bavinck M et al (2013) Managing fisheries in the blue economy: a framework for governance. Mar Policy 39:95–105. https://doi.org/10.1016/j.marpol.2012.09.001
Blended Finance and Coastal Ventures (2019) Leveraging blended finance for coastal projects. https://www.coastalventures.com/
Bloomberg New Energy Finance (2021) India offshore wind energy investment forecast. Bloomberg. https://www.bnef.com/
Central Marine Fisheries Research Institute (CMFRI) (2022) Annual report 2021–22. CMFRI, Kochi. http://www.cmfri.org.in
European Investment Bank (2020) EIB green bonds: financing offshore wind projects in emerging markets. European Investment Bank. https://www.eib.org/en/
FAO (2021) The state of fisheries and aquaculture. Food and Agriculture Organization. https://www.fao.org/state-of-fisheries-aquaculture
Global Impact Investing Network (GIIN) (2021) The state of impact investing in the blue economy. GIIN. https://thegiin.org/research/publication/blue-economy-impact-investing
GoFundMe (2020) Fundraising for coral reef restoration in the Philippines. https://www.gofundme.com/
Hughes T et al (2019) Sustainable eco-tourism and blockchain technology in the Maldives. J Tour Res 45(3):134–145. https://doi.org/10.1016/j.jtr.2019.01.002
Indian Ministry of Finance (2017) India's first green bond issuance: renew power. Ministry of Finance. https://www.finmin.nic.in/
International Finance Corporation (IFC) (2020a). Digital financial services for sustainable development. IFC. https://www.ifc.org/wps/wcm/connect/industry_ext_content/ifc_external_corporate_site/financial+institutions/publications/digital-financial-services-for-sustainable-development
International Finance Corporation (IFC) (2020b) IFC green bonds for offshore wind projects in emerging markets. https://www.ifc.org/
IRENA (2021). Renewable energy in the blue economy: opportunities for offshore wind in India. International Renewable Energy Agency. https://www.irena.org/
Kusworo A et al (2020) Sustainable aquaculture: challenges and opportunities. Aquac Res 51(5):1639–1649. https://doi.org/10.1111/are.14685
Ministry of Earth Sciences (2021) India's blue economy: a draft policy framework. Government of India. https://moes.gov.in
Ministry of New and Renewable Energy (2020a) India's offshore wind energy potential and financial strategies. Ministry of New and Renewable Energy. https://mnre.gov.in/
Ministry of New and Renewable Energy (2020b) Offshore wind energy development in India. https://mnre.gov.in/
Ministry of Shipping (2021). Sagarmala Programme: Enhancing Maritime Connectivity in India. Government of India. https://www.sagarmala.gov.in/
National Investment and Infrastructure Fund (NIIF) (2021) NIIF green bond issuance for offshore wind projects. NIIF. https://niifindia.in/
OECD (2020a) Financing a sustainable ocean economy. OECD Publishing, Paris. https://www.oecd.org
Organisation for Economic Co-operation and Development (OECD) (2020b) Financing a sustainable ocean economy. OECD Publishing, Paris. https://www.oecd.org

PwC (2021) The role of fintech in sustainable finance. PwC. https://www.pwc.com/sustainable-finance
Salagrama V (2006) Coastal livelihoods in the age of globalization. J Coast Dev 12(4):5–17. https://www.journalofcoastaldev.com/
Seychelles Government (2018) Seychelles blue bond: World's first sovereign blue bond. https://www.seychellesbluebond.com/
Tisdell C (2013) Economics of ecotourism. Edward Elgar Publishing. https://www.e-elgar.com/shop/usd/economics-of-ecotourism-9781848446911.html
UNEP (2017). The role of fisheries and aquaculture in sustainable development. United Nations Environment Programme. https://www.unep.org/
World Bank (2017) The potential of the blue economy. World Bank, Washington, D.C. https://openknowledge.worldbank.org
World Bank (2020a) Blue bonds: financing ocean sustainability projects. https://www.worldbank.org/
World Bank (2020b) Innovative financing for coastal protection. https://www.worldbank.org/
WWF (2020) Blockchain technology in sustainable fisheries. World Wildlife Fund. https://www.worldwildlife.org/
Sharma V, Gupta M, Taneja S (2024). Does FinTech adoption impact on sustainability of small businesses: mediating role of financial well-being. Global Knowledge, Memory and Communication

Empowering the Blue Economy: The Role of FinTech in Green Investment Solutions

Ewelina Idziak, Katarzyna Kachel, and Vikas Sharma

Abstract The chapter aims to highlight and explain the supporting role of the FinTech industry in issuing green financial instruments to sustain oceans, seas, and marine resources. The goal is also to show good practices for better education and encouragement to act. Currently, the aquatic environment is in the worst situation ever. That is why intensifying activities for sustainability is necessary. Standard solutions are no longer sufficient, and there is an urgent need for targeted and tailored financial instruments supported by advanced technology. There is still a gap between knowledge and implementation, so this chapter is intended to make up for this deficiency. The study combines secondary research, review of existing research, comparatives, and practical case studies to explore the usage of green financial instruments for financing the Blue Economy. Data were analyzed, and the study-case methodology was used to present examples for in-depth investigation. Visualization methods (charts, tables, schemes) were used for better understanding. The findings present the revolutionary role of FinTech solutions by using advanced technology for targeted actions. Well-structured financial, green instruments combined with advanced technology are effective and efficient in fighting for sustainability. Identified solutions may serve good examples for subsequent implementations. Practical implications are important for entities such as policymakers, financial institutions, technology developers, and businesses engaged in the Blue Economy. The need for Fintech services for oceans and seas is growing. Implications go in regulatory frameworks, adoption and scaling, education, sustainable practices, and long-term perspective directions.

E. Idziak · K. Kachel
Department of Law and Economics, Uniwersytet Kazimierza Wielkiego, Bydgoszcz, Poland
e-mail: ewelina.idziak@ukw.edu.pl

K. Kachel
e-mail: katarzyna.kachel@ukw.edu.pl

E. Idziak
Azerbaijan State University of Economics, Baku, Azerbaijan

Vilniaus Kolegija, Higher Education Institution, Vilnius, Lithuania

V. Sharma (✉)
University School of Business, Chandigarh University, Mohali, India
e-mail: vikas.sharma62@yahoo.com

V. Sharma et al. (eds.), *FinTech for Sustainable Finance and a Well-Blue Economy*, World Sustainability Series, https://doi.org/10.1007/978-3-031-92390-6_12

Keywords Fintech · Sustainability · Green investments · Blue economy · Financing

1 Introduction

Are investor's decisions a good barometer for investments in green projects? We need to consider both sides of the coin as there are many advantages and disadvantages. At least a couple of positive aspects are to be treated as investors' decisions as a barometer. First of all, the domain of investors is that they conduct the evaluation and validation of investment opportunities. In the context of investments, it means confirming that a given investment opportunity is reliable, aligned with assumptions and has real potential to achieve the intended goals. The financial market presents capital flows, which we currently observe moving towards sustainable finance. Moreover, many activities are taken through regulatory and policy influence supporting ESG regulations. Also as a strong advantage may be pointed out rapid technological shift in industry. However, there are also disadvantages for example short-term focus as many investors prioritize short-term returns. There are also highlighted negative impacts of the "greenwashing" risk causing some investments to be made based on marketing claims rather than genuine sustainable impact. There are also voices about speculative bubbles in the case of overhyped green sectors. One more problem is underinvestment in high-impact areas as for example some truly sustainable projects like ocean conservation, may struggle to attract investors due to unclear profit models.

To present the broader background of green investments, it is necessary to provide general statistics about it. In 2023, the Global Sustainable Investment Alliance released next edition of Global Sustainable Investment Review, revealing that currently 30.3 trillion USD was invested in green assets worldwide (Global Sustainable Investment Alliance 2022). As green investments are in the development phase, it is still difficult to find aggregate data that would allow comparison of this financial phenomenon. The problem may be also different methodologies of gathering and presenting data. One of the first statistics showing global commitment to sustainable investment presents growth but with some signals of a slowdown. The data about global sustainable investing assets are presented in Fig. 1.

Figure 1 presents data about the amount of money invested in sustainable initiatives across selected countries or continents in 2-year periods from 2016 to 2022 in billions of USD. Europe shows a fluctuating trend with an initial increase from 2016 to 2018, then a decrease in 2020, and another increase in 2022. Canada exhibits a steady increase in sustainable investments from 2016 to 2020, followed by a slight decrease in 2022. Australia, and New Zealand show a consistent upward trend in sustainable investments over the years. Japan demonstrates significant growth, with a notable spike between 2018 and 2022. The United States shows a significant increase from 2016 to 2020, followed by a sharp decline in 2022. The total values also show an overall increase from 2016 to 2022, with a slight decrease in 2022. The percentage

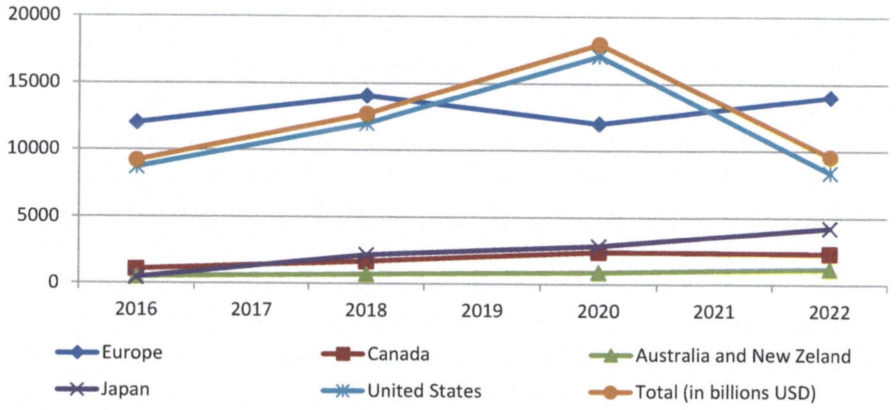

Fig. 1 Global sustainable investing assets, 2016–2022 (in billions USD) (Global Sustainable Investment Alliance 2022). *Source* Created by authors, based on Review of Global Sustainable Investment from 2022. https://www.gsi-alliance.org/members-resources/gsir2022/?utm_source=chatgpt.com (Global Sustainable Investment Alliance 2022)

change indicates a 34% increase from 2016 to 2018, a 15% increase from 2018 to 2020, and no available data for the change from 2020 to 2022 due to a change in methodology. Decreases might be caused by more rigorous asset labeling in the United States and in Europe by increased regulatory requirements following the introduction of a suite of legislation and regulations under the Sustainable Finance Action Plan and the European Green Deal. From the global perspective, it is also seen that higher requirements for getting financing in 2023 caused almost 30% of decline in international projects, what is super important for power and renewable energy infrastructure investments, so it might mean not lower interest in green investments, but overall decrease. It means that in consequence investments in Sustainable Development Goals (SDGs) fell by more than 10% (United Nations 2024a). One of many conclusions from the World Investment Report 2024, is that: "in 2023, the value of sustainable investment products, encompassing bonds and funds, reached more than $7 trillion, a 20% increase from 2022. Although the picture is nuanced, the overall positive trend in the sustainable finance market points to continued investor confidence and the resilience of sustainable investment strategies. Sustainable bonds were the main driver of growth in sustainable capital market products. Issuance climbed to $872 billion, a 3% rise from 2022, bringing the cumulative value of the market since 2018 to more than $4 trillion. Despite continued growth in number and asset value, though, sustainable funds experienced strong headwinds in 2023. Net inflows dropped from $161 billion in 2022 to $63 billion in 2023. Greenwashing remains the most significant challenge to the sustainable fund market."(United Nations, World Investment Report 2024b). Selected and estimated data presenting two main drivers of green investments (green bonds and sustainable funds) are presented in Fig. 2. For better trend visualization data are presented using different scales for green bonds and sustainable funds in Fig. 2.

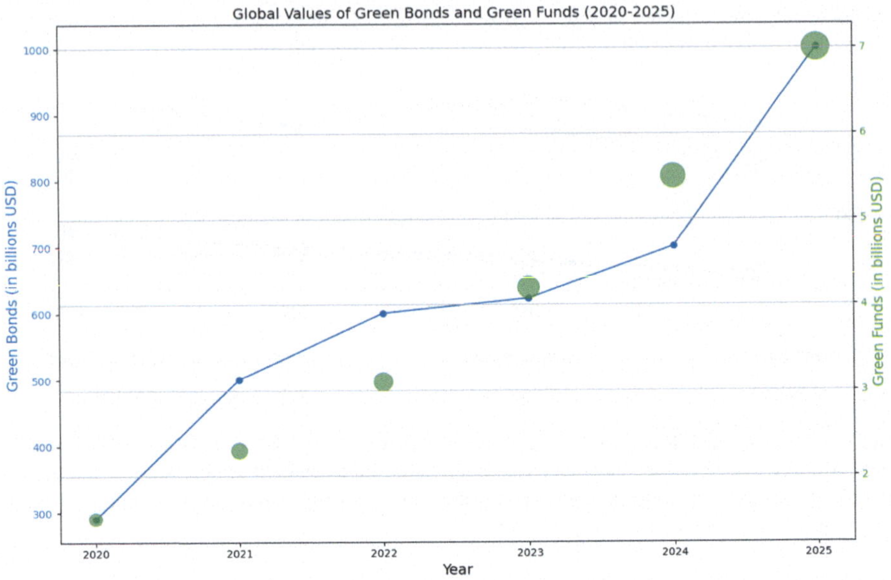

Fig. 2 Global values of green bonds and green funds (2020–2025) (https://www.statista.com/topics/9217/green-bonds-market-worldwide/; SP Global Ratings 2025; Green Climate Fund 2020b). *Source* By authors, Created by authors, based on: Statista. "Green bonds market worldwide". Available at: (https://www.statista.com/topics/9217/green-bonds-market-worldwide/), S&P Global Ratings. Sustainability Insights: Global Sustainable Bond Issuance to Hold Steady at $1 Trillion in 2025. Available at: (SP Global Ratings 2025), Green Climate Fund. Annual Results Report 2020. Available at: (Green Climate Fund 2020b)

It should be highlighted that both environmental bonds and sustainable funds present a positive correlation. The simultaneous growth of green bonds and green funds suggests a potential interdependence between the two financial instruments. The expansion of green bond markets may be driving greater investment into green funds, or vice versa. It is also observed that the global value of green bonds presents a consistent upward trajectory from 2020 to 2025. The initial value of approximately $300 billion in 2020 has increased significantly, surpassing $1 trillion by 2025. This indicates a strong and sustained interest in green financing through bond issuance and it might be treated as the main driver for green instruments. Green funds also play a significant role in building a green infrastructure because green funds have also expanded, as represented by the bubble sizes and right-hand scale in Fig. 2. Starting at a lower value in 2020, green funds investments have grown progressively, reaching approximately 7 billion USD in 2025. This trend suggests that investors are increasingly adopting sustainable investment. The rate of increase appears to accelerate significantly after 2023, particularly for green bonds. This may indicate a policy shift, increased investor confidence, or regulatory incentives supporting sustainable finance.

The observed trends highlight the increasing role of sustainable finance in global markets, but how does it go with blue finance?

It is important to emphasize that blue investments are still only a small fraction of green investments. Quantifying the exact proportion of blue investments within the broader spectrum of green investments is very challenging due to overlapping definitions and different reporting standards. However, some recent data provides insight into the growth and scale of blue economy investments, so even if the data is not comparable it is worth presenting it for the general understanding of the phenomenon.

According to the European Commission and the "BlueInvest Investor Report 2024", the scale of disclosed investments in the marine economy has significantly increased in the European Union. It's three times bigger than it was a decade ago, totaling over 13 billion EUR between 2018 and 2023 (European Commission 2024). What is more, according to the OECD Report there are huge expectations for the ocean economy to grow. It is estimated that the ocean economy will double to 3 trillion USD by 2030 (https://www.oecd.org/en/topics/sub-issues/ocean.html). Taking into consideration blended finance (public, private, and philanthropic capital) as a measurement of the scale of the aquatic economy it is crucial to emphasize the constant growth trend during the period from 2000 to 2021 which is presented in Fig. 3 (SDGInvest 2022).

It is also important to understand the spread between the ocean's needs and as it is so far. There is a significant disproportion between sectors receiving Official Development Assistance (ODA) when comes to sectors and sustainability-targeted aspects and the ocean economy. The presented data show five main receivers of both assistance types. The distribution of ODA is illuminating the strong dominance of maritime transport as a receiver, which is presented in Fig. 4. The maritime transport sector plays a crucial role with the allocation of 5.27 billion USD of sustainable

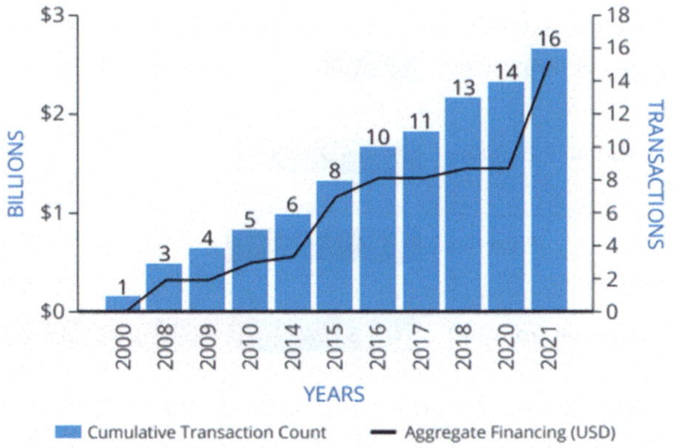

Fig. 3 Blended finance growth for the blue economy (SDGInvest 2022). *Source* By authors, SDGInvest, Report, Blended Finance & The Blue Economy, Data Brief, https://sdginvest.jointsdgfund.org/sites/default/files/2022-05/Data%20Brief%20-%20Blended%20Finance%20and%20the%20blue%20economy%20-%20Draft-2%5B94%5D.pdf?utm_source=chatgpt.com (SDGInvest 2022)

sources and about doubled (10,29) amount for other ocean economy activities in 2022. Then fisheries are the second most financed sector, similar to marine protection. Still categories like unspecified areas, disaster prevention and also preparedness receive less funding, with only 1.78 billion USD allocated in those topics. Significantly more funding than other sub-sectors received for marine transport in 2022. Investments in this sector are usually connected with enhancing infrastructure, improving safety standards, and in general promoting sustainability in shipping practices. In the case of fisheries, there is a strong need to support sustainable fishing practices and protect marine biodiversity. Analyzing data only from the perspective of sustainable ocean economy ODA the marine protection received the hugest amount about 5,79 billion USD in 2022. This allocation reflects the current situation, but at the same time shows how a small fraction of green finance is present in blue finance. It is still not enough to cover the needs.

To make a point to this introduction it is significant to illuminate synthetically how the aquatic economy complements the environmental economy supporting the same goals but in a more targeted way focusing on "water resources". Starting from the definition promoted by the United Nations Environment Programme it is said: "A green economy is one that results in improved human well-being and social equity, while significantly reducing environmental risks and ecological scarcities." (https://www.unep.org/pt-br/node/23750). Which is a wide perspective, mainly accepting all activities that reduce environmental and ecological harms. Going to the blue economy, it is worth citing the aquatic economy definition proposed by the World Bank, where the blue economy is defined as: "sustainable use of ocean resources for economic growth, improved livelihoods, and jobs while preserving the health of ocean ecosystems."

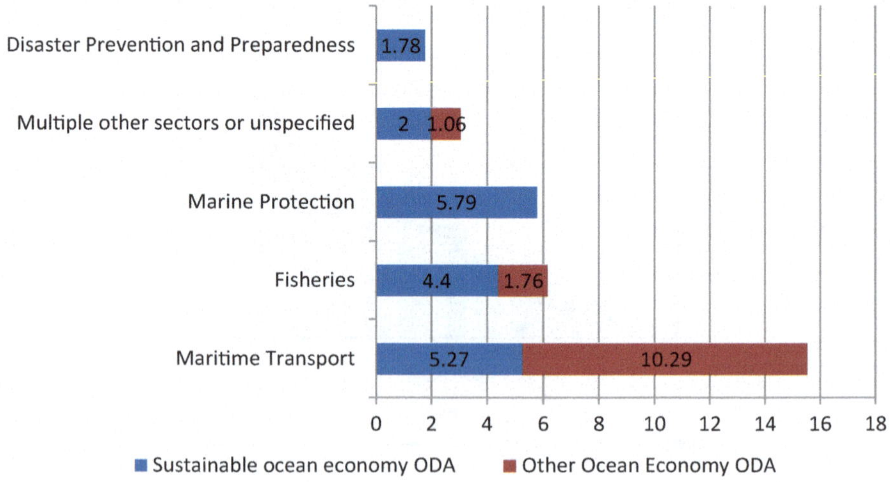

Fig. 4 Official development assistance for ocean sectors (SDGInvest 2022). *Source* Created by authors, based on: "OECD, Data Platform on Development Finance for the Sustainable Ocean Economy, (https://oecd-main.shinyapps.io/ocean/)

(World Bank Group 2017; https://openknowledge.worldbank.org/server/api/core/bitstreams/cee24b6c-2e2f-5579-b1a4-457011419425/content). This perspective is much narrower, focusing on water resources, and even just oceans in particular. It is clear that the ocean economy is a subset of the environmental economy. In this situation, it is important to set the relations between blue finance and green finance. An attempt to determine the main connections in a synthetic form is visualized in Fig. 5. A better understanding of the inclusion between blue finance and green finance (as more narrow disciplines) helps to understand the straight inclusion of blue finance into the wider category of green finance. In the Fig. 5 authors categorize seven main areas within each financial domain and highlight how they align in addressing global environmental challenges. Starting from blue finance, it is seen that according to main areas such as marine ecosystem restoration, water and wastewater treatment, tackling ocean plastic pollution, sustainable shipping practices, green tourism, and offshore renewable energy, and protecting underwater environments, there is straight support for green finance fields, which are: environmental stability, climate change mitigation, clean power projects, preventing pollution, protecting biodiversity, promoting circular economy practices, and ensuring the sustainable use of natural resources. Both areas (blue and green) have the same general goals because for both environmental stability and climate change mitigation are crucial. Moreover, offshore renewable energy aligns with renewable energy projects in the green economy. Also, eco-friendly tourism supports a circular economy from green finance. Finally, also water and wastewater care align with sustainable use of resources from green finance. The main idea for creating the Fig. 5 is to effectively and synthetically illuminate that the blue economy is a subset of green finance, addressing the same goals and concerns and supporting green initiatives as well.

In this context of introduction, it is important to present the latest research on a crossroad of green investments and FinTech that supports financing the blue economy.

2 Review of Literature

Blue economy as an emerging topic currently appears more frequently in research papers than previously, but still, it is just the beginning. The dominant for this research stage is mainly environmental sciences and economics, but according to the Web of Science database, this topic appears also in such areas as environmental studies, material science multidisciplinary, management, urban planning, green sustainable science technology, but also chemistry, and marine freshwater biology. So, blue economy is one of the multidisciplinary subjects. Filtering the Web of Science database by the filter “blue economy” gives 6804 publications connected to this topic (Fig. 6).

Surprisingly, using two filters “blue economy” and “fintech” together for the research in Web of Science the result is only one. Using those two filters together shows that there is a huge gap in research on the crossroads of blue economy and fintech. The only one paper presented in the Journal of Economic Studies is

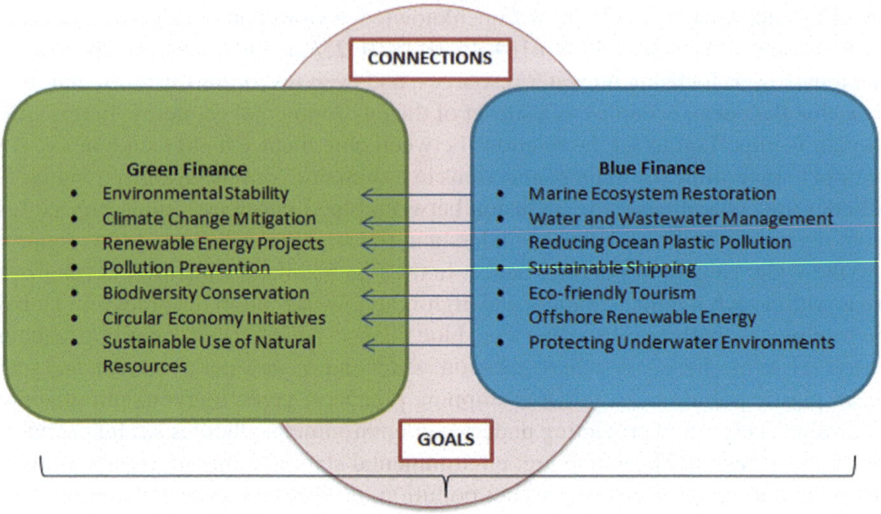

Goal	Connection
Environmental Stability	Both green and blue finance aim to achieve environmental sustainability.
Climate Change Mitigation	Blue finance contributes to climate change mitigation, which is also a key goal of green finance.
Renewable Energy Projects	Offshore renewable energy is a part of blue finance, aligning with green finance's focus on renewable energy.
Pollution Prevention	Reducing ocean plastic pollution is a specific aspect of pollution prevention under the blue finance.
Biodiversity Conservation	Marine ecosystem restoration under blue finance contributes to biodiversity conservation.
Circular Economy Initiatives	Eco-friendly tourism in blue finance aligns with circular economy initiatives in green finance.
Sustainable Use of Natural Resources	Water and wastewater management in blue finance is a part of the sustainable use of natural resources.

Fig. 5 Blue and green finance relations. *Source* By authors, Created by authors, based on: Statista. "Green bonds market worldwide". Available at: (https://www.statista.com/topics/9217/green-bonds-market-worldwide/), S&P Global Ratings. Sustainability Insights: Global Sustainable Bond Issuance to Hold Steady at $1 Trillion in 2025. Available at: https://www.spglobal.com/ratings/en/research/articles/250205-sustainability-insights-global-sustainable-bond-issuance-to-hold-steady-at-1-trillion-in-2025-13402934 (SP Global Ratings 2025), Green Climate Fund. Annual Results Report 2020. Available at: (Green Climate Fund 2020b), World Economic Forum. What is Green Finance? Available at: https://www.weforum.org/stories/2020/11/what-is-green-finance/ (Green Climate Fund 2020a), eFinance Management. Green Finance. Available at: (https://efinancemanagement.com/sources-of-finance/green-finance), IFC Blue Finance. Available at: https://www.ifc.org/en/what-we-do/sector-expertise/financial-institutions/climate-finance/blue-finance (IFC Blue Finance), IFC Guidelines for Blue Finance. Available at: https://www.ifc.org/en/insights-reports/2022/guidelines-for-blue-finance (IFC Guidelines for Blue Finance 2022)

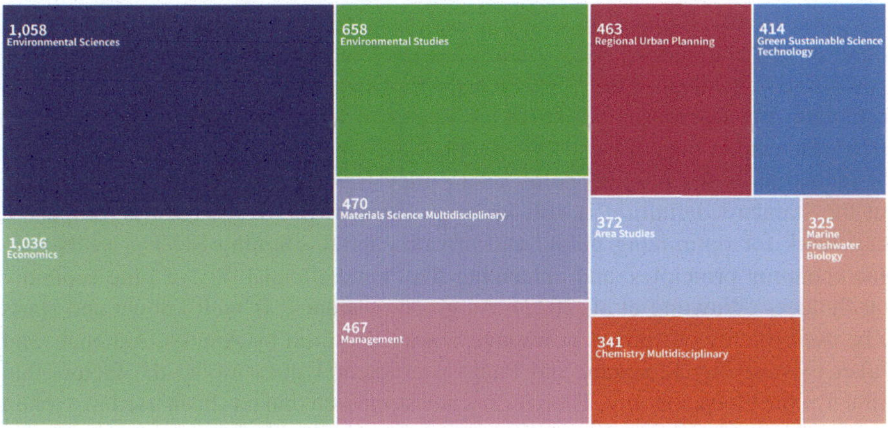

Fig. 6 Blue economy publications in the WoS. *Source* Filtering by authors, based on Web of Science, https://www.webofscience.com/wos/woscc/analyze-results/03e0f058-b0f0-41d7-a608-3a5997ca715f-014d78fdca (Web of Science database)

titled: "Gauging the dynamic interlinkage level between Fintech and the global blue economy performance" published in 2025. Le Thanh Ha analyzes connections between FinTech development and the "blue economy" from September to August 2023. Especially, the research was focused on volatility connected with blue bonds and financial technology. This is the first paper on this subject. The main findings are about two types of connections:

- the aquatic economy index is a net shock receiver, while
- most of the FinTech indexes were mainly net shock transmitters, but at the same time author claims that FinTech is crucial in promoting the marine economy (Thanh Ha 2025).

Analysis of research at the intersection of "green investments" and "blue economy" in the Web of Science results within 66 documents, which confirms that also in this area there is a scant amount of research, but much more than about the connection of "blue economy" and "fintech". This confirms also the strong connections between the blue economy and green aspects as a whole, which is presented in Fig. 5.

Tirumala and Tiwari pointed out that investment flows are much lower than expected, so the authors decided to verify whether the existing ocean initiatives are enough to the investment needs. Results of the research confirm the insufficiency of existing blue instruments and recommend establishing specialized financing facilities to gather pool low-cost funds from investors. It has also raised the importance of public–private partnerships and innovative financial structures (Tirumala and Tiwari 2022). The aim of this chapter exactly resonates with the researcher's conclusion about the need to fill the gap with innovative, FinTech instruments financing the blue economy. Moreover, the input from the blue economy to a wealth of countries is statistically important. Alharthi and Hanif proved blue economy has an impact on economic development in South Asian Association for Regional Cooperation countries and claimed aquatic economy factors have a proven impact on economic growth and help advance the UN's Sustainable Development Goal 14 to protect and sustainably manage marine resources for long-term development (Alharthi and Hanif 2020). The gap in the policies is also emphasized by Wuwung et al. They argue that the inclusion of the blue economy as a policy frameworks has its "momentum", but the standard definition is still lacking. Through this research they propose a framework for optimizing public and private capital, aligning SDG projects with blue economy principles, and enhancing the financial capability of blue economy stakeholders (Wuwung et al. 2022). As green awareness is well known and starts to be used rule by customers nowadays, it was also used by Ameko, Asiamah, and Duker to verify by R-Square and Path-Co-efficient Values model the factors that influence the blue economy. They form a new approach that has been used to explain initiatives behavior and sustainability performance in the blue economy. It is that the interest of different groups, entities, businesses, and stakeholders in increasing nowadays. They also proved that positive intention helps to realize sustainable goals. The research reveals that blue economy investments substantially mediate the relations between factors affecting its (organizational, environmental, technological, cost, risk, and environmental awareness) intention and sustainable goals (Appiah et al. 2023). Going further to dedicated, financial instruments for blue finance, Thompson presents statistics, trends, and insights about blue bonds for marine conservation. The author highlighted that blue bonds should generate both a positive environmental and/or social impact, in addition to providing a positive return. However, Thompson argues that there has been little academic attention on these early efforts to assess if the claims are valid, but drawing on his 15 years of experience, he offered valuable, synthetic insights into the blue market:

- focus more on the overall blue bond market than on individual projects financed by blue bonds;
- standardize the focus areas and avoid certain sectors for consistency for investors;
- prioritize locations where projects are most likely to succeed, focusing on low-risk projects;
- offer clear information on both impact and returns for better market understanding (Thompson 2022).

One more point of consideration is presented by Lavitt and Sageant about green infrastructure in developing economies for risk reduction and investments. They emphasize that it is wise to treat the financing of green infrastructure as a long-term investment with multiple benefits (financial and environmental). Their findings emphasize that various financial instruments like green bonds, carbon credits, and debt-for-nature swaps are underutilized in developing economies due to currency risks, political instability, negative surplus projects, and greenwashing concerns. They propose to consider risk reduction tools, such as floating rate structures, performance-linked returns, and asset backing, which can be applied to green infrastructure projects to reduce risk exposure and enhance returns for investors. So, in consequence, they propose a solution as bond combines green bonds with carbon credit and nature asset backing, floating rates tied to project milestones, biodiversity outcomes, and economic spillover effects (Lavitt and Sargeant 2024). Taking into account a variety of conditions and possibilities in different countries it might be a specialized solution for the targeted needs of blue finance and sustainable needs of oceans.

And one more insight, for summaring relations between blue and green economy. As it was written by Hamdy: "It is quite evident that green growth, fundamentally based on water resource development, is an outcome of green economy and that a green economy cannot be totally green unless it addresses water issues. In other words, a blue economy is a prerequisite of a green economy. Considering the core features of a green economy, it is quite obvious that water-use efficiency and improving crop water productivity, i.e. increasing crop by drop, are central aspects of green economy. This directly indicates the crucial role regional and locally adapted water management strategies could have in underpinning the transition towards the green economy in the water sector." (Hamdy 2014).

Based on a brief literature review it should be highlighted that the blue economy is still an emerging discipline and what is more the relations between the blue economy and FinTech still are unexplored. The analysis of the Web of Science data confirms a significant research gap in integrating financial technology with blue economy initiatives. Research papers emphasize the need for innovative financial instruments, such as aquatic bonds and specialized funding frameworks, to support sustainable ocean-based investments, the lack of legal frameworks, and policy deficiency, while the problem of supporting and protecting oceans is becoming as urgent as never before. So the answer to this need might be a more practical insight into the supporting role of FinTech, innovation, tailored financial solutions, and practical case studies to give good practices for business and educate.

Based on the literature review it is crucial to select deficiency areas which are visualised in Fig. 7 in a way of Fishbone diagram.

The main areas of blue economy and FinTech deficiency raised in literature may be grouped into main areas: policy deficiency, research gaps, financial instrument limitations, and innovative solution needs. Firstly, in the area of policy deficiency, it is emphasized the lack of legal frameworks and inadequate policy support are the main problems. Secondly, in the area of research gaps, it is seen that there is scant academic scrutiny and a lack of practical case studies to connect knowledge with practice. Thirdly, in the case of financial instrument limitations, it is raised borders such as insufficient funding facilities and underutilized blue bonds as one of the most obvious financial instruments. Moreover, in the area of innovative solutions obstacles are public–private partnerships and tailored financial solutions.

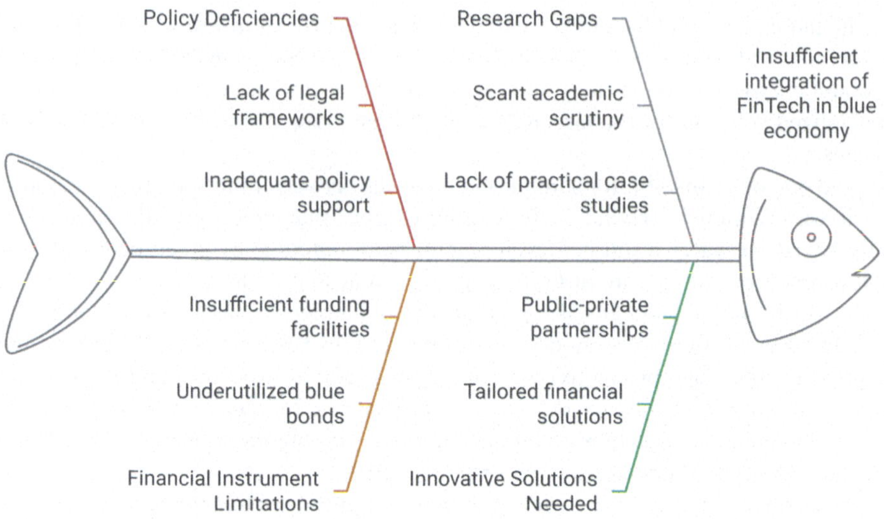

Fig. 7 Deficiency in blue economy and FinTech crossroads. *Source* Created by authors with the usage of app.napkin. Based on (Thanh Ha 2025; Tirumala and Tiwari 2022; Alharthi and Hanif 2020; Wuwung et al. 2022; Appiah et al. 2023; Thompson 2022; Lavitt and Sargeant 2024; Hamdy 2014)

3 Green Investments for Financing the Blue Economy

Tailored and targeted finance constructions play a key function in financing the narrow area of the blue economy. Types of green investments that support environmentally sustainable projects need to be more targeted to the special demands of the blue economy that supports water resources. These financial instruments direct capital towards sustainable blue initiatives. Capital flows may come from different sources like the public sectors, private sector, blended finance, and international sources. The capital may be blended to organize different stakeholders for special blue investments. Those instruments are tailored according to the needs. It is important to understand optimizing and diverse capital sources that help to scale green investments that are used to target. There is no one correct classification as this is an emerging, new, and very dynamic sphere of finance. The most frequent categorization for selecting types of green investments that support blue finance might be as follows: blue bonds, blue loans, blended finance, debt-for-nature swaps, and blue carbon credits which appear most frequently in literature and practice. The main goal is to support projects that promote aquatic ecosystems, but at the same time are economically positive. The synthetic catalogue of investment solutions that support the blue economy is presented in Table 1. The table highlights selected, main financial instruments designed to support the sustainability of marine environments. For this book selected examples build a background for understanding solutions for the aquatic economy. Each of the blue instruments plays a bit different function, but all of them are focused on supporting blue goals. Blue bonds are one of the first instruments that are very well known as debt financing for ocean sustainability. Blended finance is a combination of public and private resources for financing blue initiatives. Debt-for-nature swaps is a kind of policy agreement that lets developing countries exchange part of their foreign debt against commitments to marine conservation and sustainable coastal activities. Finally, blue carbon credits are a kind of market-based solution, for example incentivizing the protection and restoration of carbon-sequestering marine ecosystems like mangroves and seagrasses.

Depending on capital type there is a specific risk connected to the blue instruments. The gradation was done according to returns versus financial risk. Education in this matter is led by European Union through the initiative called "BlueInvest". Diverse, blue financial solutions are dedicated to different groups of investors with different levels of risk tolerance which is presented in Fig. 8. According to lower risk and returns at the beginning of the chart axis there are instruments such as blue loans and blue bonds. Going further to higher returns and a bit higher risk, the chart presents blended finance from private and public sources (debt, equity, impact capital). Going to the chart of the more risky instruments presents lower returns on public grants and funding from international financial institutions. With higher risk are presented public and private equity, investment funds, venture capital, and pension funds. Instruments are dedicated to supporting blue initiatives.

Presented in Table 1 financial solutions targeted for blue initiatives highlight the diverse range of mechanisms that may support sustainability in the blue economy. As the era of digitalization is developing dynamically, the next step for illuminating more supporting solutions for oceans is FinTech. Sustainable solutions supported by high technology implementation may design more tailored solutions for specific problems.

The practical visualization of real examples of blue financial instruments implementation is encouraging for utilization in problem-solving in the future.

A spectacular example of the very first sovereign blue bond issuance are Seychelles Blue Bonds by the Republic of Seychelles. That happens on 29th October 2018. The bond raised 15 million USD. This pioneering blue instrument was assisted by The World Bank which invited three entities: "Calvert Impact Capital", "Nuveen", and "U.S. Headquartered Prudential Financial, Inc.". Seychelles is made up of 115 islands, both granite and coral, covering an area of 455 km^2. It is surrounded by so called an "Exclusive Economic Zone" of about 1.4 million km^2. Known for its rich biodiversity, Seychelles is working to balance economic development with the protection of its natural resources (Seychelles launches World's First Sovereign Blue Bond 2018). Due to those special needs, blue bons are a more targeted solution than others. Seychelles Bonds USD (XS1885544236). The visualization of trading for Seychelles Blue Bonds in the period 29.10.2018–27.02.2025 presents in Fig. 9.

World Bank also explains the project's contribution towards climate change adaptation. This project is expected to strengthten Seychelles' boosting its resilience to climate change by expanding marine protected areas to 30% of its EEZ and promoting sustainable fisheries through better management. The project supports

Table 1 Selected types of green investments tailored to the blue economy

No	Financial instrument	Brief characteristic	Purposes	Comment
1	Blue bonds	Debt securities were issued to raise capital for projects focused on marine initiatives. Construction similar to classic bonds, but targeted for blue initiatives (World Bank, International Finance Corporation 2022)	Marine conservation Sustainable fisheries, Pollution reduction	The International Finance Corporation (IFC) has developed guidelines to assist issuers in identifying eligible blue projects, ensuring that investments align with sustainable marine and freshwater management principles (World Bank, International Finance Corporation 2022)
2	Blue loans	The type of credit is usually for entities undertaking projects with a positive impact on the marine environment	Sustainable aquaculture Wastewater management Coastal resilience	The IFC's guidelines for blue finance also encompass criteria for blue loans, providing a framework for financial institutions to support sustainable blue economy projects (World Bank, International Finance Corporation 2022)
3	Blended finance	A mix of public and private for financing positive impact on the marine environment	Coral reef restoration Sustainable tourism Blue carbon projects	"The United Nations Development Programme" highlights the importance of blended finance in addressing the investment gap in the blue economy, emphasizing the need for innovative financial instruments to attract private sector participation (investors and philantrophic) (Financing the Blue Economy 2022)

(continued)

Table 1 (continued)

No	Financial instrument	Brief characteristic	Purposes	Comment
4	Debt-for-nature swaps	Tailored financial arrangements by which a portion of a developing country's foreign debt is forgiven in exchange for local investments in environmental conservation measures (G20 2023)	Marine protected areas Sustainable coastal management	G20 countries' policy treats debt-for-nature swaps as a crucial instrument in mobilizing resources for blue economy transition (G20 2023)
5	Blue carbon credits	Tradable certificates representing the sequestration or avoidance of carbon emissions through the protecting and rebuilding marine ecosystems (https://www.unepfi.org/blue-finance/utm_source=chatgpt.com)	Mangrove restoration Seagrass conservation	The United Nations Environment Programme Finance Initiative promotes blue carbon projects as contributing to both climate mitigation and the sustainable blue economy (https://www.unepfi.org/blue-finance/utm_source=chatgpt.com)

Source By authors, Created by authors, based on: World Bank, "International Finance Corporation (2022) Guidelines for Blue Finance". https://www.ifc.org/en/insights-reports/2022/guidelines-for-blue-finance?utm_source=chatgpt.com (World Bank, International Finance Corporation 2022), United Nations Development Programme, "Financing the Blue Economy (2022)". https://www.undp.org/sites/g/files/zskgke326/files/2022-09/Financing%20the%20Blue%20Economy.pdf?utm_source=chatgpt.com (Financing the Blue Economy 2022), G20 (2023) Generating Finance for Blue Economy Transition, Task Force 6: Accelerating SDGs: Exploring New Pathways to the 2030 Agenda. https://t20ind.org/research/generating-finance-for-blue-economy-transition/?utm_source=chatgpt.com (G20 2023), United Nations, United Nations Environment Programme, Sustainable Blue Finance, Mobilising Capital for a sustainable Ocean. https://www.unepfi.org/blue-finance/?utm_source=chatgpt.com (https://www.unepfi.org/blue-finance/utm_source=chatgpt.com)

the debt-for-nature swap with The Nature Conservancy, aiming better ocean protection and climate adaptation. It will also help diversify the economy and reduce climate change risks by developing climate-smart ocean economies and expanding seafood industries (Seychelles launches World's First Sovereign Blue Bond 2018).

The last Fitch rating for Seychelles from September 16, 2024, is positive with Fitch Ratings affirming the "BB-" LT Int. Scale (foreign curr.) credit rating of Seychelles on September 13, 2024. The exact price and returns of the Seychelles Blue Bond aren't publicly available, but its design and strong backing make it a reliable investment. Since it offers a fixed 6.5% interest rate, investors can expect steady returns. Plus, with a guarantee from the International Bank for Reconstruction and Development, the risk of losing money is low. This makes it an appealing choice for investors who

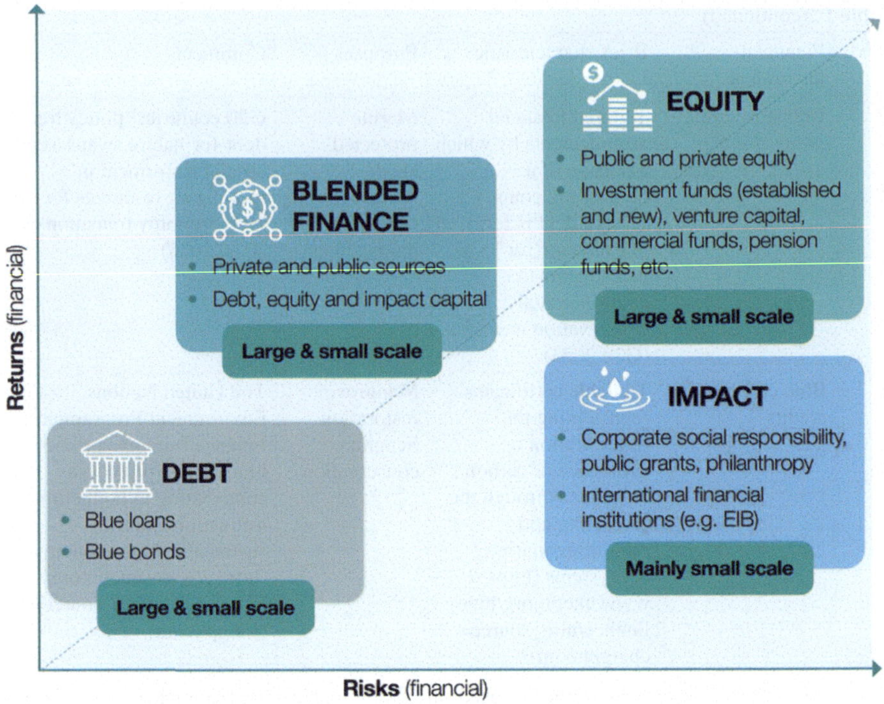

Fig. 8 Blue capital types. *Source* BlueInvest (2024) Investor report. Unlocking the potential of the Blue Economy. https://maritime-forum.ec.europa.eu/theme/investments/blueinvest_en (BlueInvest 2024)

want to support ocean conservation while earning a stable income and could be a practical example of effective utilization for supporting the blue environment.

4 FinTech for Financing the Blue Economy

Artificial Intelligence (AI) helps monitor and manage for example marine ecosystems to address the issues with the connection to climate change. "Kenya's Blue Economy" highlights the use of AI to monitor coastal environments like coral reefs, seagrass, and mangroves affected by climate change. Rising temperatures, ocean acidification, and sea-level rise are all challenging Kenya's marine ecosystems. These issues require quick and innovative solutions to protect the blue nature's future. AI plays a key role in managing climate impacts, making it essential for Kenya's sustainable development. As Kenya continues to prioritize the development of its Blue Economy sector as a key driver of economic growth and sustainable development, the integration of AI

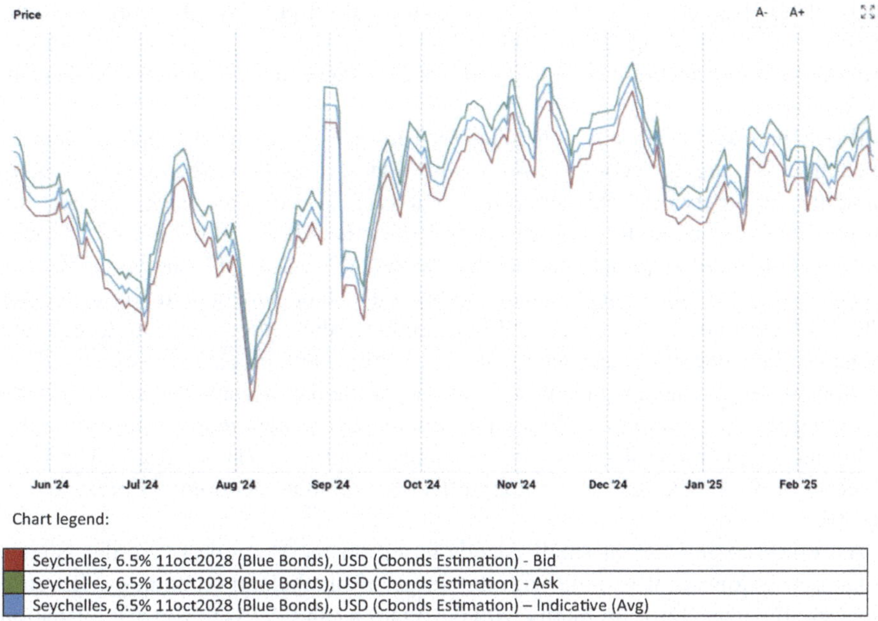

Fig. 9 Trading chart for Seychelles blue bonds. *Source* Cbonds (2025) Seychelles Bonds, 6.5% 11oct2028 (Blue Bonds), USD (XS1885544236). https://cbonds.com/bonds/542353/ (Cbonds 2025)

applications into monitoring and management practices becomes essential (Gesami and Nunoo 2024).

Cloud computing technology is used for example by Blue-Cloud, which delivers new services for marine research and the Blue Economy. It's goal is to support ocean sustainability research through five pilot projects. By building on existing resources, it creates a cyber platform that offers access to:

- multidisciplinary data from observations and models;
- analytical tools;
- computing resources for key blue science applications.

This platform consists of 2 components: the Virtual Research Environment (VRE) and the Data Discovery & Access Service, bringing together important European data systems to help users easily find and access data from different sources (European Commission 2021a). Another spectacular example is launched by the National Oceanic and Atmospheric Administration—NOAA Cloud. NOAA Fisheries Ecosystem Research uses the cloud to understand interactions between Antarctic krill and their predators. Cameras collect too much data for scientists to analyze by hand, so cloud models have been developed to automate the analysis of images. Using the cloud, the model ran in 45 s rather than the one- and two-hour standard timeframe that is typical for a traditional on-premise platform for High-Performance Computing (HPC). When broadly available to scientists, this model will provide more accurate and timely advice for fishery management (Oceanic et al. 2022).

Blockchain technology is beneficial for example for seafood supply chains. A blockchain is a digital record of transactions that is copied and shared across many computers in a network. This setup makes it very hard to alter, hack, or cheat the system. The blockchain technology is applied by the Alaska Blue Economy Center together with interdisciplinary partnerships with UAF College of Fisheries and Ocean Sciences, ACEP, College of Business and Security Management, and Alaska Center ICE. As Kevin Huo commented: "When thinking about the ocean, high-tech and trackable data doesn't immediately jump to mind, but blockchain, essentially a distributed ledger, can advance understanding of the fishing industry and help with managing resources and providing an immutable set of data that can inform nodes within the supply chain that ultimately benefit consumers." (Hugo 2022). The visualization of the blockchain technology effect for the blue economy is presented in Fig. 10.

Also the research about small islands based on the example of Seribu Islands in Jakarta, Indonesia was done with the usage of blockchain technology. Because those small islands have problems with destinations and deliver proper information for visitors in real time the blockchain technology was used to check how it may create added value. This study shows that for islands facing challenges in access and connectivity, blockchain and smart technology are essential. They help bring together resources from different stakeholders, making the development of the blue economy on these islands more effective, while also ensuring sustainability (Pranita et al. 2023).

Fig. 10 Blockchain technology for transparency (Pandiangan 2025). *Source* Pandiangan (2025) Incentro High-Tech, High Tide: How digital solutions are shaping the "blue economy". https://www.incentro.com/en-ID/blogs/high-tech-high-tide-how-digital-solutions-are-shaping-the-blue-economy (Pandiangan 2025)

A good example of using the Internet of Things (IoT) is the fishing industry, which is a major part of the economy in many coastal countries. Studies suggest that the lack of technology in this sector is a big problem. For example, fishermen use simple tools that can't provide real-time updates or alert them when it's time to collect the catch. This means they waste time and fuel by making unnecessary trips to check their gear. Using Information and Communication Technology (ICT) could help increase productivity and reduce costs. This study proposes an IoT-based smart fishing gear that counts and records the number of fish caught, showing the information on a mobile app. It also sends alerts when the required catch is reached and helps fishermen navigate to their gear in remote locations (Abdalla et al. 2022).

In case of the robotic process automation, it seems that it is a technology that may revolutionize the Blue Economy bringing lower costs and more effectiveness at the same time. This is the integration of many high-tech solutions with robots that are used for scientific research for oil and gas, robots that can interact with objects underwater, sensing and imaging oceans for data gathering, observation platforms, mobile platforms, mix the solutions, technologies, and systems for better efficiency (Robotics 2022; European Commission 2021b; Agarwala 2020; Spencer 2022).

Digital Identity Verification or Digital Twins (DT) technology is also applied in the Blue Economy. Studies show that integrating digital technologies (DT) into the blue economy can provide significant benefits, such as improving efficiency, better managing risks, and promoting sustainability. DT helps optimize marine energy systems, improve shipping operations, enhance aquaculture practices, and support sustainable fisheries (Bhati et al. 2025). It is highlighted that DT or Digital Ocean Twins will revolutionize ocean science, oceanography, and hydrography by consolidating data and bringing together data contextualization, data analytics, and finding new connections between data for better management and efficiency in the blue economy.

Another FinTech solution might be payment processing covering specific payment needs in the aquatic economy, considering the diversification of industry characteristics related to oceans, seas, coastlines, water resources, and tourism. There are many individual solutions, but the main groups are usually connected with: merchant account setup, payment gateway integration, payments orchestration, authorizations, clearing and settlement, currency conversion, compliance and security and integration, payment automation with more professional business systems, and others (https://www.globalblue.com/en/business/payment-solutions/payment-gateway?utm_source=chatgpt.com; SSC, Blue Prism 2024; https://www.aciworldwide.com/solutions/payments-hub?utm_source=chatgpt.com). For example, Global Blue offers many solutions dedicated to the blue industry as a payment gateway or omnichannel payments.

To make this full of solutions picture more synthetic, selected solutions from FinTech were chosen and synthetically characterized by finance perspective in Table 2.

Taking all those aspects into account one should remember also about the quality and true value of the implemented solution. Sharma, Jangir, Gupta, and Rupeika-Apoga proved that high-quality FinTech payment services encourage more people to

Table 2 FinTech solutions driving sustainable blue economy finance

Financial instrument	FinTech solution	Application in the blue economy	Impact	Example
Blue Bonds	Digital bond platforms and blockchain	Digital issuance of blue bonds using blockchain for transparency and smart contracts for fund allocation	Enhances investor confidence, reduces fraud, and ensures funds are used for marine conservation projects	International Finance Corporation (IFC) Blue Bonds (https://www.ifc.org/en/what-we-do/sector-expertise/financial-institutions/climate-finance/blue-finance)
Blue Loans	AI-driven credit scoring & digital lending platforms	AI-powered analysis of businesses' environmental impact to approve loans for sustainable aquaculture or wastewater management	Expands financial access to smaller entities contributing to the blue economy, and reducing risk for lenders	Blue Finance IFC Guidelines (https://www.ifc.org/en/insights-reports/2022/guidelines-for-blue-finance)
Blended Finance	Crowdfunding platforms & tokenized investment pools	Platforms like tokenized investment funds enable public–private collaboration to finance blue carbon projects and coral reef restoration	Attracts more investors by lowering entry barriers and enabling fractional investments	UNDP Blended Finance (Financing the Blue Economy 2022)

(continued)

Table 2 (continued)

Financial instrument	FinTech solution	Application in the blue economy	Impact	Example
Debt-for-Nature Swaps	Smart contracts & blockchain tracking	Automated tracking of conservation commitments using blockchain, ensuring funds are used as agreed in debt-for-nature arrangements	Increases accountability and transparency in debt forgiveness agreements linked to environmental conservation	G20 Policy on Debt-for-Nature Swaps (G20 2023)
Blue Carbon Credits	Tokenized carbon credit markets	Blockchain-based platforms allow tracking, verification, and trading of blue carbon credits for mangrove and seagrass restoration	Creates a reliable and transparent market, attracting more buyers and ensuring real impact on carbon reduction	UNEP Finance Initiative—Blue Carbon (https://www.unepfi.org/blue-finance/utm_source=chatgpt.com)

Source Created authors based on literature (IFC Blue Finance; IFC Guidelines for Blue Economy 2022; Financing the Blue Economy 2022; G20 2023; https://www.unepfi.org/blue-finance/utm_source=chatgpt.com; Gesami and Nunoo 2024; European Commission 2021a, b; Oceanic et al. 2022; Hugo 2022; Pranita et al. 2023; Abdalla et al. 2022; Robotics 2022; Agarwala 2020; Spencer 2022; Bhati et al. 2025; https://www.globalblue.com/en/business/payment-solutions/payment-gateway?utm_source=chatgpt.com; SSC, Blue Prism 2024; https://www.aciworldwide.com/solutions/payments-hub?utm_source=chatgpt.com; Pandiangan 2025)

use them. Their findings suggest that users are more influenced by how useful they perceive these services to be rather than just how easy they are to use. This challenges the common belief that ease of use is the most important factor in technology adoption, highlighting the need for FinTech companies to focus on delivering real value and reliability to drive user engagement (Sharma et al. 2024a). This is super important for FinTech companies. The next aspect in the global perspective for all FinTech adoptions is real impact on sustainability taking into account different size of the entities and scalability. For example, Sharma, Gupta, and Taneja's study shows that using FinTech can help small businesses stay sustainable. A key reason for this is financial well-being—when businesses manage their finances well, they are more likely to succeed in the long run. The research highlights that small businesses should focus on improving their financial health to get the most benefits from FinTech and ensure their long-term stability (Sharma et al. 2024b).

FinTech helps organize financing for the blue economy more efficiently, transparently, and sustainably than classical financial instruments. Technologies like AI,

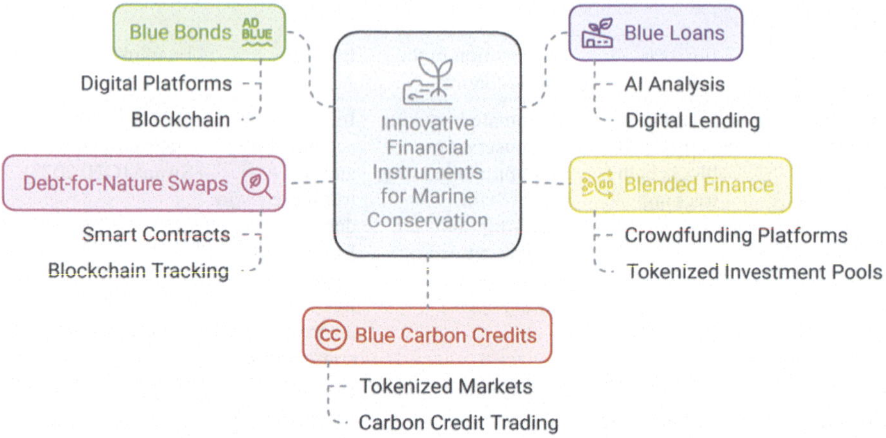

Fig. 11 Green investments and FinTech for financing the blue economy. *Source* Created by authors with the usage of the app.napkin

blockchain, cloud computing, and IoT improve tracking, monitoring, and financial processes. These tools support better marine ecosystem management and make funding ocean-related projects more effective. As digital and technology still develop, FinTech will play an even bigger role in keeping ocean industries strong and sustainable in the future.

5 Conclusion

The blue economy is at its turning point, and innovative financial solutions are its future. Traditional funding methods can not keep up with the urgent need for ocean conservation and sustainable marine industries. This is where FinTech steps in, revolutionizing blue finance with technologies like blockchain, AI, and digital platforms. These tools bring greater transparency, efficiency, and accessibility, ensuring funds reach the right projects. With solutions like blue bonds, carbon credits, and blended finance, we now have smarter and tailored investments in the ocean's health while creating economic value and higher efficiency visualized in Fig. 11.

This chapter explored how green investments and FinTech can reshape financing for the blue economy. By using financial technology, investors, businesses, and policymakers can close the gap between ideas and real impact by boosting sustainable fisheries, marine research, and infrastructure. FinTech not only makes blue finance more effective but also unlocks growth opportunities. As the blue economy expands, digital finance will be a driving force in making it more resilient and sustainable.

References

Abdalla AT, Haji N, Maiseli B, Ibwe K (2022) IoT-based smart fishing gear for sustainability of the Tanzania blue economy. Tanzania J Eng Technol 41(3):15–21. https://doi.org/10.52339/tjet.v41i3.840

ACI Worldwide, The future of unified, intelligent cloud-based payments orchestration is here. https://www.aciworldwide.com/solutions/payments-hub?utm_source=chatgpt.com

Agarwala N (2020) Monitoring the ocean environment using robotic systems: advancements, trends, and challenges. Mar Technol Soc J 54(5):42–60. https://doi.org/10.4031/MTSJ.54.5.7, https://www.researchgate.net/publication/346382738_Monitoring_the_Ocean_Environment_Using_Robotic_Systems_Advancements_Trends_and_Challenges

Alharthi M, Hanif I (2020) Impact of blue economy factors on economic growth in the SAARC countries. Marit Bus Rev 5(3):253–269. https://doi.org/10.1108/MABR-01-2020-0006

Appiah MK, Ameko E, Asiamah TA, Duker RQ (2023) Blue economy investment and sustainability of Ghana's territorial waters: an application of structural equation modeling. Int J Sustain Eng 16(1):1–15. https://doi.org/10.1080/19397038.2023.2195422

Bhati M, Goerlandt F, Pelot R (2025) Digital twin development towards integration into the blue economy: a bibliometric analysis. Ocean Eng 317:119781. https://doi.org/10.1016/j.oceaneng.2024.119781

BlueInvest (2024) Investor report. Unlocking the potential of the blue economy. https://maritime-forum.ec.europa.eu/theme/investments/blueinvest_en

Cbonds (2025) Seychelles bonds, 6.5% 11oct2028 (blue bonds), USD (XS1885544236). https://cbonds.com/bonds/542353/

eFinance Management. Green Finance. https://efinancemanagement.com/sources-of-finance/green-finance

European Commission (2021) CORDIS—EU research results, blue-cloud: piloting innovative services for marine research & the blue economy. https://cordis.europa.eu/project/id/862409/reporting

European Commission (2021) EU blue economy observatory, infrastructure and robotics. https://blue-economy-observatory.ec.europa.eu/eu-blue-economy-sectors/infrastructure-and-robotics_en

European Commission, BlueInvest investor report 2024. https://oceans-and-fisheries.ec.europa.eu/news/investments-blue-economy-are-increasing-2024-04-03_en?utm_source=chatgpt.com

G20 (2023) Generating finance for blue economy transition, task force 6: accelerating SDGs: exploring new pathways to the 2030 agenda. https://t20ind.org/research/generating-finance-for-blue-economy-transition/?utm_source=chatgpt.com

Gesami BK, Nunoo J (2024) Artificial intelligence in marine ecosystem management: addressing climate threats to Kenya's blue economy. Front Mar Sci 11:1404104. https://doi.org/10.3389/fmars.2024.1404104

Global Blue. https://www.globalblue.com/en/business/payment-solutions/payment-gateway?utm_source=chatgpt.com

Global Sustainable Investment Alliance, Global sustainable investment review (2022). https://www.gsi-alliance.org/members-resources/gsir2022/?utm_source=chatgpt.com

Green Climate Fund. Annual results report 2020. https://www.greenclimate.fund/annual-results-report-2020 [16], World Economic Forum. What is green finance? https://www.weforum.org/stories/2020/11/what-is-green-finance/

Green Climate Fund. Annual results report 2020. https://www.greenclimate.fund/annual-results-report-2020

Hamdy A (2014) Transition towards green water economies: surrounding issues and needed capacity development. Glob Environ 7(2). https://doi.org/10.3197/ge.2014.070205

Hugo K (2022) Alaska's blue economy could benefit from blockchain. Office of Intellectual Property and Commercialization, University of Alaska. https://www.uaf.edu/oipc/news/2022/january/alaskas-blue-economy-could-benefit-from-blockchain.php#:~:text=As%20researchers%20with%20the%20Alaska%20Blue%20Economy,Fisheries%20and%20Ocean%20Sciences%2C%20ACEP%2C%20College%20of

IFC Blue Finance. https://www.ifc.org/en/what-we-do/sector-expertise/financial-institutions/climate-finance/blue-finance

IFC Guidelines for Blue Finance. https://www.ifc.org/en/insights-reports/2022/guidelines-for-blue-finance

Lavitt T, Sargeant H (2024) Green infrastructure in developing economies: a unique target for risk reduction and financial investment. Clim Policy 25:1–14. https://doi.org/10.1080/14693062.2024.2409805

Nauticus Robotics (2022) How Nauticus robotics is helping shape the blue economy. https://nauticusrobotics.com/insights/how-nauticus-robotics-is-helping-shape-the-blue-economy/

National Oceanic and Atmospheric Administration, Science Council (2022) NOAA cloud action plan. https://sciencecouncil.noaa.gov/wp-content/uploads/2023/09/NOAA-Cloud-Action-Plan-V1_041922.pdf

OECD, Ocean development finance data platform. https://www.oecd.org/en/topics/sub-issues/ocean.html

OECD, Data platform on development finance for the sustainable ocean economy. https://oecd-main.shinyapps.io/ocean/

Pandiangan I (2025) Incentro high-tech, high tide: how digital solutions are shaping the "blue economy". https://www.incentro.com/en-ID/blogs/high-tech-high-tide-how-digital-solutions-are-shaping-the-blue-economy

Pranita D, Sarjana S, Musthofa BM, Kusumastuti H, Rasul MS (2023) Blockchain technology to enhance integrated blue economy: a case study in strengthening sustainable tourism on smart islands. Sustainability 15:5342. https://www.researchgate.net/publication/369346769_Blockchain_Technology_to_Enhance_Integrated_Blue_Economy_A_Case_Study_in_Strengthening_Sustainable_Tourism_on_Smart_Islands

SDGInvest, report, blended finance & the blue economy, data brief. https://sdginvest.jointsdgfund.org/sites/default/files/2022-05/Data%20Brief%20-%20Blended%20Finance%20and%20the%20blue%20economy%20-%20Draft-2%5B94%5D.pdf?utm_source=chatgpt.com

Sharma V, Jangir K, Gupta M, Rupeika-Apoga R (2024a) Does service quality matter in FinTech payment services? An integrated SERVQUAL and TAM approach. Int J Inf Manag Data Insights 4(2):100252

Sharma V, Gupta M, Taneja S (2024b) Does FinTech adoption impact on sustainability of small businesses: mediating role of financial well-being. Glob Knowl Mem Commun

S&P Global Ratings. Sustainability insights: global sustainable bond issuance to hold steady at $1 trillion in 2025. https://www.spglobal.com/ratings/en/research/articles/250205-sustainability-insights-global-sustainable-bond-issuance-to-hold-steady-at-1-trillion-in-2025-13402934

Spencer B (2022) Blue economy use cases: autonomous oyster harvesting. https://www.researchgate.net/publication/346382738_Monitoring_the_Ocean_Environment_Using_Robotic_Systems_Advancements_Trends_and_Challenges

SS&C, Blue Prism (2024) How does payment automation work? https://www.blueprism.com/resources/blog/payment-automation/?utm_source=chatgpt.com

Statista. Green bonds market worldwide. https://www.statista.com/topics/9217/green-bonds-market-worldwide/

Thanh Ha L (2025) Gauging the dynamic interlinkage level between Fintech and the global blue economy performance. J Econ Stud 52(1):55–71. https://doi.org/10.1108/JES-10-2023-0613

Thompson BS (2022) Blue bonds for marine conservation and a sustainable ocean economy: status, trends, and insights from green bonds. Mar Policy 144:105219. https://doi.org/10.1016/j.marpol.2022.105219

Tirumala RD, Tiwari P (2022) Innovative financing mechanism for blue economy projects. Mar Policy 139:104194. https://doi.org/10.1016/j.marpol.2020.104194

United Nations, World investment report 2024. https://unctad.org/publication/world-investment-report-2024

United Nations, World investment report 2024. Chapter III. https://unctad.org/publication/world-investment-report-2024

United Nations Environment Programme, UNEP green economy report. https://www.unep.org/pt-br/node/23750

United Nations, United Nations Environment Programme, Sustainable Blue Finance, Mobilising Capital for a Sustainable Ocean. https://www.unepfi.org/blue-finance/?utm_source=chatgpt.com

United Nations Development Programme, Financing the Blue Economy (2022). https://www.undp.org/sites/g/files/zskgke326/files/2022-09/Financing%20the%20Blue%20Economy.pdf?utm_source=chatgpt.com

Web of Science database. https://www.webofscience.com/wos/woscc/analyze-results/03e0f058-b0f0-41d7-a608-3a5997ca715f-014d78fdca

World Bank Group, What is the blue economy? https://www.worldbank.org/en/news/infographic/2017/06/06/blue-economy?utm_source=chatgpt.com

World Bank, International Finance Corporation (2022) Guidelines for blue finance. https://www.ifc.org/en/insights-reports/2022/guidelines-for-blue-finance?utm_source=chatgpt.com

World Bank Group, The potential of the blue economy. https://openknowledge.worldbank.org/server/api/core/bitstreams/cee24b6c-2e2f-5579-b1a4-457011419425/content

World Bank Group (2018) Seychelles launches world's first sovereign blue bond. https://www.worldbank.org/en/news/press-release/2018/10/29/seychelles-launches-worlds-first-sovereign-blue-bond

Wuwung L, Croft F, Benzaken D, Azmi K, Goodman C, Rambourg C, Voyer M (2022) Global blue economy implementation. Sec Mar Affairs Policy 9:1043881. https://doi.org/10.3389/fmars.2022.1043881

Investment and Financing in the Blue Economy to Achieving the SDGs in the EU Countries: Opportunities for Financial Technologies

Jekaterina Kuzmina, Andris Fomins, Atis Verdenhofs, and Inese Mavlutova

Abstract The topic of the blue economy is becoming increasingly relevant on a global scale due to the rapid increase in the dependence of society and the economy on ocean and coastal resources every year. The blue economy is crucial to many nations already dealing with droughts, storms, and increasing sea levels. In addition to being an environmental concern, preserving the equilibrium of our marine ecosystems is also closely related to achieving sustainable development goals (SDGs). Financial markets are experiencing exponential growth in sustainable finance and the emergence of new and diverse financial instruments, including blue bonds and blue loans. To expand the use of these instruments, it is necessary to attract modern financial technologies, such as digital payments, blockchain, AI-driven risk management, peer-to-peer finance, big data, etc. The chapter reveals the growing importance of blue finance in sustainable development, with a particular emphasis on blue bonds, and the dependence of the value added of the blue economy on several factors, including ESG risk; managing this risk can achieve the highest sustainability outcomes by attracting larger amounts of finance with modern financial technologies.

Keywords Blue economy · Blue finance · Blue bonds · SDGs · ESG risks · Financial technologies

1 Introduction

The well-known evidence is that oceans cover more than 70% of Earth's surface. Moreover, they contain about 97% of our planet's water. As a result, the blue economy is a considerable part of sustainable economic development and an urgent question within the environmental preservation agenda. Existing research highlights that the value of the global blue economy could increase up to three trillion USD by the year 2030 (The London School of Economics and Political Science 2024), which is

J. Kuzmina · A. Fomins · A. Verdenhofs · I. Mavlutova (✉)
BA School of Business and Finance, Riga, Latvia
e-mail: Inese.Mavlutova@ba.lv

V. Sharma et al. (eds.), *FinTech for Sustainable Finance and a Well-Blue Economy*, World Sustainability Series, https://doi.org/10.1007/978-3-031-92390-6_13

the domain that provides enormous opportunities for innovation, economic development, and growth. Realizing the opportunities requires considerable financial inputs, so the emergence of blue finance and the increasing role of technologies in the transformation process could help in addressing the complex challenges.

Oceans and waterways are vital to human life and the foundation of the world economy, providing the population with access to food and employment. The marine impact on climate change already affects a significant part of the world's population and this impact is expected to intensify in the coming decades. These factors will not only affect the quality of people's living conditions and places of residence from a climate change perspective but will also have a significant influence on the well-being of society and the economy, thereby achieving the sustainable development goals (SDGs 2, 6, 7, 12, 13, 14) (United Nations 2025a).

The importance of the Blue Economy in generating gross value added in Europe is growing. The Blue Economy has shown significant growth in nominal gross value added between 2015 and 2021, namely the growth of offshore wind energy by 326%, shipbuilding and repair by 54%, maritime transport by 29%, marine living resources by 27%, and port activities by 11%. Coastal tourism has been the most relevant sector, accounting for 29% of the EU's blue economy gross value added in 2021 (European Union 2024).

The Principles for Financing a Sustainable Blue Economy were launched in 2018 to promote ocean economy development as the world's first global framework of guidelines for banks, investors and insurers on financing a sustainable blue economy. SDG 14 ("Conserve and sustainably use the oceans, seas and marine resources for sustainable development") was promoted as a set of standards for integrating the financial sector into the sustainability of ocean-related sectors. The Principles (United Nations 2025b) envisage developing knowledge and data on possible risks and impacts related to the activities of financial institutions and investors, as well as promoting sustainable financing opportunities in the Blue Economy.

Blue finance is a relatively new area of climate finance that can be recognized as a specific green finance sector, attracting increasing interest from financial institutions and investors around the world. Unfortunately, other participants in the financial sector are not yet closely involved in this new direction, nor are new financial technologies useful in the development of blue finance.

The current chapter aims to analyze the role of the blue economy in sustainable development by examining the correlation of country-level ESG risks with the performance of the blue economy sector in the EU countries to substantiate the role of innovative financial technologies in relation to the achievement of SDGs in the maritime sector and the economy as a whole.

The current chapter is devoted to an important, nevertheless currently overlooked aspect of sustainable development: the relationship between ESG risks and blue economy performance. Understanding this relationship could help increase financial innovations to promote sustainable growth and achieve sustainable development goals. The findings are valuable for policymakers and investors but could also be interesting for a broad number of stakeholders and researchers.

2 The Role of the Blue Economy Role and Its Framework

The blue economy is becoming an increasingly relevant topic in the global economy to help ensure access to clean water, protect the underwater environment, and invest in a sustainable water economy. This is closely related to the fact dependence on ocean and coastal resources in society and the economy is rapidly increasing every year. Waterways and oceans are essential to human existence and a cornerstone of the global economy. Many countries, including the EU, depend on the blue economy. In the year 2000, around 1.72 billion people lived within 50 km of the shoreline. By 2018, the coastal population had increased to over 2 billion, which makes up more than 29% of the world's population. A pronounced concentration of the world's population within the 5 km coastal strip can be observed, where around 10% of the world's population is living (Cosby et al. 2024). These data indicate an extremely high level of humanity's dependence on the ocean and its resources.

The effects of climate change are already affecting a significant portion of the world's population and are expected to grow in the coming decades. These impacts will not only affect the quality of people's living conditions and places of residence from a climate change perspective but will also significantly affect the economy and the well-being of society.

Economic benefits from the use of ocean resources are increasingly contributing to the global economy. It is estimated that the value of global key ocean assets is USD 24 trillion, of which the value of goods and services produced accounts for USD 2,5 trillion, which is approximately 5% of global GDP (Facility and (GEF) Assembly 2018). In 2016, the estimated the value ocean economy was USD 1.5 trillion annually. It makes around 2.5% of global GDP and provides approximately 30 million direct jobs (OECD 2016). It is anticipated that over 40 million people will be employed in these sectors and GDP output of these sectors will increase to more than USD 3 trillion by 2030 (OECD 2024).

The blue economy plays an important role in the EU economy. According to the latest statistical data, it's turnover has increased by 21% from 513.2 billion Euros in 2020 to 623.6 billion Euros in 2021. In 2021 gross operating profit was equal to 76.4 billion Euros, which in turn means 73% increase compared to 2020. This economic sector has become a significant source of employment in Europe. In 2021 Employment has increased by 17% compared to year 2020. Blue economy has contributed 1.8% of GDP to the EU27 economy (European Union 2024). These key figures determine the increasing importance of the coastal zone in the global economy, as well as raise issues related to ensuring global sustainability.

The Blue Economy includes economic activities in various sectors which make extansive use of water resources such as fisheries, aquaculture, coastal tourism, maritime transport, ship and boat building, renewable energy and ports, etc. Marwa et al. (2024) emphasize that goal of the Blue Economy is to establish a framework that enables economic development without compromising the health of marine ecosystems. OECD estimates, that Blue Economy sectors can make a significant contribution in accelerating the green transition. Water-based renewable energy can promote the clean energy transition, aquaculture solutions can minimize coastal flood risks, and blue bioeconomy and biotechnology can reduce carbon and nutrient pollution (OECD 2016).

Authors used Scopus database as a base for literature review. Search of publications was performed on 16 January 2025. Initial search indicates that there is very limited research done in relation to blue economy and finance (technologies)—14.2% of blue economy related publications are in some way connected to finances or investments. SDG relation with blue economy was identified in 47 out of 1807 publications. Less than 1% of publications or 16 "tech" related articles were only identified among "blue economy" publications. Only three publications are to some degree related to financial technologies, others are related to technologies directly related to water, ocean, maritime and biological type of technologies. Combination of used keywords in search and result of the search is shown in Table 1.

Analyzing search results of "blue economy", authors can observe that the first two researches with keywords "blue economy" were published in 2011, and there was a stable and low increase in publication count in the following 9 years. Sharp increase in publication count started only in 2021 when 275 documents were published in comparison to 145 publications in 2020. An overview of publications indexed in Scopus database can be seen in Fig. 1.

Blue economy makes a significant contribution to balancing economic growth with environmental sustainable development of coastal regions and sustainanbility in the ocean environment. It recognizes the value of ocean resources while at the same time promoting sustainable management of those resources for the benefit of future generations (Bennett et al. 2019). Blue economy encourages the increase of productivity and resilience in coastal and marine environments and supports the health of those ecosystems, which in turn increases the sustainability of communities that rely on those resources and ecosystems.

Blue economy supports the integration of the above-mentioned aspects into the economies, at the same time emphasizing regional development and considering the needs of national and regional populations (Voyer et al. 2018).

An appropriate regulatory framework (as shown in Fig. 2) is important for the sustainable and successful development of the blue economy. Currently, its development is only at an early stage. In the authors' opinion, a broad discussion should be created among experts and policymakers, as well as in academic circles, about the

Table 1 Scopus search criteria and results

Keywords for search	Count of results
"blue economy"	1807
"blue economy" AND (financ* OR investment*)	257
"blue economy" AND sdg	47
"blue economy" AND *tech	16
"blue economy" AND fintech	3

Source Created by authors

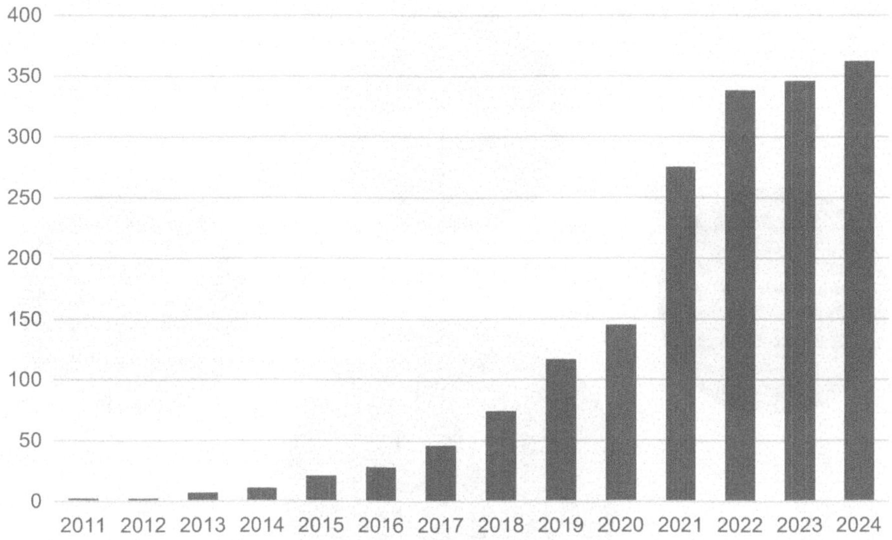

Fig. 1 Count of "blue economy" publications in Scopus by years. *Source* Created by authors

impact of various societal activities, including economic activities, on the sustainability of the ocean and the effective use of coastal resources. In the academic literature, we can observe the assessment of various local impact factors, but research is strongly focused on the Southeast Asian region. It is necessary to expand research to other regions in order to broaden a global overview of the problem and improve regulation in this area.

Authors used keyword cluster analysis to identify different directions of existing research related to "blue economy". Cluster analysis was performed by using VOSviewer and dataset containing 1807 publication information from Scopus database. Keyword clusters can be seen in Fig. 3.

Authors can identify three main clusters out of which only one is broadly related to the research topic of this paper (financial technologies or fintech). Directions of publications could be categorized in the following groups:

- Green cluster—governance, regulation, policy type of publications
- Red cluster—publications related to economic aspects of blue economy
- Blue cluster—marine-related topics like fisheries, maritime transport.

3 Blue Finance and Technologies

Blue finance as a new area of green finance that is attracting increasing interest from financial institutions and investors worldwide. The Principles for Sustainable Blue Economy Finance were jointly developed by the European Commission, the World

Fig. 2 Blue economy framework: circle of activities and benefits. *Source* Created by authors

Wildlife Fund, the World Resources Institute, and the European Investment Bank and are maintained by the UNEP FI Sustainable Blue Economy Finance Initiative. With the aim to support SDG14 ("Conserve and sustainably use the oceans, seas and marine resources for sustainable development") and other that are aimed towards good governance of the ocean (United Nations 2025b).

The Sustainable Blue Economy Finance Principles are designed to be used in developing long-term assumptions for investment decision-making, considering social, environmental and economic aspects and quantitative risks and assessing their systemic impact and development opportunities. There are also findings published by Sharma et al. (2024) indicating relations between FinTech adoption and sustainability (Sharma et al. 2024).

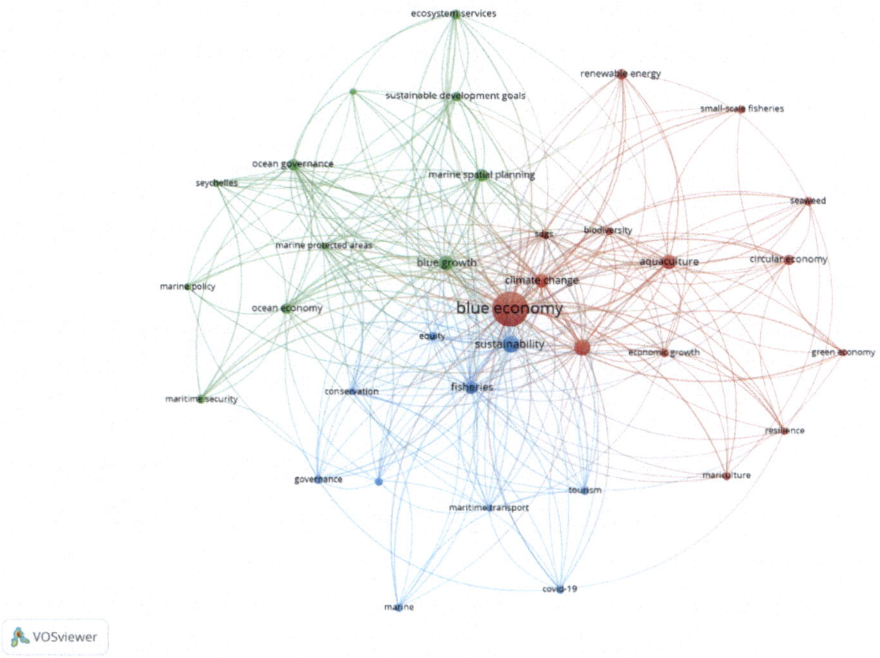

Fig. 3 Blue economy keyword clusters. *Source* Created by authors

Authors analyzed separately results of 257 publications related to either finance or investments and blue economy. Five clusters of keywords can be identified in the selected dataset. Keyword clusters can be seen in Fig. 4. Authors can categorize the following directions of research:

- Green cluster—economy related publications
- Blue cluster—governance and sustainable development goal publications
- Yellow cluster—decision making related to sustainability publications
- Red cluster—environmental aspects and nature related publications
- Purple cluster—marine and water related publications

The most cited publication from the dataset (Cisneros-Montemayor et al. 2021) describe that not natural resources but factors like national stability, corruption, and infrastructure are driving achievements in the blue economy. Fratila et al. (2021) have researched and described the relationship between pollution and economic growth, indicating a positive correlation between the two indicators, specifically focusing on maritime transportation (Fratila et al. 2021). At the same time several publications support the statement that the potential of economic activities on and around the oceans, seas, and other bodies of water are huge (Choudhary et al. 2021; Dalton et al. 2019), also after COVID-19 (Hartmann et al. 2025). Among other topics Kamberidou

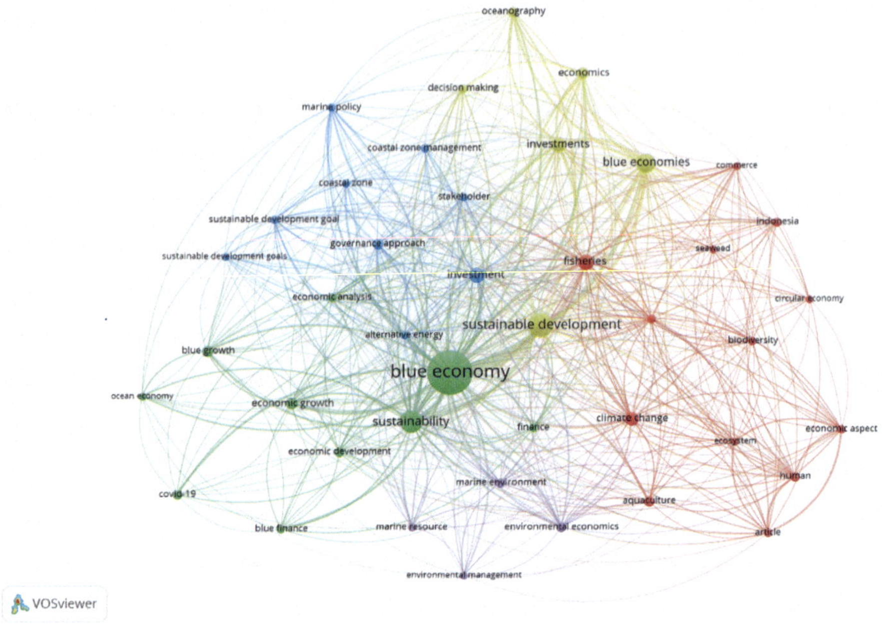

Fig. 4 Blue economy and finance or investment keyword clusters. *Source* Created by authors

(2020) has managed to describe examples of women who have become role models also in "blue economy" related entrepreneurship (Kamberidou 2020).

In relation to our publication Bennett et al. (2024) have explored and described several financing mechanisms available for territories in Carribeans (Bennett et al. 2024) still emphasizing also the need for policymakers to have an active role in the process. Diversified financial opportunities are also described in research done by Ye et al. (2024) where among other topics financing diversification is pointed out as one of three needed collective actions (Ye et al. 2024).

Sixteen publications were identified with the additional search keyword "*tech". After reviewing the results authors could create four different categories of publications as shown in Table 2.

According to the results of the Scopus database search, there are only 3 publications related directly to FinTech and the blue economy. All of them are very specific

Table 2 Tech categories identified in Scopus search

Type of technology	Count of publications
Bio-Tech	3
Fin-Tech	3
Marine-Tech	6
General-Tech	4

Source Created by authors

and narrow research. Most recent research from the identified is done by Ha (2025), and it is connected with FinTech and volatility of blue bonds where the timeframe of research is limited to just 3 years (Ha 2025). Zak et al. (2022) describe three different areas as sources of funding related to the blue economy, out of which one is categorized as "emerging forms of decentralized finance" (Zak et al. 2022). Bihorac et al. (2024) point out that digitalization also impacts the finance industry and indicates that FinTech solutions can impact the process of reaching sustainability goals (Bihorac et al. 2024).

The financial markets are witnessing the exponential growth of sustainable finance and the emergence of new and diverse financial instruments that incorporate green or social use-of-proceeds and reach sustainability targets. Currently, we are seeing a significant increase in interest in blue finance among investors as well as financial institutions, and issuers globally. Blue bonds and blue loans are innovative financial instruments. They mark the funds exclusively for ocean-friendly projects as well as protecting clean water resources and are experiencing a notable growth in interest (European Commission 2019). Contrary to green bonds and loans, industry standards and issuance guidelines for blue bonds and blue loans have not been adopted.

The existence of a generally accepted, clear and, unambiguous methodology for identifying blue financial assets is important. Currently, blue financial assets are created on the basis of green financial assets, adjusting their parameters to the needs of financing blue activities. So, in order to finance projects related to achieving the goals of the blue economy, financial instruments must, first of all, be issued in accordance with the criteria set for green bonds and green loans (see Fig. 5). Secondly, the activities to be financed must be consistent with the SDG priority areas 2, 6, 7, 12, 13, 14, namely zero hunger, clean water and sanitation, affordable and clean energy, responsible consumption and production, climate action, life below water (UN SDGs). It should be noted that goals Nr. 6, 13, and 14 are being considered as the most relevant.

Information and communication technologies play a decisive role in transforming and improving the blue economy both in managing business processes as well as in attracting capital to finance business activities. Data collection, data management, and communication within marine sectors have been highlighted as the main areas of improvement (Murray et al. 2018). Research by Berkes (2017) shows that wider use of information technologies in fisheries and aquaculture can improve decision-making as well as can increase the productivity and reduce negative environmental impacts (Berkes 2017).

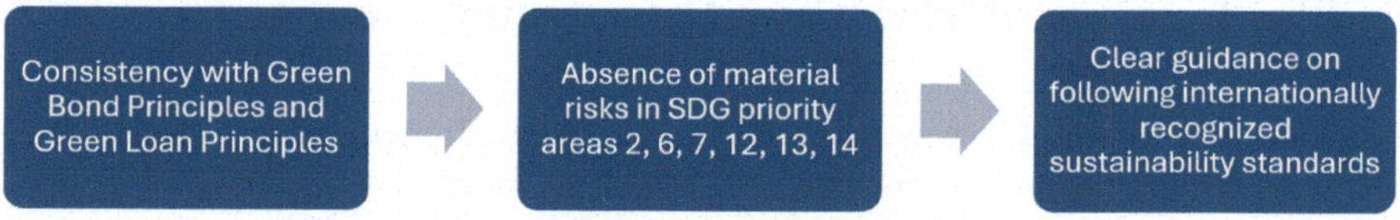

Fig. 5 Identification of blue eligible activities. *Source* Created by authors

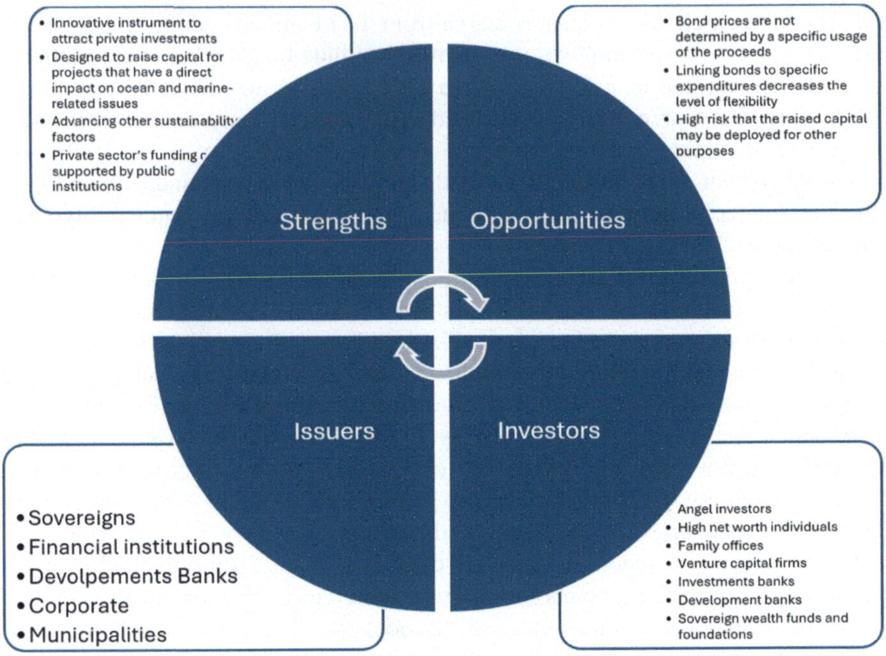

Fig. 6 Key characteristics of blue bonds. *Source* Created by authors

To expand the use of these instruments, it is necessary to attract modern financial technologies, such as electronic transactions and digital payments, blockchain and cryptocurrencies, AI-driven automated services and risk management, peer-to-peer investments and lending, big data advanced analytics to support decision making, etc. Innovative digital financial technologies will provide accessible and cost-effective financial services for blue economy, facilitate faster and inclusive blue economy growth and contribute to resilient infrastructure and industrial innovation thus thereby moving the global economic system towards conserving and increasing clean water resources, restoring ocean prosperity and biodiversity.

For example, Fig. 6 explains the set of characteristics of the most popular Blue financial instruments, such as blue bonds.

4 Blue Economy Performance Determining Factors

The following hypothesis was formulated and tested: Higher country-level overall ESG risk scores negatively correlate with blue economy performance. The definition of the hypothesis is based on the following rationality: EU member states with less advanced ESG practices demonstrated by the higher overall ESG score provided by Sustainalytics—Morningstar (2024) may experience reduced investment and development in sustainable sectors. Testing the hypothesis formulated above is critical for multiple reasons. For example, from a policy-maker's perspective, understanding this relationship could reform how governments develop and implement strategies (including strategies related to the blue economy principles). Moreover, investment and banking sector could benefit from the validation of this hypothesis would offer an important understanding of risk and portfolio management in the blue economy sectors, as investors are interested in understanding the link between ESG risk measures and blue economy outcomes in order to make more informed decisions about asset allocation alternatives.

To test the hypothesis, the authors selected an independent variable: the country's ESG overall score as determined by Sustainalytics—Morningstar in December 2024 (Sustainalytics—Morningstar 2024); a dependent variable: the blue economy gross value added by all the subsectors as determined by European Commission (2024a); and control variables: the country's GDP per capita, coastal length in km, and EU membership duration (European Commission 2024b). Table 3 demonstrates descriptive statistics of the variables mentioned above.

Based on the performed regression analysis (demonstrated in the Table 4), the authors can determine the following regression function:

$$\begin{aligned} \textit{Blue economy gross value added} = {} & 7{,}243 - 0{,}072 * \textit{ESG overall score} \\ & + 0{,}00003 * \textit{GDP per capita} \\ & + 0{,}0002 * \textit{coastal length} \\ & + 0{,}016 * \textit{EU membership duration} \end{aligned}$$

As expected, there is a negative correlation between ESG overall scores and blue economy gross value added ($r = -0.321$, $p < 0.05$), and the regression function

Table 3 Descriptive statistics

	ESG overall score	Blue economy gross value added by all the subsectors	Country's GDP per capita	Coastal length in km	EU membership duration
Mean	15.14	7,590.92	34,014.29	1,940.11	36.79
Standard Dev	3.28	11,094.48	16,158.40	3,452.99	20.01
Minimum	10.19	81.20	19,800.00	0.00	12.00
Maximum	21.40	42,868.93	86,300.00	13,676.00	68.00
25th Percentile	12.28	625.30	25,075.00	97.50	21.00
75th Percentile	16.73	4,348.61	39,525.00	3,203.75	52.00

Source Created by authors

Table 4 Regression model results

Variable	Estimate	Sig v nificance (p-value)
Intercept	12.6842	0.0127
Blue economy gross value added	-0.7234	0.0318
Blue economy gross value added ^{2}	0.0412	0.0456
Multiple R	0.7642	
Multiple R ^{2}	0.5840	
Adjusted R ^{2}	0.5240	
Standard error	0.0328	
F-statistic	9.427	0.0001

Source Created by authors

supports the hypothesis that higher ESG overall scores (demonstrating weak achievements in ESG practices) are associated with lower blue economy output. Moreover, the regression coefficient suggests that an increase in ESG risk score is related with a 7.2% decrease in blue economy gross value added, holding all other factors constant. Among the control variables examined, GDP per capita emerged as the strongest predictor, demonstrating a robust positive correlation with blue economy gross value added ($r = 0.583$, $p < 0.01$). Coastline length also showed a noteworthy positive correlation ($r = 0.412$, $p < 0.05$), while EU membership duration exhibited a weaker positive relationship ($r = 0.289$, $p > 0.05$). The statistical model demonstrated good explanatory power, with the multiple regression analysis accounting for approximately 52% of the variance in blue economy gross value added (Adjusted $R^2 = 0.524$). The model's overall statistical significance was confirmed by the F-statistic p-value falling below 0.01, indicating strong evidence against the null hypothesis. Moreover, the analysis demonstrated the following evidence: the blue economy gross value added showed substantial variation across countries. For example, landlocked nations like Austria and Luxembourg reported minimal blue economy gross value added, while the Nordic countries, consistently displaying lower overall ESG risk scores and higher blue economy gross value added results.

The statistical analysis provided evidence of the model's validity and reliability. The overall model demonstrated strong statistical significance with an F-statistic of 8.427 (p-value < 0.001) and explained a substantial portion of the variance in blue economy gross value added, as indicated by an R^2 value of 0.584 (Adjusted $R^2 = 0.524$). The ESG score coefficient appeared as a significant predictor ($p < 0.05$), along with the control variables GDP per capita and coastline length ($p < 0.05$). EU Duration showed marginal significance ($p < 0.1$), possibly influencing blue economy development.

Multiple diagnostic tests confirmed the model's robustness. The Variance Inflation Factor analysis revealed no concerning multicollinearity among predictors, with all test values remaining below 5. The Durbin-Watson statistic of 2.1034 indicated the absence of autocorrelation in the residuals. Furthermore, the Shapiro–Wilk test (p

> 0.05) defined the normality of residuals, satisfying a key assumption of linear regression.

5 Blue Economy Sector Performance Cluster Analysis

The authors performed a cluster analysis to reveal additional important insights regarding the ESG risk management performance and development of the blue economy among different European countries. The analysis identified three clusters as the most optimal grouping (k = 0.9558—demonstrating the significance of the results), each showing diverse patterns in ESG performance and economic characteristics. The visual analysis supports these findings (see Fig. 7), and the results are summarized in Table 5.

Further insights from the analysis reveal a strong correlation between overall ESG risk performance and cluster assignment. Countries with better overall ESG risk scores are consistently associated with higher-performing clusters, emphasizing the critical role of governance and sustainability in economic development. Coastal length, while contributing to the blue economy, appears less powerful in determining cluster membership than economic and ESG factors. EU membership duration also shows some limited influence.

Cluster 2 (green), "High Performers," includes countries like Germany and the United Kingdom (the length of the UK membership was calculated till Brexit).

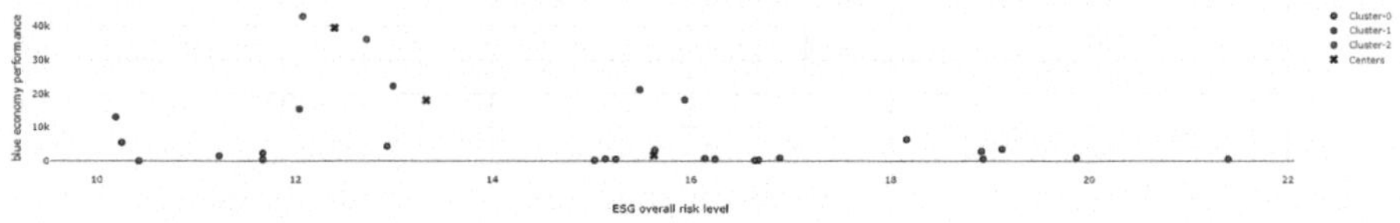

Fig. 7 Cluster analysis. *Source* Created by authors

Table 5 Cluster analysis

	Value
Number of clusters:	3
Explained variance ratio:	0.9558
Centers	[15.629, 1800.6705], [13.33, 18077.062], [12.39, 39497.125]
Maximum iterations:	10
SSE (within):	152164092.2796
SSG (between groups):	3,288851479.1599
SST (total):	3441015571.4395
SSE by group:	70135499.7895, 59290454.3693, 22738138.1209

Source Created by authors

These countries are characterized by the lowest average overall ESG risk scores (equal to 12.39), demonstrating superior governance and sustainability practices. They also reveal the highest average blue economy performance (twice as much as other countries), reflecting substantial economic contributions in maritime sectors. Furthermore, these nations possess high GDP per capita, substantial coastlines, and long-standing memberships in the EU.

Cluster 1 (blue), "Moderate Performers," comprises Denmark, France, Italy, and Spain. The countries in this cluster exhibit moderate overall ESG risk scores (equal to 13.33), while their blue economy performance is classified as medium, reflecting a balanced but not outstanding contribution to maritime economic activities. These countries display medium to high GDP per capita and variable coastline lengths, suggesting diverse geographic and economic contexts. The membership in the EU is long-standing, similar as in cluster 2.

Cluster 0 (red), "Developing Blue Economies," consists of all the other European countries (while the majority of them are newcomers in the EU in comparison to "old" nations) used in the analysis. These countries have the highest overall ESG risk scores (equal to 15.63), which points to more significant challenges in governance and sustainability. Moreover, the countries within this cluster have the lowest blue economy performance, indicating limited economic output in maritime sectors in comparison to other countries in the two clusters described above.

It is worth considering that the Baltic countries, including Estonia, Latvia, and Lithuania, are represented within the "Developing Blue Economies" cluster and are characterized by emerging economic potential and specific structural challenges. The data demonstrate that these countries exhibit remarkably consistent characteristics regarding ESG risk score and other variables. Their ESG risk scores are between 15.14 and 16.68, indicating relatively uniform environmental and governance challenges the countries are facing. The blue economy gross value added ranges from 410.29 to 782.00, reflecting modest maritime economic output. GDP per capita values between 23,500 and 29,300 and reveals the transitional economic status of all three countries.

Further analysis indicates that the Baltic countries occupy the lower percentiles within their cluster. Estonia is in the 25th percentile of blue economy gross value added, indicating limited maritime economic output. Latvia's position is at the 15th percentile, representing the most constrained maritime sector among the three nations. Lithuania performs slightly better at the 40th percentile of blue economy value added. Regarding ESG risk scores, these countries cluster around the 60th to 70th percentiles within their development group, signifying moderate environmental and governance challenges. Their GDP per capita measurements fall between the 30th and 45th percentiles, reflecting an intermediate economic development stage. This positioning underscores the ongoing economic transformation of the three countries.

Analysis reveals a critical need for specialized financial instruments and mechanisms to facilitate implementing and advancing blue economy principles in pursuit of sustainability objectives. Blue bonds represent one such innovative financial instrument that warrants particular attention.

While the European blue bond market rests in its nascent stages, this market exists within Europe's robust green bond market, which occasionally encompasses water-related initiatives. Current market data indicates that the green bond market has been valued at approximately 260 billion EUR, with approximately 15% allocated to water management and biodiversity projects that partially address blue economy requirements (European Commission 2024c). However, the adequacy of this allocation remains challenging to assess due to limitations in comprehensive and sustainable data collection and analysis.

Recent industry developments provide compelling evidence of the growing significance of blue finance. In October 2020, the Nordic Investment Bank introduced a Nordic-Baltic Blue Bond designed to support water-related investment projects. This five-year instrument, valued at SEK 1.5 billion, is a dedicated financing mechanism for water management and protection initiatives, advancing blue economy principles within the Nordic-Baltic region (Nordic Investment Bank 2020).

Furthermore, Deutsche Bank's 2023 milestone as the inaugural bank to join the #BackBlue initiative—a collaborative effort established by the Ocean Risk and Resilience Action Alliance with United Nations support—demonstrates increasing institutional commitment to ocean preservation. Through this partnership, Deutsche Bank has initiated the development of financial solutions targeting ocean protection and has successfully issued several blue bonds (Deutsche Bank 2024).

The quasi-governmental sector has also demonstrated substantial engagement in blue bond development. The International Finance Corporation (IFC), a World Bank Group member and leading global development institution focused on private sector advancement in emerging markets, has supported Banca Transilvania in establishing blue finance products. This collaboration has enabled the extension of loans to micro, small, and medium enterprises in accordance with IFC's Guidelines for Blue Finance (International Finance Cooperation 2022).

While these cases illustrate the progressive growth of the blue economy financing, the lack of a comprehensive, standardized program that also includes the use of specific financial instruments and modern financial technologies, such as electronic transactions and digital payments, blockchain and cryptocurrencies, AI-driven automated services and risk management, peer-to-peer investments and lending, big data advanced analytics remains a significant limitation.

6 Research Objectives

- To find out to what extent the global economy depends on ocean resources and what role does the blue economy has in the achievement of the SDGs. Assessing economic importance of ocean resources and coastal areas is essential in identifying factors influencing sustainable growth.
- To investigate financial instruments supporting the blue economy, especially blue bonds. Assessing their parameters is essential for determining suitability for financing blue economy objectives.

- To assess the role of new financial technologies in increasing the volume of blue finance. The reflection of opportunities and benefits provided by financial technologies is essential in the further analysis of the suitability and implementation efficiency of financial instruments.
- To analyze the correlation between country-level ESG risks and blue economy sector performance in EU countries. The analysis is used to measure the influence of elements of blue economy gross value added.
- To perform an experimental analysis of ESG risk factors and their impact on blue economy sector performance, using regression and cluster analysis. The analysis is relevant to assess the effects of governance and sustainability practices on maritime economic output.

7 Limitation

The research highlights the rising attention to the blue economy and explores financing alternatives to achieve sustainability goals that promote the blue economy principles. However, the widespread approval and effective adoption of these principles are hindered by various limitations that need to be addressed by both policymakers and stakeholders.

One significant limitation is the absence of reliable and comprehensive statistics to evaluate the blue bond market performance and their influence on achieving sustainable goals. This lack of data makes it challenging for potential bond issuers and investors to make informed investment decisions and track progress effectively.

Another notable limitation is the absence of regulation and oversight. While well-established legal frameworks govern traditional fixed-income instruments, blue bonds are relatively new to the market and lack specific regulatory structures. This absence raises concerns about transparency, accountability, and the proper use of funds, leading to a high risk of greenwashing and low acceptance among market participants.

8 Conclusion

Through GDP, employment, climate resilience, and well-being, the blue economy has an important place in attaining the UN Sustainable Development Goals, however, due to a lack of investment and financial innovation, it is still underutilized.

Although financial instruments such as blue bonds and blue loans present encouraging alternatives, the absence of standards in the market restricts investor confidence and widespread adoption. The creation of transparent regulatory frameworks for blue finance and collaboration between the public and private sectors is required.

Currently, there is no widespread consideration of the introduction of financial technologies into the market to improve financial solutions for supporting the blue economy or to assess the blue impact of financial instruments and policies. Financial markets should strive to translate general principles of blue economy financing into guidelines for the issuance of blue bonds and blue lending.

Future research should expand the geographical scope beyond the EU and emerging economies, especially Small Island Developing States, conducting interviews with policymakers and experts from financial sector, to increase appreciation of opportunities and challenges of blue finance and explore the importance of digital finance solutions for blue economy financing.

9 Future Directions and Advancements

Using the prospective of the blue economy requires introducing innovative research solutions and technological advancements. The current chapter points out several important areas for future research that scientists could address soon.

Blue Finance's Digital Innovation—This area promises new opportunities for the growth of the blue economy by integrating financial technology. The authors believe that AI-powered analytics could improve the evaluation of maritime investments, while blockchain-based tracking systems could increase transparency in blue bonds and loans.

Regional Development and Market Expansion—The authors claim that the research should be expanded outside of the EU markets, with a special emphasis on regions where the blue economy accounts for a sizeable amount of GDP. These areas provide distinctive case studies for understanding the application principles of blue finance in various economic environments.

Policy Framework and Stakeholder Engagement: The authors believe that maritime industry representatives, legislators, and financial sector professionals should be engaged and interviewed for future studies.

Impact Evaluation Techniques: The development of standardized procedures for evaluating blue financing initiatives' environmental and social impact is vital and should be addressed in future research.

These research directions will contribute to advancing both the theoretical understanding and practical implementation of blue finance techniques and solutions. Currently, there is no widespread consideration of the introduction of financial technologies into the market to propose solutions for financial support of the blue economy projects or to assess the blue impact of financial instruments and policies. Financial markets should strive to translate general principles of blue economy financing into guidelines for the issuance of blue bonds and blue lending.

Future research should expand the geographical scope beyond the EU and emerging economies, especially Small Island Developing States, conducting interviews with policymakers and experts from financial sector, to increase knowledge base of opportunities and challenges and explore the role of digital finance in blue economy financing.

Acknowledgements This research is supported by Project No.5.2.1.1.i.0/2/24/I/CFLA/007 "Internal and external consolidation of the University of Latvia" (RRF project), Project No. LU-BAPA-2024/1-0031, UL and BASBF consolidation grant project "The Role of Digital Finance Innovation in Promoting Sustainable Development Goals (DInnoFin)"

References

Bennett NJ, Cisneros-Montemayor AM, Blythe J, Silver JJ, Singh G, Andrews N, Calò A, Christie P, Di Franco A, Finkbeiner EM, Gelcich S (2019) Towards a sustainable and equitable blue economy. Nat Sustain 2(11):991–993

Bennett M, March A, Failler P (2024) Blue economy financing solutions for the fisheries and aquaculture sectors of Caribbean island states. Fish 9(8):305. https://doi.org/10.3390/fishes9080305

Berkes F (2017) Environmental governance for the anthropocene? Social-ecological systems, resilience, and collaborative learning. Sustainability 9(7):1232. https://doi.org/10.3390/su9071232

Bihorac Z, Zaimovic A, Zaimovic T (2024) Sustainable finance and FinTech: facilitating a sustainable future with the utilization of socio-economic financial services. Adv Emerg Financ Technol Digit Money 352–360

Choudhary P, Khade M, Savant S, Musale A, Chelliah MS, Dasgupta S (2021) Empowering blue economy: from underrated ecosystem to sustainable industry. J Environ Manag 291:112697. https://doi.org/10.1016/j.jenvman.2021.112697

Cisneros-Montemayor AM, Moreno-Báez M, Reygondeau G, Cheung WW, Crosman KM, González-Espinosa PC, Lam VW, Oyinlola MA, Singh GG, Swartz W, Zheng CW (2021) Enabling conditions for an equitable and sustainable blue economy. Nat 591(7850):396–401

Cosby AG, Lebakula V, Smith CN et al (2024) Accelerating growth of human coastal populations at the global and continent levels: 2000–2018. Sci Rep 14:22489. https://doi.org/10.1038/s41598-024-73287-x

Dalton G, Bardócz T, Blanch M, Campbell D, Johnson K, Lawrence G, Lilas T, Friis-Madsen E, Neumann F, Nikitas N, Ortega ST (2019) Feasibility of investment in blue growth multiple-use of space and multi-use platform projects; results of a novel assessment approach and case studies. Renew Sust Energ Rev 107:338–359. https://doi.org/10.1016/j.rser.2019.01.060

Deutsche Bank (2024) Deutsche Bank strengthens its ocean protection policies under #BackBlue initiative. https://www.db.com/news/detail/20240925-deutsche-bank-strengthens-its-ocean-protection-policies-under-backblue-initiative?language_id=1. Accessed 2 Feb 2025

European Commission (2019) Mission ocean and water service portal—Blue bond. https://projects.research-and-innovation.ec.europa.eu/en/funding/funding-opportunities/funding-programmes-and-open-calls/horizon-europe/eu-missions-horizon-europe/restore-our-ocean-and-waters/blue-bond. Accessed 21 Jan 2025

European Commission (2024a) EU blue economy observatory. https://blue-economy-observatory.ec.europa.eu/index_en. Accessed 19 Jan 2025

European Commission (2024b) Eurostat database. https://ec.europa.eu/eurostat. Accessed 19 Jan 2025

European Commission (2024c) EU continues to be a global leader on sustainable finance. https://ec.europa.eu/commission/presscorner/detail/en/ip_24_5967. Accessed 2 Feb 2025

European Union (2024) Blue economy report. https://op.europa.eu/fr/publication-detail/-/publication/ef90077b-1e82-11ef-a251-01aa75ed71a1. Accessed 30 Jan 2025

Fratila A, Gavril IA, Nita SC, Hrebenciuc A (2021) The importance of maritime transport for economic growth in the European Union: a panel data analysis. Sustain 13(14):7961. https://doi.org/10.3390/su13147961

Global Environment Facility (GEF) Assembly (2018) Blue economy. https://www.thegef.org/sites/default/files/publications/GEF%20Assembly_BlueEconomy%20Factsheet_6.19.18.pdf. Accessed 30 Jan 2025

Ha L (2025) Gauging the dynamic interlinkage level between Fintech and the global blue economy performance. J Econ Stud 52:55–71. https://doi.org/10.1108/JES-10-2023-0613

Hartmann L, Walz Y, Clelland D, Esquivel-Rodriguez C, Riganti P, Vicarelli M, Renaud FG (2025) A conceptual framework for a post-COVID-19 green recovery of the blue economy. Mar Policy 171:106446. https://doi.org/10.1016/j.marpol.2024.106446

International Finance Cooperation (2022) First blue funding backed by IFC supports sustainable use of water in Romania. https://www.ifc.org/en/pressroom/2022/first-blue-funding-backed-by-ifc-supports-sustainable-use-of-water-in-romania. Accessed 2 Feb 2025

Kamberidou I (2020) "Distinguished" women entrepreneurs in the digital economy and the multitasking whirlpool. J Innov Entrep 9(1):3. https://doi.org/10.1186/s13731-020-0114-y

Marwa T, Muizzuddin N, Bashir A, Andaiyani S, Cahyadi A (2024) Determinants of the blue economy growth in the era of sustainability: a case study of Indonesia. Econ 12(1):299. https://doi.org/10.3390/economies12110299

Murray F, Needham K, Gormley K, Rouse S, Coolen JW, Billett D, Dannheim J, Birchenough SN, Hyder K, Heard R, Ferris JS (2018) Data challenges and opportunities for environmental management of North Sea oil and gas decommissioning in an era of blue growth. Mar Policy 97:130–138. https://doi.org/10.1016/j.marpol.2018.05.021

Nordic Investment Bank (2020) NIB launches five-year SEK 1.5 billion Nordic–Baltic Blue Bond. https://www.nib.int/releases/nib-launches-five-year-sek-1-5-billion-nordic-baltic-blue-bond. Accessed 2 Feb 2025

OECD (2016) The ocean economy in 2030. https://doi.org/10.1787/9789264251724-en. Accessed 30 Jan 2025

OECD (2024) The blue economy in cities and regions: a territorial approach. https://doi.org/10.1787/bd929b7d-en. Accessed 30 Jan 2025

Sharma V, Gupta M, Taneja S (2024) Does FinTech adoption impact on sustainability of small businesses: mediating role of financial well-being. Glob Knowl Mem Commun. https://doi.org/10.1108/GKMC-04-2024-0225

Sustainalytics—Morningstar. Dataportal. Country risk rating (2024). https://dataportal.sustainalytics.com/deliverables. Accessed 19 Jan 2025

The London School of Economics and Political Science (2024) What is the blue economy? https://www.lse.ac.uk/granthaminstitute/explainers/what-is-the-blue-economy/. Accessed 8 Feb 2025

United Nations (2025) Sustainable developments. https://sdgs.un.org/goals. Accessed 21 Jan 2025

United Nations (2025) The principles. Sustainable blue finance. https://www.unepfi.org/blue-finance/the-principles/. Accessed 21 Jan 2025

Voyer M, Quirk G, McIlgorm A, Azmi K (2018) Shades of blue: what do competing interpretations of the blue economy mean for oceans governance? J Environ Policy Plan 20(5):595–616. https://doi.org/10.1080/1523908X.2018.1473153

Ye Z, Huang X, Xie Z (2024) Sustainable transformation of global ocean governance: can a global public goods approach lead the way? Mar Policy 165:106213. https://doi.org/10.1016/j.marpol.2024.106213

Zak S, Chappel B, Pilger C, Oor B, Theuretzbacher T, Truman R, von Herzen B (2022) Finance approaches supporting social enterprise, non-profits and for-purpose organisations: a marine permaculture case study. In: De Gruyter handbook of sustainable development and finance, vol 621. https://doi.org/10.1515/9783110733488-027

Fear of Missing Out and the Future of Facial Recognition Payments in the Blue Economy

Alwin Estareja and Ramona Rupeika-Apoga

Abstract This chapter explored financial technology's theoretical foundations within the Blue Economy, particularly emphasizing the user acceptance of Facial Recognition Payment Systems (FRPS). Biometric payment systems offer robust security, great convenience, and a more sustainable payment method, as they eliminate the need for physical payment cards made of PVC and other plastic materials that can drastically pollute our water resources. We extended the technology acceptance model with a cognitive bias to better understand its influence on the user acceptance of FRPS. This study aimed to develop a novel theoretical framework that identifies the key constructs impacting users' intention to use FRPS. Additionally to the components of the technology acceptance model, this study also posited the significant influence of cognitive bias in adopting this emerging technology in the burgeoning FinTech sector. The findings emphasized the role of perceived usefulness, perceived ease of use, anticipated regret, social comparison and counterfactual thinking in predicting user behavior to use a specific technology. The findings are critical for decision-makers and industry stakeholders in the Blue Economy.

Keywords Facial recognition payment systems · TAM · Fear of Missing out (FOMO) · Blue economy · FinTech

1 Introduction

The rapid evolution of financial technology has changed how people interact with digital payments, with facial recognition systems emerging as promising innovations. Simultaneously, the Blue Economy is using digital solutions more and more to raise environmental sustainability, efficiency, and financial inclusiveness. The study is significant in several domains because it examines the intersection of technology, psychology, and sustainability in digital payment adoption.

A. Estareja · R. Rupeika-Apoga (✉)
Faculty of Economics and Social Sciences, University of Latvia, Riga, Latvia
e-mail: ramona.rupeika-apoga@lu.lv

V. Sharma et al. (eds.), *FinTech for Sustainable Finance and a Well-Blue Economy*, World Sustainability Series, https://doi.org/10.1007/978-3-031-92390-6_14

Imagine a society where using a camera to purchase eliminates the need for cash, credit cards, or even a smartphone. This is the promise of a new technology called Facial Recognition Payment Systems (FRPS), which simultaneously provides sustainability, convenience, and security. Though amazing as the technology is, human psychology is equally important in deciding its acceptance. How do we feel about using our faces as a payment method? What are the reasons for our willingness to accept or reject this payment method? It's clear that the issue is bigger than just technology issues like ease of use, speed, and efficiency; it also involves worries about our privacy.

The Technology Acceptance Model has long been applied to understand why people use or avoid specific technologies. It implies that perceived usefulness and ease of use are important drivers of users' intention to use. Nevertheless, this model does not consider the emotional and social factors influencing decision-making. Cognitive biases, emotional responses, and social dynamics also affect how we perceive and adopt innovations. People can use new technology because they don't want to be left behind, or wish be modern. In today's world, where technology changes quickly, this is especially true. The fear of missing out (FOMO) is an essential concept, because it affects decision-making across multiple domains, including technology adoption. We aim to create a comprehensive framework by incorporating FOMO into TAM, addressing both emotional and intellectual motivations.

This paper explored how FOMO, combined with the well-established Technology Acceptance Model (TAM), shapes people's attitudes and behaviors toward using the payment systems that recognize faces. By understanding these factors, we can more accurately anticipate adoption trends and develop strategies to increase the accessibility and acceptance of this technology.

In addition to personal conduct, the implementation of FRPS has considerable implications for the Blue Economy, which encompasses the use of ocean resources to boom economic development, and promote ecosystem vitality. Coastal businesses, maritime trade, and tourism are all sectors that can benefit from efficient and secure digital payment solutions. FRPS can diminish dependence on plastic cards, optimize transactions, and improve financial inclusion for small-scale fishermen and coastal businesspeople lacking access to conventional banking services.

This chapter is a theoretical investigation into how psychological and technological factors intersect. By examining the potential impact of FOMO on FRPS adoption, we hoped to provide valuable insights to businesses, policymakers, and technology developers. Understanding these behavioral drivers can help shape marketing strategies, regulatory policies, and user education initiatives that increase FRPS acceptance.

In the following sections, we examined the mechanics of FRPS, the foundations of the TAM models, and the psychological aspects of FOMO more closely.

2 Facial Recognition Payment Systems

Maity et al. (2021) assert that among several biometric characteristics, such as fingerprints, palmprints, voice, and facial attributes, the face is a primary physiological trait amenable to collection and analysis by biometric systems. Big Chinese tech firms such as Baidu, Alibaba, and Tencent have developed world-leading facial recognition technologies that offer high recognition accuracy and fast speed. With its continuous progress and breakthroughs related to this emerging technology, algorithms based on facial recognition are evolving (Liu et al. 2021). Advancement in the facial recognition technologies paved way to the increasing popularity of facial recognition payment systems (FRPS). This modern digital payment system allows users to effortlessly authorize transactions using facial recognition after linking their facial data to their online banking or other payment platforms (Zhang and Kang 2019). As one of the new and innovative biometric payment methods in the world, FRPS has gained growing interest because to its effective and simple payment process (Achim 2025).

3 Technology Acceptance Model (TAM)

For mobile technologies (e.g. facial recognition payment systems), it is crucial to recognise how specific factors influence the use of such technologies (Amoroso and Lim 2013). The Technology Acceptance Model (TAM) created by Fred Davis (see Fig. 1) is a credible and resilient framework that has been utilized intensively to analyze user acceptance of a particular technology (King and He 2006). One of the suggested applications of TAM is measuring user acceptance of "emerging technology" as they surface in the market (Davis and Venkatesh 1996), making it ideal in understanding the user acceptance of FRPS. TAM is not the final stage of technology implementation (Turner et al. 2010), and its primary focus is not to measure the advantages of using a technology (Dyba et al. 2005). Its main application is to examine and understand the constructs that significantly influence users' attitudes and behavioral intention that lead them to use such technology.

3.1 Perceived Usefulness (PU)

In this chapter, the Perceived Usefulness (PU) construct signifies the degree to which an individual believes that employing FRPS would significantly enhance their payment experience. The strong correlation of perceived usefulness with the user acceptance has been studied and proven in previous studies, and it must not be overlooked while designing and implementing a specific system (Davis 1985). A research regarding mobile wallet suggests that mobile wallet services' perceived usefulness

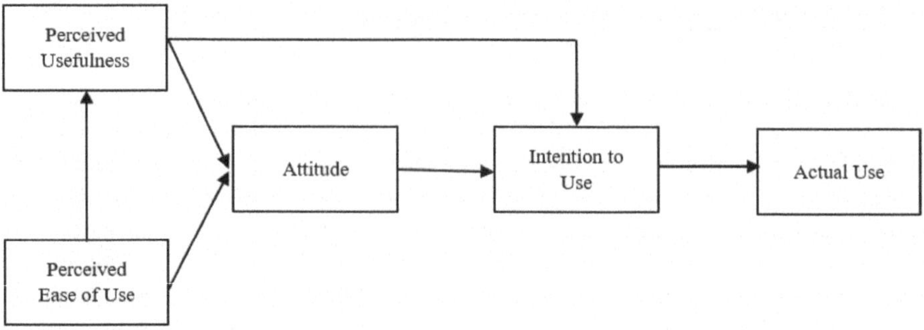

Fig. 1 Technology acceptance model. *Source* Fred Davis' theoretical model (Davis 1985)

positively influences the attitude of its users towards it (Shin 2009). A similar study about a disruptive technology in Austria indicates that perceived usefulness significantly and positively affect the usage of mobile payment systems, indicating that individuals who recognize higher advantages in utilizing mobile payments are more inclined to use them (Schmidthuber et al. 2020). In addition, recent study highlights that the perceived efficiency and convenience of financial technology solutions significantly drive consumer engagement, particularly in mobile payment systems. A study on the adoption of digital payment solutions supports the notion that perceived usefulness is a main determinant of user attitudes and behavioral intentions toward financial innovations (Sharma et al. 2024). Therefore, the researchers posit that:

H1: Perceived usefulness has a positive effect on the user's attitude toward FRPS.

H2: Perceived usefulness positively affects the intention to use FRPS.

3.2 Perceived Ease of Use (PEOU)

This chapter defines Perceived Ease of Use (PEOU) as the degree to which an individual user of FRPS believes that employing this payment method necessitates minimal cognitive effort. A previous study indicates that the perceived ease of use construct significantly positively affects attitudes toward technology acceptance (Wixom and Todd 2005). A study on the utilization of mobile money reveals that users' perceived ease of use is positively connected with their sentiments regarding its use (Gbongli et al. 2019). This aligns with the findings of a relevant study that tackles mobile wallet acceptance of consumers, indicating that users' perceived ease of use of mobile wallet products greatly affects their sentiment towards its usage (Shin 2009). Despite not being explored in the fundamental version of the TAM, the direct and positive effect of the perceived ease of use construct to the users' behavioral intention towards technology adoption was subsequently proven in previous researches, including a recent study that shows that a consumer's perceived ease of use of mobile payment systems has a positive influence to their behavioral intention to use it (Tian and Chan 2024).

Moreover, the perceived ease of use positively influences its perceived usefulness, as a system that is easier to operate is likely to enhance performance or utility for its users, everything else being equal (Davis 1985). A meta-analysis of 26 empirical investigations indicates that the association between perceived ease of use and perceived usefulness is among the most robust relationships within the Technology Acceptance Model (Ma and Liu 2004). An empirical study investigating customer perceptions and adoption of m-payment systems indicates that the perceived ease of use of mobile payment platforms in Finland, Germany, the USA, and Japan positively correlates with the perceived usefulness of this technology (Guhr et al. 2013). With these prior research findings, the researchers hypothesize that:

H3: Perceived ease of use has a positive effect on the user's attitude toward FRPS.

H4: Perceived ease of use has a positive effect on the intention to use FRPS.

H5: Perceived ease of use has a positive effect on the perceived usefulness of FRPS.

3.3 Attitude (ATT)

In this chapter, attitude (ATT) in this chapter is defined as the evaluation of users toward the consequences of using Facial Recognition Payment System (FRPS) when paying. The attitude construct has remained an important element in theoretical studies and applications based on the great assumption that it can predict social behavior (Ajzen and Fishbein 2005). According to Yang and Yoo (2004), attitude is a crucial construct that researchers should focus on because of its significant impact on how individuals and organizations use information systems. A prior study determined that attitude is the most significant factor influencing behavioral intention to utilize mobile payment systems, hence enhancing users' intention and subsequent adoption of the technology (Liébana-Cabanillas et al. 2015). This aligns with previous studies suggesting that users' attitude is positively correlated to their intention to use mobile wallet services (Shin 2009). Another study about mobile payments in China also show the significant and positive influence of its user's attitude to their intention to use one of the leading platforms for payments and digital services in the country called Alipay (Li et al. 2019). Taking these related literatures into account, the researchers suggest that:

H6: Attitude has a positive effect on the intention to use FRPS.

4 Fear of Missing Out (FOMO)

Cognitive biases are systematic, predictable, and replicable patterns of deviation from rational judgment (Kahneman and Tversky 1972). According to a prior study, cognitive biases receive very little attention and remain largely overlooked in technology acceptance research (Ciriello and Loss 2023). To extend the predictive capability of the TAM, a cognitive bias called Fear of Missing Out (FOMO) will be used. FOMO was originally defined as "a pervasive apprehension that others might be having rewarding experiences from which one is absent. It is characterized by the desire to stay continually connected with what others are doing" (Przybylski et al. 2013). In terms of technology acceptance research, the findings of a previous study about cryptocurrency investment platforms show that FOMO positively influence its user's behavioral intention to adopt such systems (Deb and Deb 2024). Another study on corporate decision makers in Austria indicates the positive correlation between FOMO and the intention of users to adopt novel technologies (Gartner et al. 2022).

In this chapter, FOMO's theoretical definition will follow the proposed definition of a prior study, stating that FOMO involves two sets of experiences: affective (Anticipated Regret) and cognitive (Social Comparison and Counterfactual Thinking). These components are based on the perceived difference between the current and possible experiences of an individual, and that between their very own experiences as well as the experiences of those around them (Neumann and Bulck 2020). It is also the basis of the suggested economic definition proposed by Kaddouhah (2024), which states that the decision-making utility of an individual can be influenced by FOMO, leading them to base their decision on their own anticipated regret and the decision of the people in their social group. Following the suggestions from related researches, this chapter will adopt the Anticipated Regret, Social Comparison, and Counterfactual Thinking as the key determinants/components of FOMO.

4.1 Anticipated Regret (AR)

Anticipated Regret (AR), also referred to as Anticipatory Regret, is the affective component of FOMO (Neumann and Bulck 2020). Based on a previous study, in case of a possible regret, an individual anticipates the unlikable effect of what they might feel, which leads them to protect themselves before making a decision (Huang et al. 2022). The primary role of anticipated regret is based on the Regret Regulation Theory (Pieters and Zeelenberg 2007), which states that "regret is an cognitive emotion that people are motivated to regulate in order to maximize outcomes in the short-term and learn to maximize them in the long run" (Neneh 2019). Taking prior studies into consideration, Anticipated Regret in this chapter is defined as the induced emotion that users experience "if they miss out on the chance to gain and enjoy the possible benefits of using FRPS when paying" (Ding 2018). A study regarding continued use of information systems in China indicates that the degree of anticipated regret of users after halting their usage of a specific mobile application is positively related to their intention to continue using it (Ding 2018). Another research shows the positive effect of anticipated regret to the behavioral intention of users to adopt mobile payment systems (Verkijika 2020). Considering the aforementioned, the researchers propose that:

H7: Anticipated regret has a positive effect on the intention to use FRPS.

4.2 Social Comparison (SC)

Social Comparison is one of the cognitive components of FOMO that refers to experiences of an individual when compared with the experiences of others (Neumann and Bulck 2020). It is based on the Social Comparison Theory that primarily states that people seek to gain accurate self-evaluations, and that in order to reduce uncertainty, people tend to compare themselves to others (Festinger 1954). In this chapter, Social Comparison is defined as the tendency of FRPS users to compare their payment method to what others are using in order to evaluate whether they confirm with their peers or social groups (Neumann and Bulck 2020). A technology acceptance research in South Korea shows that the tendency to social comparison of the users of virtual communities positively impact their intention to use Avatars on their profile (Song and Kim 2006). This is consistent with a recent study that shows that Social Comparison's informational influence and value-expressive influence significantly influence the user acceptance of electronic vehicles (Shanmugavel et al. 2022). Taking these prior studies into consideration, the researchers posit that:

H8: Social comparison has a positive effect on the intention to use FRPS.

4.3 Counterfactual Thinking (CT)

Counterfactual Thinking (also known as Counterfactual Thoughts) is the other half of the cognitive component of FOMO. It is evaluative in nature and provides alternatives that are either better or worse than the actual situation (Epstude and Roese 2008). People tend to do counterfactual thinking after making a decision and receiving feedback on the chosen alternative, often asking themselves "What could have been?" (Neumann and Bulck 2020). Counterfactual Thinking does not refer to future prospects but only to negations of the established fact (Roese 1997). Functional Theory of Counterfactual Thinking is adopted in this chapter, which states that individuals generate counterfactual thoughts (e.g. "If only…") after experiencing a failure or a negative event in order to learn from past mistakes. Counterfactual Thinking in this chapter is therefore defined as the user's view of Facial Recognition Payment System (FRPS) as an alternative to other payment methods that they have used in the past (Epstude and Roese 2008). A previous research that investigates the capability of counterfactual thinking to facilitate the behavioral intention to perform specific content-related activities suggests that counterfactual thinking is significantly correlated with behavioral intentions (Smallman and Roese 2009). This relationship is supported by the findings of another study that suggest that counterfactual thinking of users significantly influences their intention to use digital health tracking tools (Wang et al. 2021). Moreover, upward counterfactual thinking that focuses on personal choices leads to an emotion called regret (Epstude and Roese 2008). This relationship is reflected in a study in China about people's satisfaction, indicating that counterfactual thinking significantly affects anticipated regret, i.e. more counterfactual thinking induces more anticipated regret (Li et al. 2022). With these related theories and research findings, this study proposes that:

H9: Counterfactual thinking has a positive effect on the intention to use FRPS.

H10: Counterfactual thinking has a positive effect on the anticipated regret of FRPS users.

5 Theoretical Framework

In line with the aforementioned theories and hypotheses (see Table 1), the researcher proposes a corresponding theoretical framework to assess the links among the determinants of the Technology Acceptance Model and FOMO and to analyze their influence on the behavioral intention of users to adopt facial recognition payment systems (see Fig. 2).

6 Conclusion

This chapter offers a theoretical framework of intention to use Facial Recognition Payment Systems (FRPS) by integrating the Technology Adoption Model with the Fear of Missing Out (FOMO) concept. The FOMO concept represents human behavior and includes three factors: Anticipated Regret, Social Comparison, and Counterfactual Thinking. Unlike traditional TAM constructs—perceived usefulness (PU) and perceived ease of use (PEOU)—which focus on the rational decision-making process, FOMO introduces an emotional and social dimension to technology adoption, which is particularly relevant in today's digital ecosystem.

The fear of being left behind can greatly affect user behavior in a time when the digital revolution is speeding forward. The main outcome of the research is the importance of FOMO constructs in understanding an individual's intention to use emerging technologies like FRPS:

- Anticipated Regret (AR): users may begin to use facial recognition payment systems to prevent the regret of not using a quick, rapid, and creative payment option. This psychological factor pushes individuals toward usage because they

Table 1 Hypotheses

No	Hypotheses	Directions
H1	Perceived usefulness has a positive effect on the user's attitude toward FRPS	PU + → ATT
H2	Perceived usefulness has a positive effect on the intention to use FRPS	PU + → ITU
H3	Perceived ease of use has a positive effect on the user's attitude toward FRPS	PEOU + → ATT
H4	Perceived ease of use has a positive effect on the intention to use FRPS	PEOU + → ITU
H5	Perceived ease of use has a positive effect on perceived usefulness of FRPS	PEOU + → PU
H6	Attitude has a positive effect on the intention to use FRPS	ATT + → ITU
H7	Anticipated regret has a positive effect on the intention to use FRPS	AR + → ITU
H8	Social comparison has a positive effect on the intention to use FRPS	SC + → ITU
H9	Counterfactual thinking has a positive effect on the intention to use FRPS	CT + → ITU
H10	Counterfactual thinking has a positive effect on user's anticipated regret toward FRPS	CT + → AR

Source Created by the authors

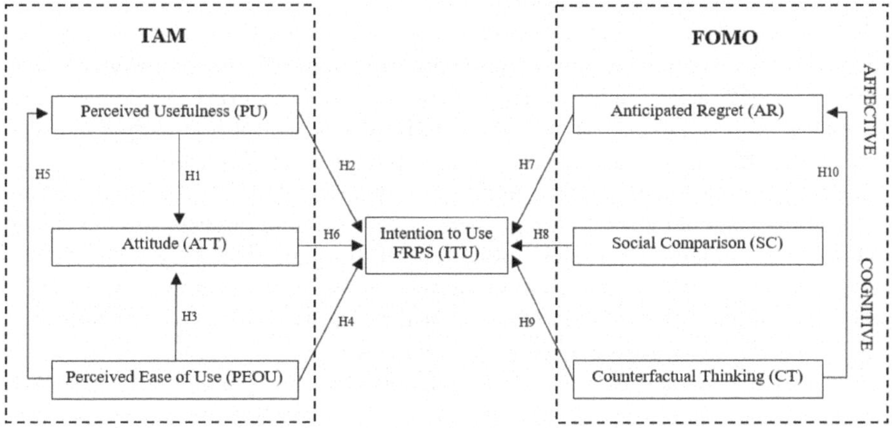

Fig. 2 Theoretical framework. *Source* Created by the authors

anticipate a loss in terms of time, convenience, or social standing if they do not embrace the technology.

- Social Comparison (SC): the influence of peers and society is critical in technology adoption. When consumers witness others employing FRPS effortlessly in routine transactions, they may experience pressure to conform to the trend, so reinforcing the perception that FRPS is an essential and advantageous innovation.

- Counterfactual Thinking (CT): Users often play "what-if" games where they compare the ways they pay now to how they might pay with FRPS in the future. If they think FRPS is a better option, they are more likely to use it. Also, thinking about what might have happened instead can make you more likely to regret what you did, which makes adoption even more likely.

To sum up, the developed theoretical framework shows that FOMO is an important concept that affects the intention to use the facial recognition payment systems. The findings of this chapter are significant not only from the theoretical point of view but also from the practical one, and they can be helpful for policymakers, industry representatives, and companies working in the area of the Blue Economy. This aligns with the insights provided by Jain et al. (2023), who emphasize the importance of understanding the evolving risk landscape in FinTech, highlighting the necessity of further empirical validation.

The environmental impact of FRPS adoption is another crucial factor to take into account, as it aligns with the sustainable objectives of the Blue Economy. The existing payment methods' dependence on cash and credit cards not only makes them inefficient, but they also contribute to plastic pollution and greenhouse gas emissions. Consequently, by eliminating the necessity for physical cards, FRPS can reduce plastic pollution and become a more environmentally responsible financial service. This transition to biometric payments is part of a broader trend of sustainable digital financial services transformation, which is designed to ensure that innovation does not compromise the environment.

Consumer behavior is an important component in the marketing process, and this study demonstrates that FOMO influences consumers' intention to use FRPS. Marketing FRPS as an exclusive and cutting-edge tool can draw in customers and establish it as a dominant force in markets in places with a high population and volume of traffic, such as coastal cities and popular tourist destinations.

Financial inclusion is another important aspect of the Blue Economy. Many communities, including small-scale fishers and coastal entrepreneurs, are excluded from traditional banking services. FRPS can fill this gap by offering a safe mobile money transfer service that does not require well-established infrastructure. Thus, biometric authentication enables financial institutions to provide new and improved digital financial services to underserved populations.

The hazards to data security and regulatory concerns cannot be overlooked (Rupeika-Apoga and Wendt 2021), even with all of FRPS's advantages. Governments and regulatory agencies must put in place robust regulations to protect users' biometric information and guarantee adherence to data protection laws as biometric payment systems proliferate (Rupeika-Apoga and Wendt 2022). The effectiveness of FRPS in the Blue Economy will be determined by the ability to balance technological advancement with ethical concerns to ensure consumers remain confident with the system.

Recognizing this study's limitations is crucial because it is entirely theoretical. The relationships between FOMO, TAM components, and FRPS adoption need further validated through empirical research, and the suggested framework has not yet been empirically tested. Future research should perform qualitative and quantitative analyses to test these theoretical claims in practical contexts.

Regional variations in regulatory frameworks, technology infrastructure, and cultural norms may affect the uptake of FRPS.

Future research should look into these contextual factors to better understand the model's applicability to different populations. Longitudinal studies could also be beneficial in tracking changes in consumer behavior over time as FRPS technology evolves and becomes more widespread.

Addressing these research gaps will allow scholars and industry professionals to improve the theoretical model and understand the psychological and technological factors influencing digital payment adoption.

References

Achim A-L (2025) Is facial recognition technology doomed to fail? https://jingdaily.com/posts/is-facial-recognition-technology-doomed-to-fail. Accessed 23 Jan 2025

Ajzen I, Fishbein M (2005) The influence of attitudes on behavior. The handbook of attitudes. Lawrence Erlbaum Associates Publishers, Mahwah, NJ, US, pp 173–221

Amoroso D, Lim R (2013) Consumer adoption of mobile technology in the Philippines. Baguio City, Philippines

Ciriello RF, Loss S (2023) The yea-paradox: cognitive bias in technology acceptance surveys. Inf Softw Technol 161:107253. https://doi.org/10.1016/j.infsof.2023.107253

Davis FD (1985) A technology acceptance model for empirically testing new end-user information systems : theory and results. Thesis, Massachusetts Institute of Technology

Davis FD, Venkatesh V (1996) A critical assessment of potential measurement biases in the technology acceptance model: three experiments. Int J Hum Comput Stud 45:19–45. https://doi.org/10.1006/ijhc.1996.0040

Deb SS, Deb DS (2024) Awareness of blockchain technology and FOMO on crypto investment: an empirical study. Educ Adm Theory Pract 30:1746–1750. https://doi.org/10.53555/kuey.v30i4.1744

Ding Y (2018) Modelling continued use of information systems from a forward-looking perspective: antecedents and consequences of hope and anticipated regret. Inf & Manag 55:461–471. https://doi.org/10.1016/j.im.2017.11.001

Dyba T, Moe NB, Arisholm E (2005) Measuring software methodology usage: challenges of conceptualization and operationalization. 2005 international symposium on empirical software engineering. IEEE, Queensland, Australia, pp 432–442

Epstude K, Roese NJ (2008) The functional theory of counterfactual thinking. Pers Soc Psychol Rev 12:168–192. https://doi.org/10.1177/1088868308316091

Festinger L (1954) A theory of social comparison processes. Hum Relat 7:117–140. https://doi.org/10.1177/001872675400700202

Gartner J, Fink M, Maresch D (2022) The role of fear of missing out and experience in the formation of SME decision makers' intentions to adopt new manufacturing technologies. Technol Forecast Soc Change 180:121723. https://doi.org/10.1016/j.techfore.2022.121723

Gbongli K, Xu Y, Amedjonekou KM (2019) Extended technology acceptance model to predict mobile-based money acceptance and sustainability: a multi-analytical structural equation modeling and neural network approach. Sustainability 11:3639. https://doi.org/10.3390/su11133639

Guhr N, Loi T, Wiegard R, Breitner M (2013) Technology readiness in customers' perception and acceptance of m(obile)-payment: an empirical study in Finland, Germany, the USA and Japan
Huang M, Li Z, Su X (2022) Anticipated regret, entrepreneurial cognition, and entrepreneurial persistence. Front Psychol 13:788694. https://doi.org/10.3389/fpsyg.2022.788694
Jain R, Kumar S, Sood K et al (2023) A systematic literature review of the risk landscape in Fintech. Risks 11:36. https://doi.org/10.3390/risks11020036
Kaddouhah M (2024) An economic definition of 'Fear of Missing Out' (FOMO). Financ Res Lett 63:105344. https://doi.org/10.1016/j.frl.2024.105344
Kahneman D, Tversky A (1972) Subjective probability: a judgment of representativeness. Cogn Psychol 3:430–454. https://doi.org/10.1016/0010-0285(72)90016-3
King WR, He J (2006) A meta-analysis of the technology acceptance model. Inf & Manag 43:740–755. https://doi.org/10.1016/j.im.2006.05.003
Li J, Wang J, Wangh S, Zhou Y (2019) Mobile payment with Alipay: an application of extended technology acceptance model. IEEE Access 7:50380–50387. https://doi.org/10.1109/ACCESS.2019.2902905
Li X, Ye J, Deng M et al (2022) Decision reversibility and satisfaction: the mediating role of counterfactual thinking and anticipated regret. PRBM 15:2193–2203. https://doi.org/10.2147/PRBM.S364548
Liébana-Cabanillas F, Ramos De Luna I, Montoro-Ríos FJ (2015) User behaviour in QR mobile payment system: the QR payment acceptance model. Technol Anal & Strat Manag 27:1031–1049. https://doi.org/10.1080/09537325.2015.1047757
Liu Y, Yan W, Hu B (2021) Resistance to facial recognition payment in China: the influence of privacy-related factors. Telecommun Policy 45:102155. https://doi.org/10.1016/j.telpol.2021.102155
Ma Q, Liu L (2004) The technology acceptance model: a meta-analysis of empirical findings. J Organ End User Comput 16:59–72. https://doi.org/10.4018/joeuc.2004010104
Maity S, Abdel-Mottaleb M, Asfour SS (2021) Multimodal low resolution face and frontal gait recognition from surveillance video. Electronics 10:1013. https://doi.org/10.3390/electronics10091013
Neneh BN (2019) From entrepreneurial intentions to behavior: the role of anticipated regret and proactive personality. J Vocat Behav 112:311–324. https://doi.org/10.1016/j.jvb.2019.04.005
Neumann D (2020) Fear of missing out. In: Bulck J (ed) The international encyclopedia of media psychology, 1st edn. Wiley, pp 1–9
Pieters R, Zeelenberg M (2007) A theory of regret regulation 1.1. J Consum Psychol 17:29–35. https://doi.org/10.1207/s15327663jcp1701_6
Przybylski AK, Murayama K, DeHaan CR, Gladwell V (2013) Motivational, emotional, and behavioral correlates of fear of missing out. Comput Hum Behav 29:1841–1848. https://doi.org/10.1016/j.chb.2013.02.014
Roese NJ (1997) Counterfactual thinking. Psychol Bull 121:133–148. https://doi.org/10.1037/0033-2909.121.1.133
Rupeika-Apoga R, Wendt S (2021) FinTech in Latvia: status quo, current developments, and challenges ahead. Risks 9:181. https://doi.org/10.3390/risks9100181
Rupeika-Apoga R, Wendt S (2022) FinTech development and regulatory scrutiny: a contradiction? The case of Latvia. Risks 10:167. https://doi.org/10.3390/risks10090167
Schmidthuber L, Maresch D, Ginner M (2020) Disruptive technologies and abundance in the service sector - toward a refined technology acceptance model. Technol Forecast Soc Chang 155:119328. https://doi.org/10.1016/j.techfore.2018.06.017
Shanmugavel N, Alagappan C, Balakrishnan J (2022) Acceptance of electric vehicles: a dual-factor approach using social comparison theory and technology acceptance model. Res Transp Bus Manag 45:100842. https://doi.org/10.1016/j.rtbm.2022.100842
Sharma V, Jangir K, Gupta M, Rupeika-Apoga R (2024) Does service quality matter in FinTech payment services? An integrated SERVQUAL and TAM approach. Int J Inf Manag Data Insights 4:100252. https://doi.org/10.1016/j.jjimei.2024.100252

Shin D-H (2009) Towards an understanding of the consumer acceptance of mobile wallet. Comput Hum Behav 25:1343–1354. https://doi.org/10.1016/j.chb.2009.06.001

Smallman R, Roese NJ (2009) Counterfactual thinking facilitates behavioral intentions. J Exp Soc Psychol 45:845–852. https://doi.org/10.1016/j.jesp.2009.03.002

Song J, Kim YJ (2006) Social influence process in the acceptance of a virtual community service. Inf Syst Front 8:241–252. https://doi.org/10.1007/s10796-006-8782-0

Tian Y, Chan TJ (2024) Predictors of mobile payment use applications from the extended technology acceptance model: does self-efficacy and trust matter? SAGE Open 14:21582440241292524. https://doi.org/10.1177/21582440241292525

Turner M, Kitchenham B, Brereton P et al (2010) Does the technology acceptance model predict actual use? A systematic literature review. Inf Softw Technol 52:463–479. https://doi.org/10.1016/j.infsof.2009.11.005

Verkijika SF (2020) An affective response model for understanding the acceptance of mobile payment systems. Electron Commer Res Appl 39:100905. https://doi.org/10.1016/j.elerap.2019.100905

Wang K-Y, Bublitz MG, Zhao G (Terry) (2021) Enhancing dieters' perseverance in adversity: How counterfactual thinking increases use of digital health tracking tools. Appetite 164:105261. https://doi.org/10.1016/j.appet.2021.105261

Wixom BH, Todd PA (2005) A theoretical integration of user satisfaction and technology acceptance. Inf Syst Res 16:85–102. https://doi.org/10.1287/isre.1050.0042

Yang H, Yoo Y (2004) It's all about attitude: revisiting the technology acceptance model. Decis Support Syst 38:19–31. https://doi.org/10.1016/S0167-9236(03)00062-9

Zhang WK, Kang MJ (2019) Factors affecting the use of facial-recognition payment: an example of Chinese consumers. IEEE Access 7:154360–154374. https://doi.org/10.1109/ACCESS.2019.2927705

Fintech Innovations to Combat Illegal Fishing

Silvio Andrae

Abstract Illegal, unreported, and unregulated (IUU) fishing endangers marine ecosystems and poses a growing threat to global food security. Modern financial technologies (fintech) offer promising solutions to create transparency and enforce existing regulations more effectively. In particular, Blockchain technologies enable tamper-proof traceability along the supply chain—from catch to consumer. Digital payment systems make it more challenging to conceal illegal financial flows and enable more precise monitoring of trade activities. Data analysis and AI-based algorithms help to uncover patterns of illegal fishing and to identify future violations at an early stage using forecasting models. This study highlights the potential of fintech innovations in the fight against IUU fishing. It examines the associated challenges regarding technological implementation, regulatory frameworks, and economic viability. The aim is to develop a comprehensive understanding of how financial technologies can support sustainable fisheries management and to use this to derive targeted recommendations for policymakers, industry, and international organizations.

Keywords Illegal fishing · Fintech · Blockchain · Data analysis · Transparency · Traceability · Digital identity

1 Introduction

Illegal, unreported, and unregulated (IUU) fishing poses an acute threat to marine ecosystems and the sustainable use of global fish stocks. It is estimated that around 20% of fish caught in the wild worldwide result from illegal fishing activities. These are increasingly being carried out on an industrial scale, with perpetrators deliberately exploiting existing control gaps. The consequences are far-reaching: a dramatic decline in biodiversity, the destruction of marine habitats, and billions of dollars in economic damage. At the same time, IUU fishing undermines international efforts

---

S. Andrae (✉)
Independent Researcher, Berlin, Germany
e-mail: andrae@zedat.fu-berlin.de

V. Sharma et al. (eds.), *FinTech for Sustainable Finance and a Well-Blue Economy*, World Sustainability Series, https://doi.org/10.1007/978-3-031-92390-6_15

to achieve sustainable fisheries and endangers the livelihoods of millions of people who depend on fishing.

IUU fishing takes several forms (Ballesteros et al. 2024; Kastoryano and Vollaard 2023), including fishing without a license, entering protected areas, using prohibited fishing methods, exceeding catch quotas, and falsifying catch documents. Due to the unofficial nature of these activities, the actual extent of these activities is difficult to quantify (Coppa et al. 2021). Nevertheless, the Food and Agriculture Organization (FAO) estimates that up to 26 million tons of fish are caught illegally annually, with an estimated market value of 23 billion US dollars (FAO 2024).

Combating IUU fishing requires multifaceted measures, including strengthening regulatory frameworks, intensifying monitoring, imposing stricter sanctions, and improving international cooperation. However, this study focuses on a technology-based approach, mainly using financial technologies (fintech) to promote transparency, traceability, and effective regulation in the fishing industry. In particular, Blockchain technologies and digital payment systems open new possibilities for securely capturing, analyzing, and accessing relevant data along the entire value chain. Advances in data analysis and the increasing availability of remote sensing technologies enable more precise monitoring of fishing activities and the identification of illegal practices. These technologies can help detect violations early, optimize compliance with existing regulations, and facilitate law enforcement.

This study aims to analyze the potential of fintech innovations in the fight against IUU fishing and illustrate specific use cases. It assesses the contribution of these technologies to increasing transparency, traceability, and regulatory efficiency and identifies challenges and limitations to their practical implementation. It also derives recommendations for action to integrate fintech effectively into fisheries control.

The study is based on a systematic analysis of scientific literature, industry reports, and relevant case studies. It is aimed at policymakers, fisheries managers, scientists, technology providers, and a broad specialist audience.

After this introduction, Chap. 2 provides an overview of existing control mechanisms for IUU fishing and the role of fintech technologies in this context. Chapter 3 is dedicated to Blockchain technology and its application in the fishing industry. Chapter 4 examines the influence of digital payment systems on the fight against illegal financial flows associated with IUU fishing. Chapter 5 presents case studies and best practices for implementing fintech solutions. Chapter 6 summarizes the key findings and provides an outlook on future developments and research perspectives.

2 Basics to Combate IUU

2.1 *Overview of Existing Control Measures*

Effectively combating IUU fishing requires a comprehensive package of measures at multiple levels. A central starting point is strengthening regulatory frameworks and their consistent implementation. It includes introducing and strictly enforcing more burdensome legal requirements, higher sanctions for violations, and more intensive cooperation between law enforcement authorities and relevant interest groups (Widjaja et al. 2023).

Another crucial factor is optimizing monitoring and control mechanisms. Modern technologies such as satellite monitoring, drone missions, and aerial reconnaissance enable more precise recording and control of fishing practices. These developments open up new possibilities for the early identification of illegal activities and increase the effectiveness of deterrence and sanctioning measures.

One innovative example is the 'Eyes on the Seas' project of the Pew Charitable Trusts, which has developed a global solution for identifying and tracking illegal fishing vessels with the Virtual Watch Room system. This system combines various data sources—including vessel monitoring systems (VMS), synthetic aperture radar (SAR), automatic identification systems (AIS), and optical satellite sensors—and links them to fishing vessel databases and oceanographic information. This technology provides an accurate, comprehensive picture of the situation, facilitating targeted law enforcement.

A central monitoring component is the vessel monitoring system (VMS), which enables continuous control and regulation of fishing activities. These devices automatically record a ship's position, speed, and direction of travel and transmit the data to the relevant authorities at fixed intervals. Since VMS devices cannot be switched off, any signal interruption triggers an alarm interpreted as a possible manipulation attempt. Data transmission is encrypted via satellite connections, with flag states primarily responsible for controlling and forwarding this information to coastal states or regional fisheries management organizations (RFMOs) (Swan 2020). In addition, VMS is increasingly being used to transmit electronic logbooks (ERS), in which captains document detailed information about the fishing trip—including start and end dates, species and quantities caught, by-catch, and stock quantities. These data are essential for quota management and monitoring fishing activities (Willette et al. 2023).

In addition to satellite-based monitoring systems, other remote sensing technologies play a central role. The automatic identification system (AIS) works similarly to VMS but transmits location data at shorter intervals and in an open format. Therefore, AIS data can be picked up by any receiver with the appropriate technology, increasing transparency. Meanwhile, web-based data providers and commercial satellite AIS services increasingly use this information, meaning ship movements can be analyzed on a large scale. While fishing companies use this data to monitor the competition, shipping companies can analyze the traffic situation in ports more efficiently.

However, AIS is not tamper-proof. Malicious actors or fishermen who want to keep their best fishing grounds secret (Welch et al. 2022) can deliberately switch off the system or send false identification and position signals—a phenomenon known as AIS spoofing (Androjna et al. 2021). Platforms such as Global Fishing Watch provide open access to global AIS data by aggregating, publishing, and analyzing satellite signals. This makes it possible to identify IUU activities on a large scale. However, coverage varies by region, and smaller fishing vessels are often not equipped with AIS technology (Willette et al. 2023).

Another innovative company in this area is Unseenlabs (https://unseenlabs.space), which specializes in space-based detection of radio frequency signals. Using its satellite constellation, Unseenlabs can locate ships worldwide by capturing and analyzing their electromagnetic signatures—even if the AIS has been disabled or tampered with. The company plans to expand its satellite fleet from 15 to 25 units by the end of 2025 and to launch a new generation of larger satellites in 2026. In the future, these could be used not only for maritime surveillance but also to monitor activities on land, in the air, and space—a potential milestone in the global surveillance of illegal fishing practices.

In addition to satellite monitoring, other innovative monitoring technologies are becoming more important. Remote Electronic Monitoring (REM) systems combine VMS with CCTV cameras and sensors attached to winches, conveyor belts, and freezers. While AIS and VMS only provide positional data, REM captures visual and physical data on the actual species caught and the quantities landed. This technology is crucial for monitoring by-catches, often not declared correctly. However, less than one percent of global fisheries are currently subject to REM monitoring. Artificial intelligence is expected to revolutionize the automated analysis of video material and thus significantly increase the efficiency of monitoring measures (Monkman et al. 2019).

Other technologies for identifying and tracking illegal fishing activities include Long-Range Identification and Tracking (LRIT), an international tracking system for ocean-going vessels, and the use of unmanned aerial vehicles (drones). Drones with high-resolution cameras and sensors enable targeted monitoring of remote coastal regions and protected marine areas where satellite imagery alone is insufficient (Harris et al. 2019; Toonen and Bush 2020).

Electronic catch documentation systems (eCDT) contribute to the traceability of fisheries products and help to identify illegal catches at an early stage. In this context, Willette et al. emphasize:

'Traceability alone does not deliver improved fish stocks, healthier ecosystems, nor sustained benefits for people. It does improve our ability to understand whether sustainable practices are adopted, identify fraudulent activities, and aid enforcement rules aimed to support greater transparency of the supply chains' (Willette et al. 2023, p. 2).

Port state controls are also adequate in identifying illegal catches and ensuring compliance.

Since IUU fishing is a transnational problem, increased international cooperation is essential. This includes developing and implementing international agreements, exchanging best practices, and helping developing countries strengthen their fisheries management capacities.

In addition, informed consumer decisions play a central role. Certification programs such as the Marine Stewardship Council (MSC) or the Aquaculture Stewardship Council (ASC) increase transparency in the supply chain and strengthen consumer trust in sustainably caught seafood.

Technological innovations, including Blockchain-based traceability systems and DNA analysis, could play an even greater role in combating IUU fishing in the future by creating tamper-proof documentation and more sustainable supply chains (Chap. 3).

2.2 Introduction to the World of Financial Technology (Fintech)

The rapid development of financial technology (fintech) fundamentally changes the financial sector and offers promising approaches to promoting transparency and efficiency in the fishing industry. Innovative technologies such as Blockchain, big data, cloud computing, and mobile applications are giving rise to new financial products and services that are often more efficient, cost-effective, and user-friendly than traditional solutions (Frost 2020).

There is no legal definition of fintech, but the term essentially describes the application of digital technologies in the financial services sector. Companies are often classified as fintech if they have existed for no longer than ten years. Thus, fintech encompasses both emerging start-ups and established financial companies that integrate technological innovations into their services.

Fintech companies operate globally, although market structures vary significantly from region to region. While in some regions, fintechs occupy a complementary niche role, they dominate large parts of the financial sector in other markets. Fintech solutions have become particularly widespread in the area of digital payments. A prime example is M-Pesa, the mobile money transfer system launched by the Kenyan telecommunications provider Safaricom in 2008 (see Sect. 5.1).

In the fishing industry, fintech offers the potential to ensure greater transparency, traceability, and regulatory control. Blockchain technologies enable tamper-proof documentation of fisheries products along the entire supply chain. At the same time, digital payment systems can improve the monitoring of financial flows and detect suspicious transactions related to illegal fishing. At the same time, big data enables in-depth analysis of patterns and trends, allowing more accurate risk assessments to be carried out.

The use of fintech in the fisheries industry offers several key advantages:

- **Increased transparency**: Digital platforms enable better traceability of catch quotas, trade flows, and financial transactions.

- **Improved traceability**: Blockchain-based systems provide tamper-proof proof of origin for fisheries products.
- **More efficient monitoring**: Automated systems, combined with real-time data, make monitoring fishing activities and financial flows easier.
- **Stronger regulation**: The ability to identify and prosecute illegal activities more quickly helps to improve the enforcement of existing regulations.

Despite these advantages, implementing fintech solutions presents significant challenges. In many regions, the necessary technological infrastructure is lacking to implement these systems on a large scale. Data protection remains critical, as the collected data must be protected from misuse. Furthermore, acceptance within the industry is crucial, as many fishermen and companies prefer traditional methods and view new technologies skeptically. Another important aspect is the high implementation and operating costs, which require long-term investments and sustainable financing models.

Despite these challenges, fintech offers a promising approach to effectively combating IUU fishing. By identifying and overcoming existing obstacles and developing suitable solution strategies, fintech technologies can be successfully and sustainably integrated into the fishing industry. The rest of this study analyzes the specific technologies and their use cases in more detail.

3 Blockchain Technology in Fisheries

This section explores the potential applications of Blockchain technology in the fishing industry. Blockchain (BC) provides complete traceability of fisheries products from the moment they are caught, which is critical in preventing illegally caught or unreported fish from entering the market. More importantly, it catalyzes promoting sustainable and responsible fishing practices, thus contributing to the industry's and the environment's well-being (Widjaja et al. 2023).

3.1 How Blockchain Works and Its Advantages

Blockchain-based industry solutions for seafood traceability have been introduced commercially (Hang et al. 2020). Large technology companies and start-ups like IBM, SAP, Oracle, and Microsoft have developed these solutions. Provenance was one of the first Blockchain-based solutions, launched in 2013. It enables consumers to trace fish products back to the fisherman. Fishcoin is another Blockchain-based traceability system for seafood developed in 2018 to establish a trusted data exchange between stakeholders (Ferreira et al. 2021). IBM Food Trust is a Blockchain-based supply chain food solution launched in 2018.

Blockchain technology is based on a decentralized, distributed, immutable digital ledger that records transactions securely and transparently. Unlike centralized databases, which are managed by a single authority, Blockchain is operated by a network of computers (nodes). Each network participant has a complete copy of the Blockchain, ensuring high transparency and resilience.

For example, let us say a fisherman catches a tuna. This catch is recorded as a 'transaction' in the Blockchain.

- Block creation: The information about the catch (e.g., date, location, species, and fisherman ID) is stored in a 'block.'
- Concatenation: This block is given a unique digital 'fingerprint' (hash) and linked to the previous block in the chain. This creates a 'chain' of blocks—the Blockchain.
- Validation: Before this new block is finally added to the Blockchain, several computers in the network (the 'nodes') must verify and confirm the information through a 'consensus mechanism.'
- Immutability: Once the block is confirmed and added to the Blockchain, it cannot be changed or deleted. Any attempt at manipulation would be immediately apparent, as the block's digital fingerprint would change, and the chain would be broken.

Each step in the tuna supply chain (e.g., transport to the processing plant, sale to the distributor, export) is recorded as a transaction in a new block. The Blockchain enables all parties involved (e.g., authorities, distributors, consumers) to trace the entire history of the fish—from catch to sale in the supermarket (Larissa and Parung 2021). Illegal activities, such as fishing in protected areas or falsifying catch documents, can be more easily detected because the Blockchain's information is transparent, tamper-proof, and accessible to all parties, giving a sense of control. Consumers can scan a QR code on the fish packaging to verify the fish's origin and journey, ensuring they are not buying illegal fish. The technology's security features, such as the cryptographic linking of blocks, make tampering nearly impossible, ensuring data integrity and reducing the risk of fraud and counterfeiting, thereby increasing trust.

The decentralized nature of Blockchain also reduces the risk of censorship or tampering, as no single entity has complete control. At the same time, the technology increases efficiency by streamlining processes in the fishing industry. For example, catch quotas can be managed automatically. This ensures that fishermen comply with their limits and reduces the risk of overfishing. Certificates can be issued digitally, reducing costs and processing times.

Furthermore, Blockchain can streamline the process of tracking and verifying the sustainability of a catch, making it easier for consumers to make informed choices about the fish they buy. Private Blockchains are particularly suitable for applications that contain sensitive business data or require specific access and participation restrictions (Kumperšcak et al. 2019). By implementing access control lists, granular data access rights can be defined based on the respective roles of the system's participants (Qin et al. 2020).

In the context of combating IUU fishing, different types of Blockchains can be used, each with its advantages and disadvantages (Ismail et al. 2023):

Public Blockchains. These Blockchains are open to everyone and are transparent. Anyone can view transactions and participate in the consensus process. This provides high transparency and traceability, which is beneficial for tracking fisheries products and verifying catch data. However, public Blockchains can be slower and less scalable than other types.

Private Blockchains. These Blockchains are restricted to specific participants and require approval to join. They offer more control and privacy but can be less transparent than public Blockchains. Private Blockchains could be used for internal supply chain management systems or collaboration between specific players in the fishing industry.

Consortium Blockchains are hybrids of public and private Blockchains. They are maintained by a group of organizations or institutions that control the consensus process. Consortium Blockchains can balance transparency and control well and may be well-suited for collaboration between governments, fishing companies, and other stakeholders.

Public and consortium Blockchains increase transparency along the supply chain by providing an immutable record of catch data, transport information, and trade activities. This makes smuggling illegal catches into the legal supply chain more complex. Consortium Blockchains can promote collaboration between different players in the fishing industry by providing a common platform for data exchange and collaboration (Ismail et al. 2023).

Another important point concerns data storage and processing. The choice of data storage affects system performance, latency, integrity, and costs. The following forms of storage are distinguished (Ismail et al. 2023):

- On-chain storage involves storing data and metadata directly in the Blockchain. This approach is secure, but it can lead to performance issues when storing large amounts of data.
- Off-chain storage: This approach stores only essential data and metadata in the Blockchain, while the actual data is stored externally (e.g., in cloud storage) (Reno and Haque 2023). This improves performance but requires mechanisms to ensure data integrity.
- Hybrid models combine on-chain and off-chain storage to leverage the advantages of both approaches.

Blockchain technology relies on fundamental cryptographic techniques, most notably hashing and asymmetric encryption.

Hashing. A hash results from a mathematical function that converts any input into a unique fixed value. If the input changes slightly, the hash value will be completely different. In the Blockchain, hashing is used to ensure data integrity: any manipulation would be immediately apparent because the hash produced would no longer match the previous value.

Asymmetric encryption uses a public key and a private key. The public key encrypts data that can only be decrypted with the corresponding private key. This process plays an important role in Blockchain systems for authenticating transactions and ensuring that only authorized users can make changes.

Finally, consensus protocols are an essential feature of BC. Consensus protocols are mechanisms by which Blockchain network participants agree on the validity of transactions and the state of the Blockchain. Examples include Proof of Work (PoW), Proof of Stake (PoS), Proof of Authority (PoA), and others. The Proof of Work mechanism is a widespread consensus mechanism. In a decentralized network without a central authority, a mechanism must exist to agree on the state of the Blockchain. The proof-of-work mechanism solves this problem by forcing the participants (so-called miners) to use computing power to solve mathematical problems. These tasks consist of calculating a suitable hash value that meets specific requirements. Only when a miner has solved this problem can he add a new block to the Blockchain and confirm the transactions. The difficulty of this computational task ensures that it is virtually impossible to manipulate the Blockchain. An attacker would need immense computing resources to outsmart the network—which is hardly feasible economically or technically. The choice of consensus protocol influences the system's security, performance, scalability, and energy consumption (Ismail et al. 2023).

Besides these three primary considerations regarding the type of BC, data storage, and consensus protocols, other quality attributes must also be taken into account when designing Blockchain-based systems:

- Traceability: the ability to track the movement of a product through the SC;
- Performance: efficiency of the system in terms of throughput and response time;
- Security: protection of the system and data from threats;
- Privacy: protection of sensitive information of the parties involved;
- Interoperability: the system's ability to communicate with other systems;
- Scalability: the system's ability to handle a growing number of users and data;
- Latency: delay in data transmission;
- Integrity: ensuring the immutability and accuracy of data and
- Usability: the system's user-friendliness.

3.2 Use Cases for Blockchain in Fisheries

Blockchain technology can be used in various areas of the fishing industry to promote transparency and sustainability. It follows a multi-level architecture model that enables seamless documentation of fish products along the entire value chain (Ismail et al. 2023). One key area of application is supply chain traceability. Here, integrating Internet of Things (IoT) devices with Blockchain technology can be a promising solution for secure data collection and registration in a distributed immutable database system. IoT devices such as wireless sensors, RFID chips, GPS trackers, and other devices collect data along the supply chain. These devices monitor physical and chemical parameters (e.g., temperature, humidity, location) and collect product information. The collected data is used to improve traceability, monitor product quality, and combat IUU activities.

Ismail et al. present a three-layer IoT architecture consisting of IoT devices, edge servers, and the cloud (Ismail et al. 2023). Edge servers process data close to the IoT devices, reducing latency and improving efficiency. Blockchain is deployed in the cloud to process time—and energy-intensive data.

'In this system, fish can be processed, packaged, and shipped several times during its journey, while Blockchain records all data transactions in an immutable ledger shared between stakeholders. Blockchain assures system users that the relayed information is unalterable, trusted, and accurate' (13, p.20).

Another essential area of application is the management of fishing quotas. Overfishing is one of the greatest threats to global fish stocks. Blockchain can provide an efficient solution by registering the amount of fish caught in real-time. This lets authorities and fishermen see immediately whether quotas have already been exhausted, making it easier to enforce sustainable fishing practices.

Blockchain can also improve the certification of sustainable fisheries products. Organizations such as the Marine Stewardship Council (MSC) and the Aquaculture Stewardship Council (ASC) award sustainability seals to ensure that fisheries products are produced in an environmentally friendly manner. Blockchain can help to store these certifications in a tamper-proof manner and make their authenticity verifiable at any time.

In combination with identity solutions, Blockchain enables decentralized user and producer data management, ensuring tamper-proof identification of actors in the supply chain (Sect. 4.2).

3.3 Challenges and Limitations of Blockchain Implementation

Despite the promising possibilities, various challenges stand in the way of a broad implementation of Blockchain technology in the fishing industry (Ismail et al. 2023):

- High implementation costs: Building a Blockchain-based infrastructure requires significant hardware, software, and specialized personnel investments.
- Technical complexity: The technology requires cryptography and software development expertise, which can be a hurdle for smaller fishing companies.
- Lack of standardization: There are currently no uniform standards for Blockchain in fisheries, so interoperability between different systems can be problematic.

A key problem is the so-called Blockchain trilemma, which states that it is challenging to ensure scalability, security, and decentralization at the same time:

- Scalability: The ability to process large amounts of transactions efficiently.
- Security: Protection against cyber attacks and manipulation.
- Decentralization: Avoiding a central control instance to ensure independence and trust.

Finally, the technology's acceptance is a crucial factor. Many smaller players in the fishing industry do not have the necessary resources to implement Blockchain applications. Therefore, training and a step-by-step introduction are essential. Despite these challenges, Blockchain technology offers enormous potential for improving transparency, traceability, and efficiency in the fishing industry. Overcoming existing hurdles will enable the sustainable and future-oriented integration of the technology (Ismail et al. 2023).

4 Digital Payment Systems and Financial Analysis

IUU fishing is closely linked to other forms of organized crime, including drug trafficking, money laundering, and tax evasion. Therefore, targeted monitoring of financial flows is an essential strategy to combat these activities effectively. Digital payment systems and modern financial analysis offer innovative ways to identify suspicious transactions, prevent money laundering, and facilitate the prosecution of those responsible. This chapter examines the role of these technologies in combating IUU fishing, analyzes their potential, and discusses existing challenges.

4.1 Significance of Tracking Financial Flows

The systematic tracing of financial transactions has proven to be a particularly effective way of dismantling criminal networks and disrupting their business models. Illicit revenues from the fishing industry are often laundered to conceal their origin. A detailed analysis of these financial flows makes it possible to uncover links between the various actors along the value chain—from fishermen to traders, processors, and exporters. Furthermore, identifying and confiscating illegal funds can help destroy the financial basis of these activities and make it more difficult to continue them.

Various methods are used to track illicit financial flows. Analysis of banking transactions can reveal unusual movements of money while monitoring trade flows can show whether illegally caught fish are entering the global market. Financial investigators use advanced analytical techniques to decode disguised transactions and expose criminal networks. Since illegal financial flows usually operate across borders, close international cooperation between financial regulators, banks, and law enforcement agencies is essential.

Despite the effectiveness of these approaches, significant challenges remain. Criminal organizations are developing increasingly sophisticated money laundering techniques and exploiting regulatory gaps in certain countries. In particular, regions with weak financial supervision or opaque legislation make it difficult to trace illegal financial flows. Nevertheless, the analysis of financial transactions remains an indispensable tool in the fight against IUU fishing, mainly through modern technologies and increased international cooperation.

4.2 Digital Payment Systems to Combat Money Laundering

Digital payment systems offer a promising approach to combating money laundering in connection with illegal fishing. Their greatest strength lies in the increased transparency: every transaction is recorded electronically and can be monitored by the relevant authorities. This makes it much more difficult for criminal networks to launder illegal proceeds into the legal, financial system, as the origin and use of the funds can be traced at any time.

Seamless digital documentation provides financial transactions with time stamps and unique identifiers. This enables law enforcement authorities to monitor suspicious money movements in real-time and initiate targeted investigations. Modern analysis tools use algorithms and artificial intelligence to detect patterns and anomalies in transactions automatically. This makes it possible to identify suspicious financial flows early and take countermeasures.

Another advantage of digital payment systems is the way they promote international cooperation. Since transaction data is recorded digitally, it can be exchanged efficiently between supervisory authorities in different countries. This is particularly relevant given that IUU fishing often operates across borders and money laundering networks are active in several jurisdictions.

Various approaches for using digital payment systems in the fight against money laundering already exist today. The increased shift from cash to electronic payments makes it more difficult for criminals to transfer illegal proceeds unnoticed. Mobile payment systems enable real-time monitoring of financial flows and immediate detection of suspicious transactions. The use of Blockchain technology, which enables a tamper-proof and transparent record of financial transactions, is auspicious. This makes it much more challenging to conceal illegal money flows.

Despite these advantages, challenges remain. Some digital payment methods still allow anonymous transactions, which can limit traceability. Furthermore, the introduction and operation of such systems require specialized technical knowledge, which is not sufficiently available in all countries. Developing countries, in particular, often face high implementation costs and a lack of infrastructure. Nevertheless, digital payment systems offer enormous potential for increasing transparency in the financial sector and more effectively controlling illicit financial flows.

4.3 Data Analysis to Detect Suspicious Transactions

Data analysis is a central component of modern financial investigations and a powerful tool for detecting suspicious transactions related to illegal fishing. By systematically examining large amounts of data, patterns, and anomalies that indicate criminal activity can be identified.

Various analytical techniques are used to identify suspicious financial flows. One proven method is supervised machine learning (SML). This approach is called supervised learning because the outcome of interest is determined a priori. An SML algorithm is fed a set of training data. This data set contains the input–output pairs and the labels for the observations. The goal is to classify individual data points by identifying, among multiple classes (i.e., categories of observations), the one to which a new observation belongs. This is inferred from analyzing a sample of past observations (the training dataset) for which the group (category) is known. The algorithm's objective is to predict the category of a new observation given its independent variables.

Examples:

Transaction classification: A supervised learning model can be trained to recognize suspicious financial transactions related to illegal fishing. The model would be trained with labeled data sets of known legal and illegal transactions to recognize patterns and classify new transactions as potentially suspicious or non-suspicious.

Risk assessment of companies: An algorithm could be developed to assess the risk of fishing companies regarding money laundering. Based on historical data on company profiles, transaction histories, and known cases of money laundering, the model could provide risk assessments for new or existing companies in the fishing industry.

Alternatively, unsupervised learning (UML) can be used. Where SML assumes that each input in the training set has an associated set of output targets, the UML techniques examine how systems can derive a function to describe a hidden structure from unmarked data. In UML, the data is not specified in advance as input or output variables. Instead, the UML algorithm is given all the variables and is tasked with finding structure in the entire data set. The algorithm must identify the clusters by regrouping the observations for which it finds similar features or patterns. As such, there is no immediate measure of success.

Examples:

Anomaly detection in financial data: Unsupervised learning algorithms can identify unusual patterns in financial data from fishing companies without needing pre-labeled examples. This could help uncover new and previously unknown methods of money laundering associated with illegal fishing.

Clustering of transaction networks: By applying clustering algorithms to transaction networks, suspicious groups of companies or individuals potentially involved in money laundering from illegal fishing could be identified. This could reveal complex relationships and patterns that are difficult to see using conventional methods.

Another approach is network analysis, which can reveal links between actors within criminal structures. Structural network analysis is of particular interest. It examines the overall structure of a network. There are three main approaches:

1. Affinity approach: This approach analyses direct and indirect contacts as well as measures such as density (ratio of realized to possible connections). It can be used to assess the cohesion within a network, e.g., criminal organizations in the fishing industry.
2. The positional approach identifies roles and positions based on sociological concepts. This approach helps identify key actors in a network, e.g., agents who distribute corruption money.
3. Inferential statistical approach: This approach provides explanatory models for network structures and can be used to understand patterns in criminal activities.

Relational network analysis focuses on the relationships of individual nodes (actors) in the network. It is beneficial for analyzing transaction networks, or communication flows between actors.

Examples:

Uncovering corporate networks: Network analysis can reveal complex corporate structures that conceal illegal activities. One example is the corporate network identified by Greenpeace in 2020 around a Greek shipowner to whom 20 freezer ships with suspicious maneuvers were assigned.

Identifying key figures: Central criminal network actors can be identified by analyzing transactions and business relationships.

Exposing transshipment patterns: Network analyses can reveal patterns in transshipment operations at sea by visualizing links between fishing vessels and reefer vessels. This helps to identify vessels that are frequently involved in suspicious transshipment activities.

These analytical methods have proven to be highly effective in practice. Abnormally large or repeated transactions can indicate money laundering, especially when offshore accounts or companies in high-risk regions are involved. Modern AI-based systems make it possible to automatically detect conspicuous patterns in financial transactions and generate alerts promptly.

Nevertheless, there are also challenges in using data analysis to combat money laundering. One of the main difficulties is the availability of high-quality data since much information is inaccessible or only accessible to a limited extent. Furthermore, there is a risk of false positives, in which legal transactions are wrongly classified as suspicious. This can lead to unnecessary investigations and additional work. Another challenge is the need to consider data protection aspects. Analyzing sensitive financial data poses potential risks to the privacy of the individuals concerned, and it is crucial to balance the need for effective crime prevention with respect for the individual's privacy.

Despite these challenges, data analysis remains essential for combating money laundering in illegal fishing. Combining data analysis with modern technologies such as digital payment systems and Blockchain can create a comprehensive monitoring system that makes financial flows transparent, detects illegal activities, and helps dismantle criminal networks.

4.4 Challenges and Limits of Digital Payment Systems

Digital payment systems, including cryptocurrencies, are crucial in combating money laundering and present various challenges. A significant problem is the anonymity that some of these systems allow. Because transactions are often pseudonymous or anonymous, tracing financial flows is more complex and provides criminals with loopholes to disguise illegal funds.

Another major challenge arises from the transnational nature of illegal fishing. Effective prosecution of financial flows across different countries requires close cooperation between national authorities. However, the lack of uniformity in the legal frameworks and regulatory standards in different countries presents a hurdle that complicates the exchange of information and law enforcement.

The technical complexity of digital payment systems also presents a hurdle, particularly for developing countries. Implementation and operation require expertise and advanced technical infrastructure that are not equally available everywhere. Furthermore, the high costs can be a significant burden, especially for countries with limited resources.

In addition to these technical and operational challenges, there are tory and structural weaknesses. The regulation of digital payment systems is still in its infancy in many countries, allowing criminals to exploit existing gaps to launder illegal funds. Furthermore, international cooperation in the fight against money laundering is often inadequate, as differing interests, data protection regulations, and bureaucratic hurdles make coordination difficult.

Another problem is the lack of awareness of the risks of money laundering. Many people are unaware that they can unknowingly be involved in illegal financial transactions, for example, by investing in complex financial products with unclear ownership structures or using specific digital payment systems that offer a high degree of anonymity.

Despite these challenges, digital payment systems and modern financial analytics have the potential to significantly contribute to the fight against money laundering related to illegal fishing. These technologies can increase the transparency of financial flows, detect illegal activities, and improve the efficiency of law enforcement. A comprehensive approach combining technological innovation with a sound legal framework and international cooperation is a critical step in the fight against illegal fishing and the associated money laundering.

5 Case Studies and Best Practices

The application of fintech innovations in fisheries, an issue of increasing importance, is a relatively new area. This chapter presents four case studies and best practices crucial to understanding how these technologies can be successfully implemented.

By analyzing successful and failed projects, we can identify the key success factors and challenges in applying fintech solutions in fisheries.

5.1 Transformative Success Stories of Fintech in the Fishing Industry

Blockchain-Based traceability of tuna in Indonesia

Nasution et al. analyze the traceability systems and supply chains of the tuna fishery in two different types of fishing ports in Indonesia (Nasution et al. 2023). The study also shows that the Indonesian government's traceability system is document-oriented. Indonesian tuna exporters must complete documents from the pre-catch to the post-catch to the processing plant. The traceability system for tuna fishing is already in place when the tuna is caught and is an integral part of the tuna supply chain even before the tuna is landed at the fishing port. At each stage of the supply chain, the traceability documents must be completed so that the tuna products can be appropriately recorded and certified. However, paper-based traceability is too slow for today's growing information needs. Therefore, the NWO project was launched in 2015 to implement further and develop electronic traceability and catch documentation (eCDT) in Indonesia and the Asia–Pacific region. In 2018, the TraceTales platform was introduced to introduce Blockchain technology. Integrating Blockchain into the TraceTales system ensures that data is tamper-proof as it moves from processing units to the production line, thus avoiding the risk of incorrect labeling. It also enables supply chains to maintain market access by visibly committing to tackling IUU fisheries products entering supply chains and mislabeled products. This project improves consumer transparency and builds trust in the supply chain. It enables the identification and sorting of illegal catches, thus contributing to the sustainability of tuna stocks. In addition, fishermen can command higher prices if they can prove the legal origin of their catches (IPNLF 2019).

Electronic fisheries monitoring system in Saudi Arabia

FAO and Saudi Arabia have developed a fully digitized fisheries management system. Cameras installed on fishing vessels, landing sites, and auction halls allow fisheries officers to access data and information online. The data is then processed using deep learning algorithms, and the results are automatically analyzed using statistical software. This expands the data collected at the landing sites and overcomes problems caused by extreme weather conditions, remote areas, and difficulties in identifying species. Since the end of 2021, the data has been collected in the new digital system using mobile apps and instant updates from the sector via interactive maps, graphs, and tables. Since 2023, cameras and deep learning algorithms have been used to automate data collection, thus completing the autonomous operation of the entire system (FAO 2024).

Satellite-based surveillance system in Palau

The challenges of policing such a vast territory, which covers approximately 1.5 million square miles, persist due to a lack of resources, funding, and capacity. Technological advances enable remote monitoring and control via satellites and drones (Brown et al. 2021). In 2020, Palau was the first country to test a new, state-of-the-art radar system to detect and track illegal fishing boats in its territorial waters. The virtual surveillance area in Chap. 4 covers the Pacific island nation of Palau. Such systems' deterrent effect helps prevent illegal fishing in the first place (Fujii et al. 2021; PICRC, COS 2019).

Mobile payment system for fishermen in Kenya

There are about 150 fintech companies in Kenya. Only a few fintechs deal with digital payment processing and online banking. The best-known project is the one that has been causing a stir for over 15 years: the globally praised payment system M-Pesa from the Kenyan provider Safaricom. M-Pesa, a money transfer platform via mobile phone, has contributed to significant financial inclusion and provided the impetus for the founding of other fintechs. M-Pesa enables Kenyans to send and receive money via SMS without having a bank account (Burns 2018; Tyce 2020). This system improves the financial inclusion of fishermen, who often lack access to traditional banking services. Safer and more efficient payments reduce the risk of theft and fraud. At the same time, fishermen benefit from faster access to their income, which they can use flexibly for investments or daily needs. In this way, the system not only contributes to the financial stability of the fishermen but also strengthens their economic independence and prospects.

One key factor for successful projects is close cooperation with local communities and authorities. Their understanding of the specific needs and challenges of the respective region is invaluable. In addition, these initiatives use innovative technologies that are specifically adapted to local conditions. These examples show how fintech innovations can significantly contribute to fisheries' sustainable development. They increase transparency and promote the responsible use of resources while also strengthening the economic situation of fishermen.

5.2 *Gaining Insights from Failed Projects*

Analyzing failed fintech projects in the fisheries sector provides valuable insights that can make future initiatives more successful. A common problem with failed projects is the lack of involvement of fishers, the most important stakeholders in the fishing industry. Their needs, knowledge, and experiences must be incorporated into the planning and implementation of fintech projects. Projects designed 'from the top down' and do not actively involve fishers often face resistance and are not adopted. Therefore, it is crucial to pursue participatory approaches that ensure the active involvement of fishers in all project phases and make them feel engaged and part of the solution.

Another obstacle is inadequate technical infrastructure. Fintech solutions require a stable technical foundation, including internet access, power supply, and suitable hardware. However, this infrastructure is often inadequate or non-existent in many coastal communities, especially in developing countries. In such cases, it is crucial to consider local conditions and select technologies that work even under challenging conditions to ensure effective implementation.

Another common problem with failed projects is a lack of sustainability. Many fintech projects are time-limited and end after the project period has expired without achieving a sustainable impact. For fintech innovations to be successful in the long term, they must be integrated into existing structures and processes. Equally important is building local capacity so fishers and other actors can independently use and further develop the technologies.

Other reasons for project failure include lack of financial viability, insufficient training and user support, and user acceptance. Without the necessary training and support, fishers may be unable to use the technologies efficiently, leading to low adoption and project failure.

To substantiate the model for combating IUU fishing with fintech, both theoretical and practical approaches could be pursued using quantitative and qualitative methods.

(a) Quantitative analysis: data-based model tests

- Transaction analyses could examine payment flows in fishing hotspots and identify patterns of illegal financial flows.
- Blockchain prototypes and simulations could be developed to verify the efficiency of tamper-proof supply chains.
- The satellite and AIS data analysis combined with machine learning could help identify IUU patterns by comparing them with transaction data.

(b) Qualitative methods: stakeholder surveys and case studies

- Interviews with fishers and supply chain actors could be conducted to assess the acceptance of digital payment systems and Blockchain-based traceability solutions.
- A regulatory analysis could compare various fintech-based anti-IUU strategies in countries with different legal frameworks.
- We could examine case studies of successful implementations, such as in Norway or Indonesia, where technological innovations are already being used against IUU fishing.

Finally, a deeper analysis of the impact of fintech integration on the competitiveness of legal fishing businesses, global seafood markets, and sustainable supply chains, as well as an assessment of the regulatory landscape and an analysis of compliance costs, could help to develop informed policy recommendations.

6 Conclusions

Almost two decades after the adoption of the International Plan of Action to Prevent, Deter, and Eliminate Illegal, Unreported, and Unregulated (IPOA-IUU) Fishing, IUU fishing in its various forms remains one of the most critical challenges facing global fisheries. IUU fishing poses a significant threat to marine biodiversity, the sustainability and balance of marine ecosystems, and global fish stocks. Illegal fishing practices are a complex problem that requires a multifaceted approach combining monitoring, enforcement, and economic incentives. This paper examined the potential of fintech innovations to combat IUU fishing. It showed that Blockchain and digital payment systems are crucial in improving transparency, traceability, and regulation in the fisheries sector (Ismail et al. 2023). The certainty offered by these technologies in terms of improving transparency and traceability is an important step in the fight against IUU fishing. Combining these technologies with other measures, such as strengthening regulation and international cooperation, can create a comprehensive system for combating IUU fishing.

'Technologies cannot solve all the issues of IUU fishing and fish fraudulent practices, but they have now demonstrated the capacity to detect IUU fishing activities at sea, which was nearly impossible 20 years ago. Progressive digital transition and more integrated data exchanges have also made it possible to improve fisheries management, reduce the catch documentation analysis workload, produce more reliable risk analysis based on vessel history, and allow much faster verification prior to landing.' (Willette et al. 2023, p. 6).

Fintech solutions have the potential to fundamentally change the fisheries industry and make an important contribution to combating IUU fishing. They can help improve fisheries products' transparency and traceability and facilitate fishing activity monitoring and enforcement. However, implementing fintech innovations presents high costs, technical complexity, and data privacy concerns. The development of fintech solutions for the fisheries sector is still in its infancy, and further research is essential to understand their full impact. Further advances can be expected in Blockchain and digital payment systems, which improve effectiveness. Continuous research and development will be key to realizing the full potential of fintech solutions in the fight against IUU fishing. Combining fintech solutions with other measures, such as strengthening regulation and international cooperation, will create a comprehensive system for combating IUU fishing.

References

Androjna A, Perkovič M, Pavic I, Mišković J (2021) AIS data vulnerability indicated by a spoofing case-study. Appl Sci 11:5015. https://doi.org/10.3390/app11115015

Ballesteros HM, Sánchez-Llamas E, Rodríguez-Rodríguez G (2024) Estimating illegal catches in data-poor S-fisheries: insights from multispecies shellfish poaching in Galician small scale fisheries. Mar Pol 163:106084. https://doi.org/10.1016/j.marpol.2024.106084

Brown CJ, Desbiens A, Campbell MD, Game ET, Gilman E, Hamilton RJ (2021) Electronic monitoring for improved accountability in western Pacific tuna longline fisheries. Mar Pol 132:104664. https://doi.org/10.1016/j.marpol.2021.104664

Burns S (2018) M-Pesa and the 'market-led' approach to financial inclusion. Econ Affairs 38(3):406–421. https://doi.org/10.1111/ecaf.12321

Coppa S, Pronti A, Massaro G, Brundu R, Camedda A, Palazzo L et al (2021) Fishery management in a marine protected area with compliance gaps: socio-economic and biological insights as a first step on the path of sustainability. J Environ Manag 280:111754. https://doi.org/10.1016/j.jenvman.2020.111754

FAO (2024) The state of world fisheries and aquaculture (SOFIA): blue transformation in action. https://openknowledge.fao.org/server/api/core/bitstreams/f985caed-cc7a-457e-8107-7ce16c6ef209/content Accessed 13 Mar 2024

Ferreira JC, Martins AL, Tokkozhina U, Helgheim BI (2021) Fish control process and traceability for value creation using Blockchain technology. In: Proceedings of the international conference on innovations in bio-inspired computing and applications, Seattle, WA, USA, 16–18 December 2021, pp 761–773

Frost J (2020) The economic forces driving fintech adoption across countries. BIS Working Papers No 838. https://www.bis.org/publ/work838.htm Accessed 11 July 2021

Fujii I, Okochi Y, Kawamura H (2021) Promoting cooperation of monitoring, control, and surveillance of IUU fishing in the Asia-Pacific. Sustainability 13:10231. https://doi.org/10.3390/su131810231

Hang L, Ullah I, Kim DH (2020) A secure fish farm platform based on blockchain for agriculture data integrity. Comput Electron Agric 170:105251. https://doi.org/10.1016/j.compag.2020.105251

Harris JM, Nelson JA, Rieucau G, Broussard WP III (2019) Use of drones in fishery science. Trans Am Fish Soc 148:687–697

IPNLF (2019) Implementing electronic traceability – the journey of Anova Food USA and the Indonesian handline tuna fishery, 13th November 2019. https://ipnlf.org/implementing-electronic-traceability-the-journey-of-anova-food-usa-and-the-indonesian-handline-tuna-fishery/ Accessed 11 Aug 2020

Ismail S, Reza H, Salameh K, Kashani Zadeh H, Vasefi F (2023) Toward an intelligent blockchain IoT-enabled fish supply chain: a review and conceptual framework. Sensors 23:5136. https://doi.org/10.3390/s23115136

Kastoryano S, Vollaard B (2023) Unseen annihilation: illegal fishing practices and nautical patrol. J Env Econ Man 122:102881. https://doi.org/10.1016/j.jeem.2023.102881

Kumperšcak S, Medved M, Terglav M, Wrzalik A, Obrecht M (2019) Traceability systems and technologies for better food supply chain management. Qual Prod Improv 1:567–574. https://doi.org/10.2478/cqpi-2019-0076

Larissa S, Parung J (2021) Designing supply chain models with Blockchain technology in the fishing industry in Indonesia. IOP Conf Ser Mater Sci Eng 1072:012020. https://doi.org/10.1088/1757-899X/1072/1/012020

Monkman GG, Hyder K, Kaiser MJ, Vidal FP (2019) Using machine vision to estimate fish length from images using regional convolutional neural networks. Meth Eco Evo 10:2045–2056. https://doi.org/10.1111/2041-210X.13282

Nasution AM, Nugraheni RD, Atmaja NN (2023) Traceability schemes and supply chains of tuna fisheries in Indonesian fishing ports: case study of Bitung Ocean Fishing Port and Pondok Dadap Beach Fishing Port, Indonesia. AACL Bioflux 16(4):1985–2001. http://www.bioflux.com.ro/aacl

PICRC, COS (2019) Palau's national marine sanctuary: managing ocean change and supporting food security. http://picrc.org/picrcpage/palau-national-marine-sanctuary Accessed 14 Sept 2020

Qin J, Vasefi F, Hellberg RS, Akhbardeh A, Isaacs RB, Yilmaz AG et al (2020) Detection of fish fillet substitution and mislabeling using multimode hyperspectral imaging techniques. Food Control 114:107234. https://doi.org/10.1016/j.foodcont.2020.107234

Reno S, Haque MM (2023) Solving blockchain trilemma using off-chain storage protocol. IET Inform Secur 17:681–702. https://doi.org/10.1049/ise2.12124

Swan J (2020) Illegal, unreported, and unregulated fishing: are RFMOs effectively addressing the problem? Ocean Yearb Online 34(1):297–336. https://doi.org/10.1163/9789004426214_013

Toonen HM, Bush SR (2020) The digital frontiers of fisheries governance: fish attraction devices, drones and satellites. J Env Pol & Plan 22(1):125–137. https://doi.org/10.1080/1523908X.2018.1461084

Tyce M (2020) Beyond the neoliberal-statist divide on the drivers of innovation: a political settlements reading of Kenya's M-Pesa success story. Word Dev 125:104621. https://doi.org/10.1016/j.worlddev.2019.104621

Welch H, Clavelle T, White TD, Cimino MA, Van Osdel J, Hochberg T et al (2022) Hot spots of unseen fishing vessels. Sci Adv 8:eabq2109. https://doi.org/10.1126/sciadv.abq2109

Widjaja S, Long T, Wirajuda H, As H Van, Bergh PE, Brett A et al (2023) In: Lubchenco J, Haugan PM (eds) The blue compendium. Springer, pp 553–591. https://doi.org/10.1007/978-3-031-16277-0_15

Willette DA, Ababouch L, Barber PH, Bunje PME, Cauzac JP, Conchon A et al (2023) Emerging monitoring technologies to reduce illegal fishing activities at sea and prevent entry of fraudulent fish into markets. Front Sust Food Syst 7:1–7. https://doi.org/10.3389/fsufs.2023.1166131

The Future of Fintech-Driven Sustainable Finance in Marine Industries

Pawan Pant, Kaushal Kishore Mishra, Azmee Zaheer, Lucius J. S. Botes, and Wynand C. J. Grobler

Abstract Marine industries include fisheries, shipping, aquaculture, and offshore energy, among others, which are an important component of the global economy. These industries raise severe sustainability challenges, including overfishing, pollution, habitat destruction, and climate-related risks. Financial technology, or Fintech, may be the key to transforming these sectors through innovative, efficient, and scalable financial solutions. This chapter explores where Fintech and sustainable finance intersect in marine industries, discussing how blockchain technology, artificial intelligence, and decentralised finance can be used to provide transparency, resilience, and inclusivity. The chapter discusses how Fintech is promoting economic empowerment and environmental conservation in coastal and marine-dependent communities by examining the role of blue bonds, crowdfunding platforms, and digital payment systems. Marine industries involve fishing, shipping, aquaculture, and offshore energy. Other challenging sectors of the world economy include overfishing, pollution, habitat destruction, and risk due to climate change. Fintech has enormous potential for changing everything mentioned above by using innovative, efficient, and scalable financial solutions. Ultimately, this sees the future of Fintech in marine sustainability as well. It has newer trends, such as digital twins, IoT integration, and

P. Pant · K. K. Mishra (✉) · A. Zaheer
University School of Business, Chandigarh University, Mohali, Punjab, India
e-mail: kkmsitmbbk@gmail.com

P. Pant
e-mail: pawanpant04@gmail.com

A. Zaheer
e-mail: azmee.zaheer@gmail.com

L. J. S. Botes
Development Studies and Director Research Development, North-West University, Potchefstroom, South Africa
e-mail: Lucius.Botes@nwu.ac.za

W. C. J. Grobler
School of Economic Sciences, North-West University, Potchefstroom, South Africa
e-mail: wynand.grobler@nwu.ac.za

V. Sharma et al. (eds.), *FinTech for Sustainable Finance and a Well-Blue Economy*, World Sustainability Series, https://doi.org/10.1007/978-3-031-92390-6_16

AI-enhanced financial tools. Fintech shall help the marine industries evolve towards a sustainable blue economy.

Keywords Fintech · Sustainable finance · Blue economy · Marine industries · Decentralized Finance (DeFi) · Blue bonds · Marine sustainability · Artificial Intelligence (AI) · Crowdfunding · Digital payment systems · Climate resilience · Ocean conservation

1 Introduction

The blue economy, or ocean economy, is a pillar of global economic activity, valued at around $2.5 trillion annually (FAO 2022). It sustains the livelihoods of over three billion people, food security, and industries such as fisheries, aquaculture, shipping, and offshore energy (World Bank 2021a). However, growing pressures of overfishing, habitat loss, pollution, and climate change risk the long-term ocean ecosystem and economic welfare of reliant communities (World Bank 2018). There has never been greater urgency to achieve a blue economy transition toward sustainable management as it requires new forms of financial arrangements supporting environmental conservation without compromising economic progress and social fairness (Nordic Investment Bank (NIB) 2020).

In this context, financial technology (Fintech) emerges as a revolutionary force, offering innovative solutions to overcome the age-old problems of maritime industries (UNEP 2023). By using technologies like blockchain, artificial intelligence (AI), decentralised finance (DeFi), and mobile banking, Fintech can make financial transactions transparent, efficient, and accessible in maritime-based businesses (OECD 2020b). Blockchain, in contrast, enables traceable and transparent supply chains, preventing illegal, unreported, and unregulated (IUU) fishing and promoting sustainable resource management (FAO 2022). Meanwhile, DeFi platforms democratise capital access, empowering small-scale fishers and marine start-ups, and AI-driven analytics optimise investments in climate-resilient projects (OECD 2020a).

In spite of these encouraging developments, significant obstacles stand in the way of Fintech's mass application in marine sectors. Digital infrastructure shortages, regulatory ambiguity, and conservatism in established industries are intense obstacles to the World Bank (2021b). Moreover, coastal and island societies—frequently leading the way in marine-based livelihoods—also do not possess the financial knowledge and technological aptitude to utilise these innovations optimally (UNCTAD 2019). Fighting such challenges is accomplished through coordination by governments, banks, and technology suppliers in attempting to create enabling environments with policy stimulus, capacity-building programs, and global regulatory convergence (WEF 2022).

This chapter delves into the fundamental role Fintech plays in transforming sustainable marine industry finance. It examines the means through which financial technologies are applied to deepen economic resilience, improve inclusivity, and strengthen ocean conservation (World Bank 2021c). Through a critical review of actual case studies and evaluating the effect of innovations like blue bonds and artificial intelligence-powered financial modelling, the discussion plots a course for leveraging Fintech towards attaining an integrated convergence of economic growth, social justice, and ocean sustainability. Finally, this chapter acknowledges that Fintech holds the potential for bridging economic growth with marine conservation, and thus, the blue economy can become an equitable and sustainable future (World Bank 2021a).

2 The Role of Fintech in Marine Industries

Fintech is transforming marine industries to become more efficient, sustainable, and transparent. Perhaps the most compelling application is blockchain technology, which allows seafood supply chains to be traced. This ensures marine products are harvested responsibly while preventing IUU fishing, which overexploits the ocean's resources and ruins the livelihoods of rightful fishers. Blockchain, which allows for real-time tracking of seafood from catch through to the consumer, assists in creating trust among stakeholders and ensuring the long-term sustainability of ocean ecosystems (OECD 2020a).

Furthermore, blockchain enhances the addition of smart contracts, whereby compliance with marine environmental rules can be automated with reduced administrative costs, and there is enhanced responsibility in the marine industries (World Bank 2021b). The second Fintech technology is DeFi. There is now a shift in the way capital mobilisation and dispensation happen in marine activities. DeFi-based platforms have eliminated intermediaries to banks. It provides peer-to-peer financing options to small-scale fishers, marine startups, and coastal communities for sustainable ventures and business growth (UNEP 2023). Mobile banking and electronic payment systems have revolutionised financial inclusion in blue economies, particularly in developing coastal communities without the usual banking institutions. These technologies allow for cross-border payments, reduce transaction costs, and increase micro-financing possibilities, which allow coastal communities to access financial resources more effectively. With increased economic resilience, such innovations give power to small-scale fishers, aquaculture producers, and marine entrepreneurs, ensuring their sustainable participation in the blue economy (FAO 2022).

Artificial intelligence and data analytics are significant in maritime industries because they improve decision-making and risk management (Lloyd's Register 2020). They will predict fish population trends, assess environmental risks, maximise supply chain efficiency, and lead to more environmentally friendly fishing techniques and resource allocation (WEF 2022). Big data analysis also assists the stakeholders in keeping track of the health of oceans in real-time. Thus, they make investment decisions wisely and improve compliance regulations. All these technological improvements boost the sustainability of the oceanic industries and advance the risk appraisal of the investors and stakeholders (World Bank 2021b).

3 Financial Innovations Empowering Blue Economies

Blue economy growth takes centre stage when it comes to new financial tools, enabling the application of oceanic resources for financial growth, sustenance, and conservation of nature. Fintech technologies are transforming access to capital, improving transparency, and connecting ocean investments with economic returns and sustainable surroundings.

One significant innovation is the application of blue bonds, which have a similar function to green bonds but are specifically dedicated to securing oceans and ocean-friendly initiatives. In 2018, for instance, Seychelles issued its first blue bond and raised $15 million to support marine conservation and sustainable fishery projects (Thomas and Benjamin 2020).

Since then, it has served as a financial instrument, allowing small island developing states (SIDS) and other coastal nations to raise needed capital for sustainable marine projects, according to the (World Bank 2021c). Another major innovation in blue finance is decentralised finance (DeFi), which enables direct investment and financing in marine industries without traditional intermediaries like banks or insurance companies. DeFi platforms enable peer-to-peer lending, allowing small-scale fishers, coastal businesses, and marine startups access to capital, particularly in underbanked areas where financial institutions are in short supply. Through blockchain technology and smart contracts, DeFi lowers transaction costs, improves security, and inspires financial inclusivity, fostering a just distribution of resources for sustainable oceanic enterprises (UNEP 2023).

Crowdfunding websites play a critical role in aiding blue economies by facilitating the mobilisation of funds by individuals and organisations for marine conservation, eco-tourism, and aquaculture innovation. Crowdfunding websites democratise access to capital, making small-scale projects comparable to larger ones. Crowdfunding taps a worldwide network of investors, making sure that significant social and environmental projects receive the resources to succeed and facilitating sustainable ocean industry development (WEF 2022).

The intersection of sustainable finance and impact investing is driving blue economy sector innovation. The increasing pool of impact investors, who are more concerned with ESG outcomes than profitability, is gaining momentum. Impact investments are increasingly supporting blue economy projects, and the tools of sustainable finance include ESG scores, carbon credits, and sustainability-linked loans. These new offerings offer cheap capital, encourage responsible behaviour, and guarantee profitability in harmony with ocean conservation (OECD 2020a).

Electronic payment systems and mobile payments are broadening coastal ocean economies' financial access, especially for communities located far off from the mainland. Coastal communities can make payments, earn insurance and savings, and receive microloans via cell phones. By deepening financial inclusion, such innovations allow small-scale fishers, aquaculture farmers, and businesspeople to access the blue economy, export their products to global markets, and maintain their livelihoods (FAO 2022).

4 Case Studies and Best Practices

4.1 Seychelles' Sovereign Blue Bond

Background and Motivation

Seychelles is an island nation in the Indian Ocean that is considerably dependent on its marine resources for economic development, mainly through fisheries and tourism. Issues such as overfishing, habitat destruction, and climate change are grave threats to marine biodiversity and economic stability (Thomas and Benjamin 2020). Understanding these challenges, the Seychelles government sought out innovative financing options to back sustainable fisheries and marine conservation, leading to the launch of the world's first sovereign blue bond (Seychelles Conservation and Climate Adaptation Trust (Conservation et al. 2018; UNDP 2019).

Key Features of the Bond

1. Issue Amount and Tenor: Seychelles had issued a $15 million blue bond with a tenor of 10 years.
2. Guarantees and Risk Mitigation: The World Bank provided a partial guarantee of $5 million to mitigate the risk for investors, and the Global Environment Facility provided a concessional loan of $5 million to support the bond (Thomas and Benjamin 2020; UNDP 2019).
3. Objective: To promote sustainable fisheries, protect marine ecosystems, and stimulate the blue economy.
4. Governance Structure: Proceeds will be disbursed through a Blue Grants Fund and a Blue Investment Fund.

The Seychelles Conservation and Climate Adaptation Trust managed those funds but ensured transparency and accountability in implementing those projects (World Bank 2018).

- **Marine Protected Areas (MPAs)**: A large part of the funds allocated went into expanding marine protected areas, which now cover 30% of Seychelles' EEZ, some 410,000 km^2.
- **Sustainable Fisheries**: These funds supported the local fishers' shift to sustainable practices, such as using eco-friendly fishing gear and training in stock management techniques.
- **Community Development**: Small-scale loans and grants were made to the coastal communities to diversify into aquaculture, eco-tourism, and other sustainable enterprises.

Key Outcomes

1. Environmental Impact: Creation of the Indian Ocean's largest network of marine protected areas. Conservation of coral reefs and biodiversity is essential for both the local and global environment.

2. Economic Impact: Increased productivity and long-term sustainability of Seychelles' fisheries sector. Generated alternative income sources for coastal communities, reducing dependence on overfishing (UNDP 2019)
3. Policy and Governance Impact: Enhanced marine spatial planning frameworks to harmonise conservation efforts with economic activities. Positioned Seychelles as a global frontrunner in blue finance and sustainable development (World Bank 2018).

Challenges

- **Market Size**: The island nation of Seychelles, therefore, was limited in its capacity to attract large-scale private-sector investment.
- **Implementation Capacity**: Ensuring compliance with sustainable practices requires significant capacity-building efforts for local communities and institutions.
- **Replicability**: The bond was a success, but replicating it in other regions may require comparable international support and guarantees, which are not always available.

Global Recognition and Replicability Seychelles' blue bond proved the concept of using financial instruments for marine conservation (United Nations Environment Programme (UNEP) 2020). It has inspired other nations and organisations, such as Fiji and the Nordic Investment Bank, to consider blue bonds as a tool to address climate change and marine sustainability challenges (Seychelles Conservation and Climate Adaptation Trust (Conservation et al. 2018; Thomas and Benjamin 2020).

Key Takeaways

It details how innovative financial vehicles, such as the Sovereign Blue Bond of Seychelles, can be used to meet environmental and economic objectives. Thus, small islands can raise funds engaged with all stakeholder concerns and partnerships for the above aims, creating an essential capital investment lifeline for sustainable developmental purposes and participation in global conservation.

4.2 Nordic-Baltic Blue Bond Initiative

Background and Motivation

This is the Nordic-Baltic Blue Bond Initiative, launched to meet the urgent demand for financing projects that improve marine sustainability and environmental protection in the Baltic Sea and surrounding regions (Plan 2021). The Baltic Sea is one of the most polluted marine ecosystems in the world, and eutrophication, overfishing, and impacts of climate change are at issue (Olesen and Hansen 2021). The Nordic Investment Bank (NIB) supported the initiative, with a strong interest in using the blue bond model to finance innovative solutions related to protecting marine biodiversity and reducing pollution in the sea, as well as supporting the blue economy (Nordic Investment Bank (NIB) 2020).

Key Features of the Initiative

1. **Issuer and Structure**: The Nordic Investment Bank, a multilateral financial institution owned by the Nordic and Baltic countries, issued the blue bond. It garnered €200 million, making it one of the most significant blue bond initiatives in Europe (Olesen and Hansen 2021).
2. **Utilisation of the Proceeds**: It targeted financing projects that aimed at the sustainable use and preservation of marine resources. The key focus areas remained wastewater treatment, sustainable aquaculture, efficient maritime transport, and restoration of marine biodiversity.
3. **Innovations**: Outcome-based financing models, wherein funding remained linked to measurable environmental improvements. Environmental, social, and governance criteria were incorporated into the selection and assessment of projects WWF (Plan 2021).

Fund Allocation The money was distributed to several projects along the Baltic region:

1. Quality Improvement of Water Capital investment in high-tech wastewater treatment and reduction of nutrient runoff in the Baltic Sea. Farm infrastructure is upgraded to limit fertiliser and pesticide leakage into the sea.
2. Sustainable Development of Aquaculture Financial support for environment-friendly aquaculture with the introduction of recirculating aquaculture systems (RAS). Training programs for aquaculture producers to adopt eco-friendly farm practices.
3. Environment-Friendly Maritime Transport Financing low-emission shipping technologies and alternative fuels. Modernising port infrastructure with an environmental impact (World Economic Forum 2021).
4. Marine Biodiversity and Habitat Restoration: Restoration of seagrass meadows and wetlands, which are critical habitats for marine species. Improvement of fish stock management and prevention of overfishing.

Outcomes and Impact

1. Environmental Impact: Substantial reduction of nitrogen and phosphorus run-off into the Baltic Sea, reducing eutrophication. Enhanced marine biodiversity through habitat restoration and sustainable aquaculture practices.
2. Economic Impact: The blue economy has been strengthened in the region through jobs in sustainable aquaculture, eco-tourism, and green shipping sectors. Sustainable practices, such as the (Plan 2021), have also enhanced the competitiveness of Nordic and Baltic industries.
3. Social and Policy Impact: Regional cooperation among Nordic and Baltic countries was facilitated to address common marine issues and influenced policy frameworks by integrating marine conservation goals into national economic strategies (United Nations Environment Programme (UNEP) 2020).

Challenges and Limitations

1. Sophisticated Stakeholder Coordination Involving multiple countries requires extensive collaboration and harmonisation of environmental standards and regulations.
2. Data Gaps: Long-term monitoring of a project's impact on marine ecosystems requires environmentally robust and reliable monitoring systems.
3. Scaling and Replicability: Although it was successful in the Baltic region, the initiative called for similar efforts in other vulnerable marine ecosystems.

Future Directions

The success of the Nordic-Baltic Blue Bond Initiative has opened up the avenue for discussions on scaling similar projects globally by Olesen and Hansen (2021). NIB is exploring partnerships with international partners to replicate this model in other regions, including North Sea and Arctic waters, focusing on innovative financing mechanisms such as blended finance and public–private partnerships by the Global Environment Facility (2018).

4.3 WavePay in Myanmar

Background and Context

It belongs to a popular mobile financial service under Wave Money, which came about as a result of its strategic partners, a collaboration between the Telenor Group and Yoma Bank, and has been operational since the year 2016 in Myanmar—a country known for almost minimal banking structures and, more importantly, none of most of its peoples banked (Telenor Group 2020; Digital Financial Services in Myanmar: A Roadmap for Growth 2020). WavePay uses mobile technology to provide accessible, affordable, and secure financial services to under-banked communities, primarily rural and remote areas (GSMA 2020; Times 2020).

WavePay Services and Features

1. Mobile Wallet: WavePay allows consumers to save money in a mobile wallet that is accessible on mobile and feature phones. They can send and receive money, pay bills, recharge mobile phones, and even make in-store payments using QR codes.
2. Agent Network: WavePay is an agent network of over 68,000 agents across the country. Agents allow cash-in and cash-out, meaning that most people who would not have been able to open a bank account can now access it (Telenor Group 2020).
3. Financial Inclusion Reaching out, particularly to women and marginalised peoples. Access to finances is highly barred socially and economically, especially for women and marginalised individuals—partnership with NGOs and

microfinance institutions to offer microfinance products and insurance products (Digital Financial Services in Myanmar: A Roadmap for Growth 2020).

4. Disaster and Pandemic Response During COVID-19, Wave Pay conducted cash transfers to marginalised persons through government and non-government organisations' relief programs.

Impact and Outcomes

1. Financial Inclusion: Over 5 million active users by 2020, with a majority from the rural and underserved. Enabled women to be empowered through financial independence since more female entrepreneurs started using WavePay for small business transactions (Money 2020).
2. Economic Development: WavePay improved smoother and faster remittances from urban to rural communities, thereby lessening reliance on informal channels. It also Boosted local economies by improving payment systems for micro and small businesses.
3. Digital Literacy: WavePay programs included training to improve users' digital skills and allow them to use financial services safely.
4. COVID-19 Relief: WavePay, in collaboration with NGOs and the government, provided relief funds in excess of $13 million during COVID-19.

Challenges and Barriers

1. Regulatory environment: Myanmar's financial regulatory landscape is immature, creating scalability issues for fintech applications, such as WavePay. Implementing AML KYC processes for rural facilities was laborious and enormously tricky in the early stages of the United Nations Capital Development Fund (Digital Financial Services in Myanmar: A Roadmap for Growth 2020).
2. Connectivity and infrastructure: The limited reach of the Internet and mobile networks prevented wavePay from reaching more distant villages. This excluded the most vulnerable groups from participation: those dependent on smartphones and feature phones (Digital Financial Services in Myanmar: A Roadmap for Growth 2020).
3. Trust and Adoption: Gaining trust from people who did not use digital financial products earlier is time-consuming and painstakingly difficult. Initial fears of fraud and misuse of funds were common barriers to the adoption of Money (2020).
4. Political Instability: The political instability and military coup in 2021 threw WavePay and the overall fintech ecosystem off track.

Lessons Learned and Best Practices

1. Contextualization: WavePay worked well because it tailored its services to Myanmar's specific socio-economic context, offering multilingual support and low fees (Money 2020).
2. Trust through Agents: This network of agents has been helpful in gaining users' trust because they are more likely to do a transaction through a known local intermediary.

3. Partnerships for Impact WavePay expanded its service offerings and successfully leveraged social impact by working with banks, NGOs, and the government of the United Nations Capital Development Fund (Digital Financial Services in Myanmar: A Roadmap for Growth 2020).
4. Technology as a Leveler: WavePay showed that mobile technology could actually defeat the barriers of traditional banking and be a tool for economic empowerment.

Future Directions

1. Higher Incomes Launch new financial products, such as savings accounts, small business loans, and insurance plans, to increase penetration.
2. Increased Access: Employ satellite internet and engage with telecommunication service providers to expand to more rural locations.
3. Improving Cyber Security Expand security controls to prevent cases of fraud and protect information for clients.
4. Regulatory Coordination Collaborate closely with regulators to formulate policies to promote fintech, taking care not to destabilise finance or threaten consumers.

5 Challenges and Barriers

Although Fintech-driven sustainable finance has tremendous transformative potential for marine industries, several challenges and barriers must be overcome in order to unlock its full impact. Challenges are spread across technical, regulatory, socio-economic, and environmental domains; hence, a coordinated approach from governments, financial institutions, technology providers, and local communities is required to overcome them.

5.1 Digital Infrastructure Gaps

The most significant risk to Fintech uptake by marine businesses is poor digital infrastructure, especially in far-flung coastal and island regions. Poor access to high-speed internet, mobile network coverage, and reliable electricity slows the use of digital financial products like mobile payments, blockchain, and DeFi platforms. Small-scale fishers, coastal businesspersons, and marine conservation initiatives have limited access to the technology requirements, limiting them from joining the blue economy at full capacity by the United Nations Environment Programme (United Nations Environment Programme (UNEP) 2020). Closing the infrastructure gap requires increased investment in connectivity-satellite internet and solar power energy solutions so that remote marine spaces are not left behind in digital transformation (FAO 2022; UNEP 2023).

5.2 *Regulatory and Legal Uncertainty*

Regulation of fintech in the shipping sector is in a developing phase, and hence, there are legal uncertainties that discourage investment in green maritime businesses. The lack of harmonised regulation in nations regarding digital financial products, data protection, and blockchain prevents domestic and international regulatory harmonisation. Such a regulatory lack makes it challenging to launch blue bonds, DeFi initiatives, and other digital financial innovations in ocean conservation. Unclear rules can lead to litigation, minimal cross-border investment, and retarded innovation. A globally harmonised regulatory framework is needed to offer legal, secure, and scalable Fintech services for the blue economy (OECD 2020a; World Bank 2021a).

5.3 *Financial Literacy and Capacity Building*

Financial capacity building is a critical issue in the context of small-scale sea enterprises since the stakeholders have yet to clearly comprehend how they would be empowered through Fintech innovations such as blue bonds, blockchain, and DeFi. Their technicality also discourages them, mainly where little exposure has been made to digital finance. Such ignorance complicates access to sustainable finance by limiting funding of initiatives that underpin the blue economy. Targeted education and capacity-building initiatives are necessary to empower local communities to utilise Fintech for economic growth and sustainability (UNEP 2023; WEF 2022).

5.4 *High Transaction Costs*

In spite of assertions that Fintech reduces transaction costs, excessive charges for blockchain transactions, cross-border remittances, and digital financial services continue to be a considerable impediment, particularly to micro-enterprises and small-scale fishers. It is too expensive to join blockchain platforms or mobile payment networks, and the costs are out of reach for such stakeholders, hence restricting access to microtransactions and sustainable financing solutions such as crowdfunding, impact investing, or carbon credits. Reduction of these costs is paramount in order to promote level participation and expand the cost–benefit of online solutions to coastal unbanked communities (OECD 2020b).

5.5 Environmental and Social Risks

As Fintech drives innovation in the blue economy, it is, however, faced with environmental and social risks. Blockchain networks, particularly those based on proof-of-work, operate at high energy use, which could destroy sustainability objectives. Moreover, if new digital financial instruments do not extend to poor coastal communities, they may exacerbate socio-economic disparities. In order to minimise these risks, Fintech solutions should have energy efficiency, inclusiveness, and genuine social impact, which will ensure sustainable finance reaches all the stakeholders of the marine economy (UNEP 2023; World Bank 2021b).

6 Regulatory Frameworks for Scaling Fintech

This graphic offers an in-depth view of the regulatory structures guiding fintech growth concerning eco-friendly marine businesses. The essential components of this framework concerning regulations include ensuring that Fintech solutions are sustainable environmentally and financially (Fig. 1).

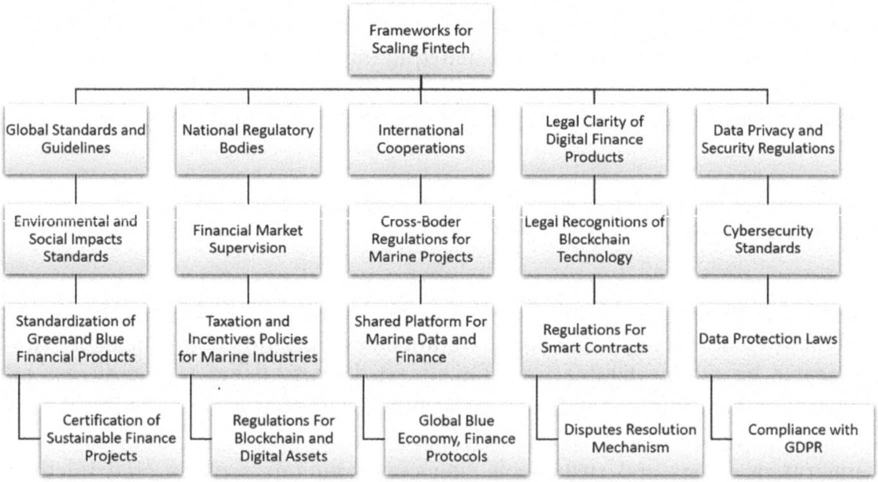

Fig. 1 Framework for scaling Fintech. *Source* Author's creation

6.1 Global Standards and Guidelines

Standards worldwide are required so that Fintech solutions within the marine industries run according to principles accepted everywhere around the world, especially those involving sustainable finance. This category includes:

- Environmental and Social Impact Standards: These standards ensure that marine-based projects funded through Fintech platforms positively contribute to the environment and local communities. They guide financial institutions in assessing the potential impacts of investments on marine ecosystems and coastal communities (Fi 2019).
- Standardization of Green and Blue Financial Products: This would mean standardising financial products funding ocean-related initiatives, including sustainable marine tourism, fisheries, and renewable marine energy. Standardisation brings transparency and confidence to both investors and regulators (Mikulova 2021).
- Certification of Sustainable Finance Projects: The certification processes ensure that the certified projects adhere to the established sustainability criteria. This will allow funds to be utilised to finance marine industries in accordance with environmental preservation and the achievement of climate goals (World Bank 2021b).

6.2 National Regulatory Bodies

National regulators ensure the integrity of financial markets while, at the same time, promoting sustainable investments. The main elements in this section are:

- Financial Market Supervision: The regulatory bodies must supervise the financial markets to ensure they are fair, transparent, and accountable. This is crucial to the growth of Fintech in marine industries as trust is built among investors and stakeholders (OECD 2020b).
- Taxation and Incentive Policies for Marine Industries: This refers to giving tax breaks and other forms of subsidy to governments that allow investment in the marine sector. The incentive programs reduce risk when financing a project in the challenging marine sector (UNCTAD 2019).
- Regulations for Blockchain and Digital Assets: As blockchain technology and digital assets, such as cryptocurrencies and tokenised assets, grow, regulations are needed to govern the use of such applications in marine industries. Such defined rules reduce the risk associated with fraud, market manipulation, and data privacy (Zohar 2021).

6.3 International Cooperation

It gives the nations the opportunity to cooperate and achieve a common goal while formulating regulations for easy integration of global financial systems. It includes:

- Cross-Border Regulation for Marine Project: Marine industries often have operations in several countries that require a legal framework to enable cross-border financial activities. This involves creating international agreements on environmental standards and investment frameworks (Gollin 2020).
- Shared platforms for marine data and finance: Shared platforms that integrate marine data with financial systems can drive better decision-making in marine finance. Such shared platforms allow stakeholders to monitor the performance and sustainability of the funded projects, ensuring accountability (Liu 2022).
- Global Blue Economy Finance Protocols: The UN's Blue Economy finance protocols lay down a guideline on financing for marine projects according to the UN's Blue Economy initiatives. The global protocols promote a consistent application of sustainable finance mechanisms, thereby enhancing investment in ocean-based solutions (OECD 2020a).

6.4 Legal Clarity for Digital Finance Products

Fintech products and services, especially those utilising blockchain and other digital technologies, require clear legal frameworks so that they can be enforced and fit within established legal systems. These include:

- Legal Recognition of Blockchain Technology: For Fintech to grow, blockchain must be legally recognised as a legitimate means of conducting transactions in the marine finance sector. This legal clarity surrounding blockchain ensures that investments are secure and compliant, whether they are in sustainable marine energy projects or ocean conservation efforts (Risius 2020).
- Regulation for Smart Contracts: Smart contracts automate financial agreements, and regulation is important to ensure that such contracts are fair and enforceable in the context of marine projects. Such contracts may simplify the financing of marine projects by saving time and reducing costs.
- Dispute Resolution Mechanisms: The dispute resolution frameworks for marine finance help avoid and resolve conflicts that can arise in international projects and provide a structured mechanism to resolve issues efficiently (UNCTAD 2021)

6.5 Data Privacy and Security Regulations

Since the Fintech solutions are, by nature, digital, data privacy and security laws must be more than sufficient to prevent an erosion of investor and consumer confidence.

- Cybersecurity Standards: Due to the increased use of digital tools in marine finance, cybersecurity standards must be established to ensure the protection of sensitive financial information. The same shift is further reducing cyberattacks that can disrupt the integrity of financial platforms (OECD 2020b).
- Data Protection Laws: The marine industries generally involve large amounts of personal and corporate data, especially concerning monetary transactions; hence, robust data protection legislation must prevail. With such provisions, this law attempts to safeguard privacy rights about sensitive information, and it was observed in Europe when GDPR was implemented (Janssen 2021).
- Compliance with GDPR: If a Fintech company conducts business in Europe or provides its services to customers in the European Union, it must observe GDPR. That would include gathering, storing, and processing personal data in a way that does not violate privacy rights, therefore creating trust with Fintech solution users at the European Commission (European Commission 2021).

7 Future Directions and Emerging Trends

As the blue economy moves further in that direction, Fintech will play an even more central role in the evolution of maritime industries, one that will drive innovation, ensure sustainability, and improve financial inclusion (FNV GL 2021; Sharma et al. 2024). Some of these trends and future frontiers will likely disrupt the landscape, with the potential to address issues such as resource management, the protection of the marine environment, the responsible utilisation of natural capital and socioeconomic development. These trends are closely tied to technological advancements, the regulatory environment, and the market demand for sustainable solutions.

7.1 Artificial Intelligence (AI) and Big Data Fintech

AI and big data are key propellers for the blue economy. AI gathers enormous quantities of ocean data via satellites, drones, and IoT sensors to provide feedback on the health of oceans, fish stock levels, and ecosystems in real-time. AI drives efficient supply chains, predicts fish stocks, and secures fishing at a sustainable rate by making informed decisions based on data. When blockchain and IoT are coupled with big data, big data makes ocean supply chains more transparent and ensures environmental compliance. Blockchain and IoT maximise the effectiveness of risk management, asset use, and investment to optimise financial development without sacrificing ocean sustainability (Mishra and Pant 2025; Sharma et al. 2024; WEF 2022).

7.2 *Rise of Green and Blue Digital Assets*

Increasing demand for blue and green sustainable financial products is fueling expansion in digital assets like carbon credits on the blockchain and marine conservation projects in tokenised formats. Such digital assets that possess the real-world value of environmental protection can be exchanged in digital exchange, unlocking new financing for marine conservation projects. For instance, blue carbon credits that measure the ability of mangroves, seagrasses, and salt marshes to store carbon can be tokenised and traded on blockchain exchanges to finance the restoration and conservation of ecosystems. All this is in the broader shift toward digitised sustainable finance with novel economic incentives to protect the oceans (OECD 2020a; Pant and Mishra 2025; United Nations Environment Programme (UNEP) 2020; ; ; ; ; ; ; ;).

7.3 *Decentralised Finance or DeFi and Crowdfunding of Marine Projects*

The development of Decentralized Finance (DeFi) is revolutionising shipping companies through new intermediary-free models of financing. With advancing blockchain technologies, DeFi platforms are emerging as go-to channels for green maritime projects' capital mobilisation. Through the use of smart contracts, DeFi offers disclosure, cost reduction, and safe exchange, facilitating peer-to-peer lending and staking to fund environmentally friendly fishing enterprises, aquaculture technology, and blue tourism activities. In addition, blockchain crowdfunding and cryptocurrencies enable direct investment in small-scale marine businesses, providing new avenues of capital to fund the blue economy and support ocean conservation efforts (Mishra et al. 2024; UNEP 2023; Zaheer et al. 2024).

7.4 *Climate Finance and Resilience Bonds*

Climate resilience financing is becoming more focused on the marine ecosystem since it is faced with an increasing chain of environmental stresses. Climate adaptation bonds and resilience bonds are slowly coming to be seen as important finance tools to help local communities deal with sea level rise, coast erosion, and intensified weather. They connect finance to quantifiable environmental and social results, including coastal protection and the rehabilitation of marine ecosystems. Combined with blockchain technology, resilience bonds also present transparency, accountability, and effective tracing of money, making them appealing to public and private investors who are interested in green investment with real-world environmental impact (World Bank 2021c).

7.5 Expansion of Sustainable Investment Products

Investment in blue economy industries that are green is also forecast to increase by a significant extent, fueled by the need for returns aligned with environmental, social, and governance (ESG) objectives. Impact investment vehicles, green bonds, and blue bonds are progressively focusing on marketplaces like green fisheries, sea renewable energy, ocean conservation, and green tourism. Such financial products hold strong appeal for institutional investors who want both profitability and positive environmental consequences. As sustainable finance goes mainstream, financial institutions will create more advanced mechanisms to track and report the environmental and social effects of marine investments, increasing transparency and driving investment in the blue economy (OECD 2020a; WEF 2022).

8 Conclusion

The incorporation of fintech into marine industries might become a solution for sustainable finance and solve some of the problems that have long been bothering the blue economy, the most stubbornly entrenched ones. Fintech solutions like blockchain, AI-driven financial tools, and digital payment platforms have empowered stakeholders by improving transparency, streamlining operations, and granting access to global capital markets. These innovations can thus drive the shift towards a sustainable and resilient marine economy, promoting investment in renewable energy, biodiversity conservation, and sustainable fisheries. However, there are still significant challenges: inconsistencies in regulatory frameworks, data privacy, and technological gaps in emerging economies. All these issues require the development of comprehensive regulatory frameworks that will balance innovation with financial and environmental security. International cooperation will be critical to harmonise cross-border regulations and standardise financial instruments for blue economy projects. Additionally, future breakthroughs in technology will likely reshape the fintech landscape with advancements in quantum computing and new blockchain protocols, thereby providing new avenues for marine sustainability. The future of fintech-driven sustainable finance in marine industries lies in a collaborative approach from regulators to private stakeholders to international organisations. By aligning technological innovation with environmental and social objectives, fintech can play a critical role in achieving well-blue economy goals while preserving marine ecosystems by unlocking sustainable economic growth.

References

Baltic Sea Action Plan (2021) Reducing eutrophication and protecting biodiversity in the Baltic Sea. HELCOM Website

DNV GL (2021) Sustainability and risk management in maritime industries
European Commission (2021) EU blue economy report: sustainable development in marine sectors
FAO (2022) Food and agriculture organization of the United Nations. The state of world fisheries and aquaculture 2022: Towards blue transformation. Rome
Global Environment Facility (2018) Innovative financing for sustainable marine practices
Gollin L (2020). Marine finance and cross-border regulations. Ocean Policy Institute
GSMA (2020) State of the mobile money industry report: WavePay case study
Janssen M (2021) Digital finance and data privacy: a European perspective. Oxford University Press
Liu Y (2022) Marine data platforms for sustainable finance. J Ocean Econ 45(3):134–148
Lloyd's Register (2020) AI and big data in shipping: enhancing risk management
Mikulova E (2021) Standardising blue economy financial products. Mar Sustain Rev 23(2):78–92
Mishra KK, Pant P (2025) Subconscious factors affecting consumer preferences toward green investments. Sustain Innov Consum Preference 119–148
Mishra KK, Pant P, Singh H, Mishra SK (2024) The philosopher's stone: applications of data alchemy—customer personalization, profiling, and retention. In: Data alchemy in the insurance industry. Emerald Publishing Limited, pp 77–95
Myanmar Times (2020) WavePay's role in enabling cashless transactions in rural Myanmar
Nordic Investment Bank (NIB) (2020) Nordic-Baltic blue bond: sustainable financing for the Baltic Sea. NIB Website
OECD (2020a) Regulatory frameworks for sustainable finance in the blue economy. OECD Publishing
OECD (2020b) Organisation for economic co-operation and development. Blockchain technologies as a tool for sustainability in the fisheries sector. Paris
Olesen M, Hansen T (2021) Evaluating the Nordic-Baltic blue bond initiative. Mar Policy 124:104309
Pant P, Mishra KK (2025) Behavioural economics and the rise of sustainable investments: urgency for a new paradigm in financial decision-making. In: Sustainability, innovation, and consumer preference. IGI Global Scientific Publishing, pp 329–360
Risius M (2020) Smart contracts and legal clarity in fintech. J Legal Innov 29(4):56–71
Sharma V, Gupta M, Taneja S (2024) Does FinTech adoption impact on sustainability of small businesses: mediating role of financial well-being. Glob Knowl Mem Commun
Seychelles Conservation and Climate Adaptation Trust (SeyCCAT) (2018) Blue bond: financing sustainability in Seychelles
Telenor Group (2020) The role of WavePay in Myanmar's financial ecosystem
Thomas A, Benjamin N (2020) Scaling blue bonds: lessons from Seychelles. Mar Policy J
UNCTAD (2019) Taxation policies for financing marine sustainable development. UNCTAD Publications
UNCTAD (2021) Digitalisation in maritime transport: opportunities and challenges. UNCTAD maritime report
UNDP (2019) Blue economy case study: Seychelles. UNDP Website
UNEP FI (2019). Environmental and social impact standards for marine finance. United Nations Environment Programme Finance Initiative
United Nations Environment Programme (UNEP) (2020) Innovative financing for marine conservation: case studies
United Nations Capital Development Fund (UNCDF) (2020) Digital financial services in Myanmar: a roadmap for growth
UNEP (2023) United Nations Environment Programme. Financing the blue economy: investing in sustainable use of ocean resources. Nairobi
Wave Money (2020) Annual report: financial inclusion in Myanmar. Wave Money Official Website
WEF (2022) World Economic Forum. Harnessing the fourth industrial revolution for sustainable development: ocean solutions. Geneva
World Bank (2018) Seychelles launches world's first sovereign blue bond. World Bank Website
World Economic Forum (2021) The future of shipping: harnessing AI for sustainable operations

World Bank (2021a) World Bank. The potential of the blue economy: increasing long-term benefits of the sustainable use of marine resources for small island developing states and coastal least developed countries. Washington, DC

World Bank (2021b) Certification mechanisms for marine finance projects. World Bank Reports

World Bank (2021c) Financial inclusion in Myanmar: challenges and opportunities

Zaheer A, Singh A, Mishra KK, Vardari L (2024) The future of alchemy: emerging trends and technologies metaverse in insurance–a virtual customer experience. Data Alchemy Insur Ind 133–152

Zohar A (2021) Blockchain regulations and legal frameworks for Fintech. Harvard Business

Financial Market Dynamics of the Blue Economy: An Econometric Analysis of Sectoral Performance and Investment Implications

Vikas Sharma, Tejinder Singh, and Ewelina Idziak

Abstract The Blue Economy, which assimilates the sustainable use of ocean resources with economic growth and environmental preservation, has attracted global importance. An innovative Blue Economy Index has been introduced to examine the financial dynamics of ocean-based industries, including metals and mining, ports and logistics, offshore energy, shipping, fishing, and marine engineering. Using daily time-series data from May 2022 to July 2024, the study examines the dynamic relationships between future maturity prices and open interest and trading volume. This study uses the Vector Auto-Regressive model, Granger-causality tests, impulse response functions, and variance decomposition analysis, highlighting market sentiment, price discovery mechanisms, and investor behaviour within the Blue Economy sector. The findings show the extent to which open interest and trading volume influence future maturity prices, helping various market stakeholders. This study fills the gap between financial markets and sustainability with a structured methodology to measure the economic significance of ocean-based industries.

Keywords VAR: Vector autoregression · IRFs: Impulse response functions · OECD: Organisation for Economic Co-operation and Development · UNCTAD: United Nations Conference on Trade and Development · GPP: Gujarat Pipavav Port · CS: Cochin Shipyard

---

V. Sharma (✉)
University School of Business, Chandigarh University, Mohali, Punjab, India
e-mail: vikas.shrarma62@yahoo.com

T. Singh
Department of Law and Economics, Uniwersytet Kazimierza Wielkiego, Bydgoszcz, Poland

E. Idziak
Women Researchers Council, Azerbaijan State University of Economics, Baku, Azerbaijan

V. Sharma et al. (eds.), *FinTech for Sustainable Finance and a Well-Blue Economy*, World Sustainability Series, https://doi.org/10.1007/978-3-031-92390-6_17

1 Introduction

The global economy is segregated into sectors that drive economic activities and development across primary, secondary, and tertiary sectors, each forming an essential role in determining global industries (Du et al. 2022). The primary sector covers extracting and harvesting natural resources, including agriculture and allied sectors such as fishing, forestry, mining, and oil and gas extraction (Wellmer and Becker-Platen 2002). It forms the basis for developed as well as developing countries. Raw materials are processed into finished goods under the secondary sector, which covers industrial production and manufacturing. Construction, automobile manufacturing, steel production, and textile industries are part of the secondary sector, forming an important component of economic growth.

The tertiary sector includes banking, finance, tourism, hospitality, retail trade, and transportation. This sector has expanded quickly in developed economies and contributes the highest share of GDP and employment (Ashley et al. 2007).

The blue economy is an essential part of the global economy, referring to economic activities related to oceans, seas, and coastal regions (Martinez-Vázquez et al. 2021). Fisheries, marine tourism, maritime transport, and offshore renewable energy industries are covered under the blue economy. The blue economy is a significant supplier to global GDP, predominantly in coastal and island nations. As per World Bank estimates, blue economy industries contribute between 3.5 and 5% of global GDP, with leading sectors such as shipping, fisheries, and marine tourism.

With more than 80% of global trade by volume, the shipping and maritime industry contributes the most to the blue economy (Spalding 2016). Furthermore, $400 billion is generated annually from the fisheries and aquaculture sector globally (Boyd et al. 2022). Overall, the blue economy signifies a leading section of the universal economic system by providing huge opportunities for economic development while underlining the need for sustainable practices to safeguard marine ecosystems and ensure sustainability (Geng et al. 2024).

The Blue Economy, which involves the sustainable deployment of ocean resources for economic growth, better livelihoods, and a healthy ocean ecosystem, has attracted global importance in recent years (Shah and Niles 2016). Keeping in view the blue economy's importance and its pivotal role in addressing environmental challenges, this study has used an innovative Blue Economy Index to ascertain the economic activities of ocean-based industries. Blue Economy Index comprises stocks from sectors such as metals and mining, ports and logistics, offshore energy, shipping, fishing, and marine engineering. The study aims to study the dynamic relations between future maturity prices, a dependent variable, and the independent variables, open interest and option volume. Time series observation from May 2022 to July 2024 has been used to apply econometric techniques to show insights into the movements and causal relationships in the blue sector. This initiative fills the gap of a critical tool to track Blue Economy performance for policymakers, investors, and stakeholders in the marine sciences and sustainability fields.

The Blue Economy represents a juncture between economic development and environmental sustainability, making it essential to design analytical tools for measuring its growth and efficiency (Shah and Niles 2016; Burmaoglu et al. 2023). This study has used a quantitative model to examine market sentiment and price discovery mechanisms in ocean-based industries. Open interest and option volume as variables throw a fresh viewpoint on how market participants react to developments in the blue economy sector. They serve as proxies for investor sentiment, liquidity, and risk perception, providing insights into market behavior. Vector autoregression (VAR) and Granger causality tests have been used to study the lead-lag relationship between future maturity prices and the independent variables. This study increases understanding of speculative and hedging activities in Blue Economy-related stocks and can help investors and researchers develop informed trading strategies and risk management mechanisms. Financial analysis combined with sustainability metrics can lead to resilient economic models that justify the fundamentals of the circular economy.

The blue economy, which encompasses shipping, coastal tourism, and fisheries industries, plays a crucial role in economic growth (Martínez-Vázquez et al. 2021). However, its sustainability is susceptible to risk owing to climate change and sea pollution. It, therefore, becomes inevitable to study blue finance within the financial market landscape of the blue economy to measure the impact of capital flows, investment styles, and risk management techniques and their impact on the long-term growth of the blue economy.

An analysis of sectoral performance within the blue economy and investment implications can provide significant data insights into financial market reactions to market opportunities in the blue economy. Financial indicators such as market volatility, investment inflows, and optimal portfolio theories can help research measure the economic flexibility of ocean-based industries. Furthermore, evaluating the validity of financial instruments such as blue bonds and sustainability-linked fundraising can augment measuring their efficacy in activating capital for ocean preservation and sustainable development.

Integrating blue finance into dynamic global financial markets showcases the need for state-of-the-art financial solutions and governing support to fasten sustainable ocean-based investments. Overall, this study contributes to the literature on the blue economy by associating the gap between financial markets and sustainability. The study's structured methodology for evaluating the economic importance of ocean-based industries helps in-depth understanding of the role of the blue sector in economic development. The paper is organized as follows. The remainder of the paper is as follows: Sect. 2 covers the literature review. Section 3 discusses Research Methodology. Section 4 presents Findings. Section 5 discusses the Conclusion.

2 Literature Review

The Blue Economy forms an important frontier in sustainable economic development. Recent studies have recommended the use of sector-specific financial instruments to quantify and analyze the economic impact of marine industries.

2.1 *Blue Economy and Financial Markets*

The Blue Economy covers shipping, fisheries, offshore energy, and marine infrastructure industries which play a crucial role in global trade and economic stability Juneja et al. (2021) By studying investment instruments such as maritime bonds and exchange-traded funds, we examined how financial markets help promote sustainable ocean-based industries. These instruments are in the research trend for economic analysis and investment decisions.

Stock market volatility in the maritime sector was evaluated due to COVID-19 pandemics and other major geopolitical events as a part of equity performance. The findings show that ocean-based industries are more sensitive to exterior shocks, necessitating solid risk management systems (Kamal et al. 2022).

Pragmatic evidence was offered on integrating marine logistics and coastal energy markets with financial market indexes (Ahmad et al. 2024). They emphasized the need for financial tools to calculate the economic importance of ocean industries, underlining the need for a Blue Economy Index (Jolliffe et al. 2021). A well-structured Blue Economy Index can help increase market efficiency by studying sector-specific dependencies.

2.2 *Use of Sector-Specific Index in Financial Markets*

The importance of sectoral indexes in analyzing market trends has been realized by the researchers. Fama and French used multi-factor asset pricing financial models to show that industry-specific indexes show better risk-adjusted returns and increase portfolio diversification (Lalwani and Chakraborty 2020). The importance of industry-based indexes in capital allocation, proving that sectoral indices may act as leading indicators for economic growth was explored (Hrazdil and Scott 2013). Green energy indices was examined, illustrating matches between sustainable investment and ocean-based financial models (Bei and Wang 2023). These studies confirm the importance of sector-specific indices in transparent market efficiency and portfolio management underlining the need for a Blue Economy Index.

2.3 Role of Vector Autoregression VAR in Financial Analysis

VAR model is an influential econometric tool for studying the alliance between financial variables. VAR model introduced by Sims 1980 supported the fact that traditional macroeconomic models were futile in studying dynamic relationships between variables in context of financial markets. This gap was by filled using VAR which integrating both present and past values in a multivariate system. VAR model was utilized in predicting commodity prices and the trends in financial markets where the usability of VAR in analysing the futures markets, trading volume, and open interest was emphasized which fits well in context of Blue Economy Indices (Füss et al. 2010; Labys 2017). The utility of impulse response functions (IRFs) and variance decomposition inside VAR models was explored, which are used to enumerate the effects of shocks on financial market variables (Uddin 2023; Shi 2023).

These existing literatures validate VAR modeling in financial market analysis, making it a suitable methodology for reviewing the Blue Economy Index.

2.4 Ensuring Stability and Stationarity in VAR Models

The inverse roots of the AR characteristic polynomial need to fall in the unit circle to approve model stationarity (Christiano and Eichenbaum 1990). Johansen (1991) encompasses this concept by presenting cointegration tests that help determine whether financial variables preserve long-term equilibrium relationships (Hjalmarsson and Österholm 2010). The stability test results established that the VAR model is well-worked and appropriate for predicting the trends in Blue Economy.

2.5 Implications for the Blue Economy Index

Ocean-based industries based financial market indices have grown recently. OECD (2016) showcases the role of financial innovation in the Blue Economy underlining the importance of well-designed indices in attracting sustainable investments (Mega and Mega 2016).

UNCTAD (2021) stresses that financial market contribution is vital for developing marine industries such as ports, logistics, and offshore energy (Garcia-Alonso et al. 2025). These studies highlight the utility of the Blue Economy Index as a trending financial instrument with practical insights.

3 Research Methodology

The research methodology section summarizes the methodology to examine the economic impact of ocean-based industries in the financial markets landscape. A Blue Economy Index is constructed using econometric techniques to investigate the influence of independent variables on future maturity prices.

Data Collection and Sample Selection: Daily values of the Blue Economy Index were obtained for 26 months from May 2022 to July 2024 with 520 observations with Future maturity prices as the Dependent variable and Open interest and trading volume were the proxy for Independent variables. The Blue Economy Index was comprised of stocks from six major sectors of the Blue Economy, warranting comprehensive industry coverage: Metals and Mining Industry: Vedanta Ltd, Hindalco Industries Ltd, NMDC Ltd. Logistics and Ports: Adani Port Limited, Gujarat Pipavav Port (GPP), Cochin Shipyard (CS). Offshore Energy and Marine Services: Reliance Industries, ONGC, Tata Power, Petronet LNG, Bharat Petroleum Corporation Limited (BPCL).

Shipping and Maritime Transport: Shipping Corporation of India (SCI), Great Eastern Shipping Company Ltd. (GESCL). Fishing and Aquaculture: Avanti Feeds Ltd. (AFL), Coastal Corporation Ltd. (CCL). Marine Engineering and Infrastructure: Garden Reach Shipbuilders and Engineers Ltd. (GRSE), Mazagon Dock Shipbuilders Ltd. (MDSL).

These stocks were selected based on their market relevance, sectoral representation, and data availability over the study period.

Dependent Variable: Future maturity prices of the Blue Economy Index, representing the anticipated value of the index based on its underlying stocks. Independent Variables: Open interest measures the total number of outstanding derivative contracts, reflecting market participation and liquidity. Trading volume represents the total number of shares or contracts traded in a given period, indicating investor activity and market sentiment.

To examine the interdependencies and dynamic relationships among these variables, the VAR Model was employed. Econometric Model and Analytical Techniques: Vector Auto-Regressive (VAR) Model: The VAR model is used to analyze time series data that involve multiple interdependent variables. This technique allows for the examination of both current and lagged relationships among variables, capturing the dynamic interactions between future maturity prices, open interest, and trading volume.

Granger-Causality Tests: To determine the causal relationship between variables, Granger-Causality Tests were applied. These tests assess whether past values of open interest and trading volume can predict future changes in the future maturity prices of the Blue Economy Index. H_1: Open interest does not Granger-cause future maturity prices. H_2: Trading volume does not Granger-cause future maturity prices.

Impulse Response Functions were used to analyze how future maturity prices react over time to unexpected shocks in open interest and trading volume. This technique helps understand the short-term and long-term effects of market movements on price dynamics. Variance Decomposition Analysis was used to quantify the contribution of each independent variable to the fluctuations in future maturity prices. This technique breaks down the percentage of variation in the dependent variable that can be attributed to each explanatory factor over different time horizons. Data Processing: The data was processed using advanced econometric tools to ensure accuracy and reliability. EViews software was used to check the robustness checks and additional econometric testing.

This study provides a novel perspective on the financial dynamics of the Blue Economy, contributing to both academic research and industry insights. The findings are expected to increase investor understanding of Blue Economy stocks and their market behaviour and Support policymakers in evaluating economic sustainability in ocean-based industries. The study also offers a framework for future research on sector-specific indices and financial market interactions. Applying a robust econometric methodology, this study aims to uncover key relationships between open interest, trading volume, and future maturity prices within the Blue Economy Index. Using VAR modeling, Granger-causality, impulse response functions, and variance decomposition ensures a comprehensive dataset analysis, providing valuable insights into the financial sustainability of ocean-based industries.

4 Findings

The Vector Autoregression (VAR) model examines the dynamic relationships between Future Maturity Prices (FMP), Open Interest (COI), and Trading Volume (CVP) by incorporating past values (lags) of each variable. It is clearly evitable from Table 1 that FMP(−1) has the strongest impact on its future values (0.8537, t = 14.44), indicating strong persistence.

COI(−1) has a moderate positive effect on itself (0.3153, t = 2.62), meaning past open interest influences future open interest. CVP(−1) significantly affects itself (0.4112, t = 3.37), suggesting strong momentum in trading volume. COI(−1) and CVP(−1) have weak effects on FMP, implying that past open interest and trading volume do not strongly predict future maturity prices. COI(−2) and CVP(−2) show weaker influences, suggesting that older lags have a diminishing impact. FMP is highly autoregressive, meaning its past values largely determine its future movements. COI and CVP influence themselves but have limited direct impact on FMP in the short term.

Autoregressive Inverse Root Characteristic Polynomial plot is used to evaluate the consistency of a VAR model to examine whether the estimated VAR system is stationary or not.

Interpretation of the Plot: The plot shows a unit circle (radius = 1) in the complex plane, with blue dots representing the inverse roots of the characteristic polynomial.

Stability Condition:

For the VAR model to be stable (stationary), all inverse roots must lie within the unit circle (i.e., their absolute values should be less than 1).

Table 1 Vector autoregression estimates

	FMP	COI	CVP
FMP (−1)	0.853662	0.333123	0.276184
	(0.05912)	(0.09330)	(0.09431)
	[14.4397]	[3.57046]	[2.92855]
FMP (−2)	0.032930	−0.128755	−0.169206
	(0.05979)	(0.09435)	(0.09537)
	[0.55079]	[−1.36461]	[−1.77417]
COI (−1)	0.054856	0.315272	0.130339
	(0.07628)	(0.12038)	(0.12168)
	[0.71915]	[2.61898]	[1.07117]
COI (−2)	−0.037729	−0.073999	−0.113638
	(0.07604)	(0.12001)	(0.12130)
	[−0.49616]	[−0.61662]	[−0.93680]

Source Author's work (obtained from Eviews)

Table 2 VAR lag order selection criteria

Lag	LogL	LR	FPE	AIC	SC	HQ	Lag	LogL
0	−1867.150	NA	0.298688	7.305275	7.330109	7.315010	0	−1867.150
1	−1308.718	1108.140	0.034924	5.159053	5.258389*	5.197993	1	−1308.718
2	−1281.359	53.97006	0.032507	5.087338	5.261176	5.155482*	2	−1281.359
3	−1277.365	7.831238	0.033150	5.106894	5.355234	5.204243	3	−1277.365
4	−1273.479	7.574698	0.033819	5.126871	5.449712	5.253425	4	−1273.479
5	−1252.127	41.36855	0.032227	5.078623	5.475966	5.234381	5	−1252.127
6	−1240.016	23.32461	0.031839	5.066467	5.538312	5.251431	6	−1240.016
7	−1221.754	34.95395*	0.030709*	5.030289*	5.576636	5.244457	7	−1221.754
8	−1215.353	12.17780	0.031025	5.040440	5.661288	5.283812	8	−1215.353

Source Author's work (obtained from Eviews)
LR: sequential modified LR test statistic (each test at 5% level), FPE: Final prediction error, AIC: Akaike information criterion, SC: Schwarz information criterion, HQ: Hannan-Quinn information criterion

Table 3 Root and modulus range

Root	Modulus
0.874838	0.874838
0.797993	0.797993
0.290134	0.290134
−0.215258	0.215258
−0.083785 − 0.015981i	0.085296
−0.083785 + 0.015981i	0.085296

Source Author's work (obtained from Eviews)

If any root lies on or outside the unit circle, the system is non-stationary, meaning the VAR estimates may not be valid.

All blue dots appear inside the unit circle, indicating that the estimated VAR model is stable.

This suggests that the future maturity prices, open interest, and trading volume relationships are well-behaved and can be reliably analyzed using the VAR framework.

The VAR model is correctly specified and stable, meaning that shocks in one variable will not cause an explosive effect in the system, and predictions based on this model are statistically sound. As per above discussion, the VAR model for Blue Economy Index has found to be stable.In the above figure, where future maturity prices, open interest and volume were the endogenous variables, no roots were outside the unit circle and VAR satisfied the stability condition. The correlogram results are mentioned in Fig. 2 ensure the stationarity.

Figure 2 contains multiple correlograms displaying autocorrelation of variables in different lags. Each plot shows blue bars showing correlation coefficients for these lags, with dashed lines highlighting confidence intervals. Substantial correlations outside the confidence bands advocate possible patterns in the data.

Figures 3, 4 and 5 describe significant results that provide insights into the relation within the Blue Economy Index framework. Figure 3 provides the Cholesky response matrix, which explains the impulse response of future maturity prices to surprises in open interest and option volume, stressing both short- and long-term effects.

Figure 4 shows the variance composition results, highlighting the proportion of forecast error variance credited to each independent variable, highlighting their explanatory power over time.

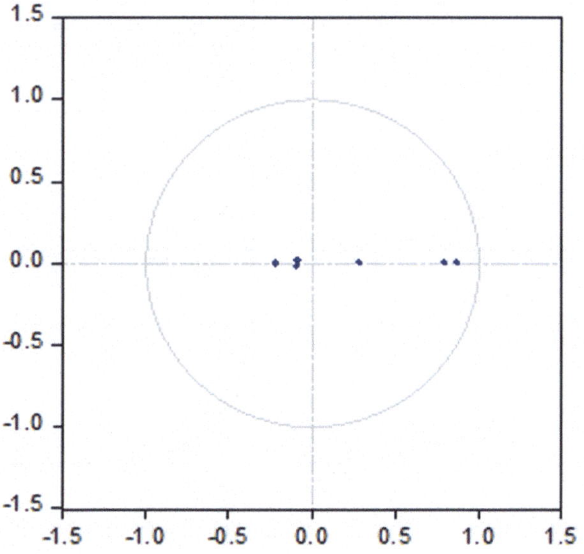

Fig. 1 Inverse roots of AR characteristics. *Source* Author's work (obtained from Eviews)

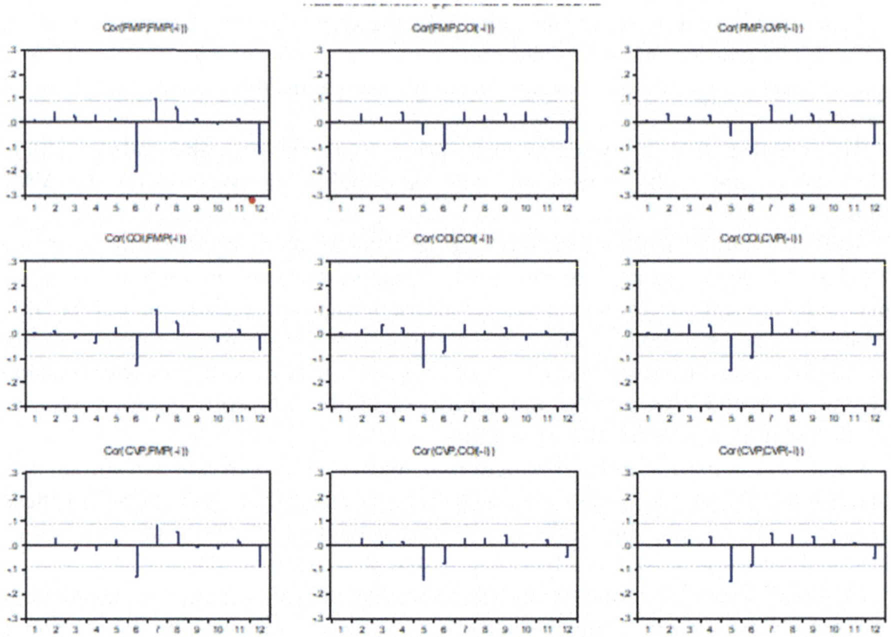

Fig. 2 Auto correlation results. *Source* Author's work (obtained from Eviews)

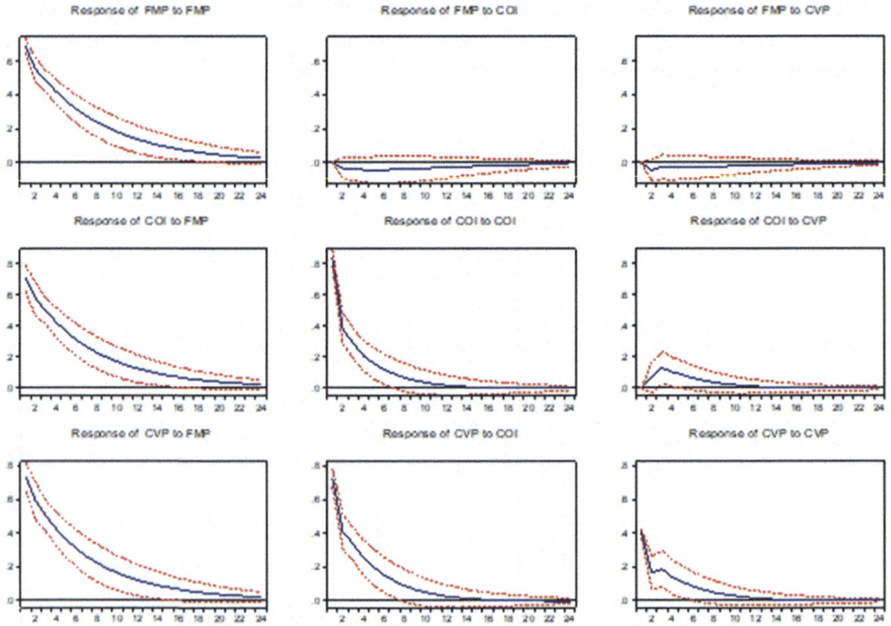

Fig. 3 Cholesky response matrix results. *Source* Author's work (obtained from Eviews)

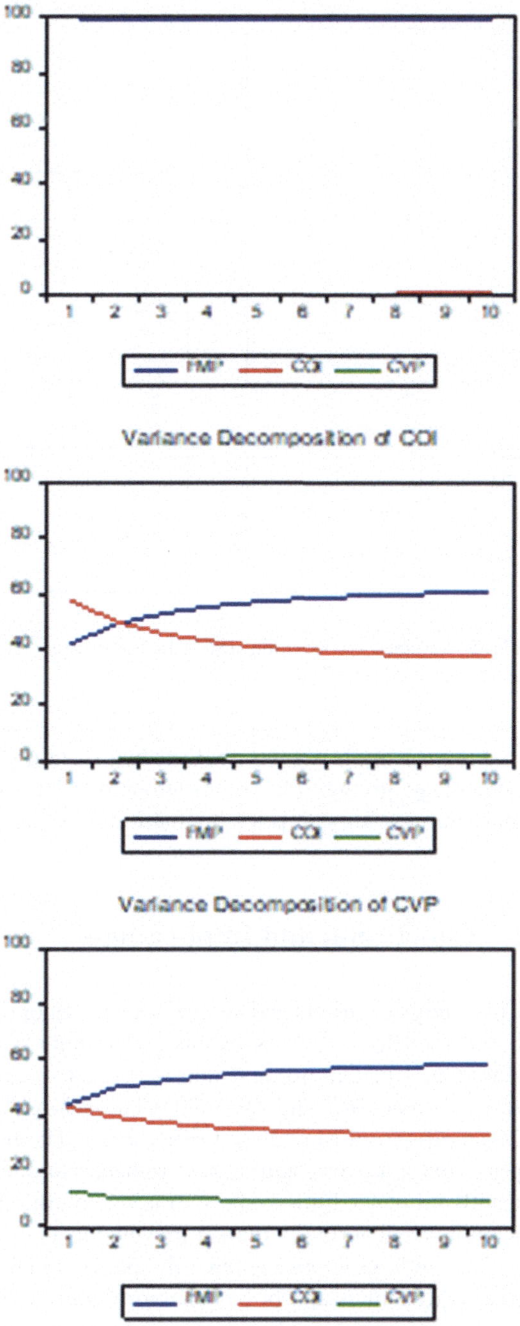

Fig. 4 Variance composition results: Cholesky response matrix. *Source* Author's work (obtained from Eviews)

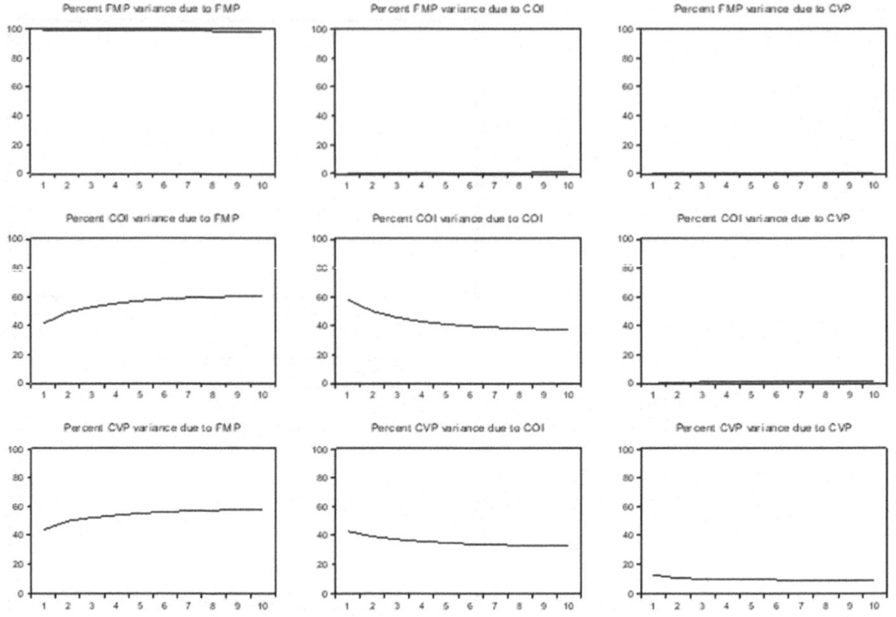

Fig. 5 Variance composition results in percentage: Cholesky response matrix. *Source* Author's work (obtained from Eviews)

Figure 5 improves this analysis by exhibiting the variance composition results in percentage, giving a clearer visualization of the relative contribution of open interest and option volume to price fluctuations.

5 Conclusion and Implications

This study constructs and analyzes a Blue Economy Index by integrating key ocean-based industries such as metals and mining, ports and logistics, offshore energy, shipping, aquaculture, and marine infrastructure. By applying a VAR model, the research examines the dynamic relationships between FMP, COI, and CVP over 26 months from May 2022 to July 2024. The findings suggest that future maturity prices exhibit strong autoregressive behavior, indicating that past price trends significantly influence future values. Trading volume shows consistency over time proving its more critical role in market movements than open interest.

The findings expose noteworthy patterns in the Blue Economy Index, concerning market dynamics and price behavior. Future maturity prices display autoregressive inclinations, implying historical price trends impact future trends. This underlines the consistency of price momentum in ocean-based industries. Trading volume outshines open interest in driving market movements emphasizing the importance of liquidity and trading activity in influencing price fluctuations, offering inputs for investors and policymakers engrossed on financial stability and strategic decision-making within the blue economy.

The study makes important implications in context of financial markets for market stakeholders Firstly, Blue Economy Index increases explicity in ocean-based industries setting a sector-specific benchmark index to measure market performance. Investors may use this index for portfolio management, risk measurement, and finding out trading opportunities in blue economy-based sectors. Secondly, the findings show the importance of liquidity measures in affecting the price movements. These are important for policymakers at the decision-making level for investments in the Blue Economy. Thirdly, the study provides the platform for future research in blue economy, inspiring the application of alternative modeling techniques, such as blockchain, algorithmic trading machine learning for enhancing predictive capabilities. Including macroeconomic variables such as oil prices and climate risk factors may increase the strength of future models.

This study is a novel contribution because it connects financial market analytics with the Blue Economy, providing inputs into sector-specific financial integration. The research shows how future maturity prices, open interest, and trading volume deal within these sectors highlighting market efficiency, price dynamics, and investment risks. By assimilating financial insights with blue economy sectors, the study increases the ability of investors, policymakers, and industry stakeholders to change targeted strategies for sustainable growth, risk management, and economic elasticity in ocean-based markets.

References

Ahmad MS, Fei W, Shoaib M, Ali H (2024) Identification of key drivers for performance measurement in sustainable humanitarian relief logistics: an integrated fuzzy Delphi-DEMATEL approach. Sustainability 16:4412

Ashley C, De Brine P, Lehr A, Wilde H (2007) The role of the tourism sector in expanding economic opportunity. John F. Kennedy School of Government, Harvard University Cambridge, MA

Bei J, Wang C (2023) Renewable energy resources and sustainable development goals: evidence based on green finance, clean energy and environmentally friendly investment. Resour Policy 80:103194

Boyd CE, McNevin AA, Davis RP (2022) The contribution of fisheries and aquaculture to the global protein supply. Food Secur 14:805–827

Burmaoglu S, Ozdemir Gungor D, Kirbac A, Saritas O (2023) Future research avenues at the nexus of circular economy and digitalization. Int J Product Perform Manag 72:2247–2269

Christiano LJ, Eichenbaum M (1990) Unit roots in real GNP: Do we know, and do we care? In: Carnegie-Rochester conference series on public policy, pp 7–61

Du Y, Liu Y, Hossain MA, Chen S (2022) The decoupling relationship between China's economic growth and carbon emissions from the perspective of industrial structure. Chin J Popul Resour Environ 20:49–58

Füss R, Adams Z, Kaiser DG (2010) The predictive power of value-at-risk models in commodity futures markets. J Asset Manag 11:261–285

Garcia-Alonso L, González-Laxe F, Sanchez RJ (2025) Sustainability, externalities and ocean grabbing: pressing challenges for maritime transport. Mar Policy 174:106598

Geng B, Wu D, Zhang C, Xie W, Mahmood MA, Ali Q (2024) How can the blue economy contribute to inclusive growth and ecosystem resources in Asia? A comparative analysis. Sustainability 16:429

Hjalmarsson E, Österholm P (2010) Testing for cointegration using the Johansen methodology when variables are near-integrated: size distortions and partial remedies. Empir Econ 39:51–76

Hrazdil K, Scott T (2013) The role of industry classification in estimating discretionary accruals. Rev Quant Financ Account 40:15–39

Jolliffe J, Jolly C, Stevens B (2021) Blueprint for improved measurement of the international ocean economy: an exploration of satellite accounting for ocean economic activity

Juneja M, De Souza C, Giriyan AL, Ganeshan S (2021) Contextualising blue economy in Asia-Pacific region. HD Lohano MB Maqbool.

Kamal MR, Chowdhury MAF, Hosain MM (2022) Stock market reactions of maritime shipping industry in the time of COVID-19 pandemic crisis: an empirical investigation. Marit Policy & Manag 49:1184–1199

Labys WC (2017) Modeling and forecasting primary commodity prices. Routledge

Lalwani V, Chakraborty M (2020) Multi-factor asset pricing models in emerging and developed markets. Manag Financ 46:360–380

Martínez-Vázquez RM, Milán-García J, de Pablo Valenciano J (2021) Challenges of the blue economy: evidence and research trends. Environ Sci Eur 33:61. https://doi.org/10.1186/s12302-021-00502-1

Martinez-Vázquez RM, Milán-Garcia J, de Pablo Valenciano J (2021) Challenges of the blue economy: evidence and research trends. Environ Sci Eur 33:61

Mega VP, Mega VP (2016) Resourceful and resilient cities and marine ecosystem services. Conscious Coast Cities Sustain Blue Green Growth Polit Imagin 63–106

Shah KU, Niles K (2016) Energy policy in the Caribbean green economy context and the Institutional Analysis and Design (IAD) framework as a proposed tool for its development. Energy Policy 98:768–777

Shi Y (2023) A new unique impulse response function in linear vector autoregressive models. Int Rev Financ 23:460–468

Spalding MJ (2016) The new blue economy: the future of sustainability. J Ocean Coast Econ 2:8

Uddin MJ (2023) Investigating the impulse responses of renewable energy in the context of China: a Bayesian VAR approach. Renew Energy 219:119485

Wellmer F-W, Becker-Platen J (2002) Sustainable development and the exploitation of mineral and energy resources: a review. Int J Earth Sci 91:723–745

Water Sports, Sustainability, and FinTech

Amitava Pal, Amruta Deshpande, and Kavita Mathad

Abstract The Global Market for water sports valued at approximately USD 40 Billion in the year 2023 would reflect a highly exponential growth in this forecast period ending 2032 at a CAGR of over 5%, mainly due to the popularity of this particular activity to enjoy as either for fitness purposes or as recreation for stress reduction. It explains how the water sports industry development under the sustainable growth of the blue economy through financial technologies works. Financed with the applications of blockchain and predictive analytics-based FinTechs and digital crowdfunding platforms, emerging eco-practices in the sports sector foster the development of initiatives for both sustainable sport and marine conservation activities. By involving sustainable finance within water sports, this sector would open the field of economic growth and contribute toward pressing global issues such as the health of oceans and, more importantly, environmental sustainability. This chapter covers FinTech's contribution to the widening opportunities in the water sports industry, mainly in ensuring ecological and Financial Sustainability while democratizing access to more water sports. With a perfect intersection of FinTech, Sustainability, and water sports, this chapter adds unique insights into comprehending resilient and inclusive strategies that benefit people and the planet.

Keywords Water sports · Blue economy · Marine conservation · Sustainability · Digital transformation · Blockchain · Fintech · Crowdfunding · Green finance

A. Pal (✉)
ICFAI Management School, ICFAI University Jharkhand and Associate Dean International Institute of Sports and Management, Mumbai, India
e-mail: research.amitava@gmail.com

A. Deshpande
Director Research, Indira School of Business Studies, Indira Group of Institutes, Pune, India

K. Mathad
Dean Presidency University, Bangalore, India

V. Sharma et al. (eds.), *FinTech for Sustainable Finance and a Well-Blue Economy*, World Sustainability Series, https://doi.org/10.1007/978-3-031-92390-6_18

1 Introduction

The global water sports products market is expected to grow due to increased interest in surfing, kayaking, paddle boarding, and snorkeling, as well as boosting demands for related gear and safety products. North America's immense water sports kingdom generates approximately $13 billion annually and is estimated to grow past $20 billion by 2032. Sizeable commercial water athletes wield roughly two-thirds control of this opportunity, organizing competitive contests and training programs that consistently lure substantial spectators and recruits. Recreational activities such as kayaking, sailing, and jet skiing have caught significant popularity's currents in recent years, granting physical exercise and stress-relieving escapes from daily drudgeries to novice dabblers and dedicated athletes across the vast dominion. This industry's momentum shows no indication of slowing as emerging technologies unveil novel ways to experience water's delights and its enormous potential for enjoyment, invitations neither the hemisphere nor its inhabitants show signs of declining. The seas remain open and welcoming to all who would experience water's timeless pleasures and restorative powers. Its primary contributors are Airhead Sports Group, Aqua Lung International, and Body Glove (Global Market Insights Inc. 2032). One can claim measurable benefits to mental health from engaging in water sports like swimming, kayaking, or paddleboarding. A 2022 Frontiers in Public Health study shows that aquatic exercise improves mood and decreases anxiety symptomatology (Tang et al. 2022). Involvement in group-based aquatic activities promotes community bonding. Studies in BMC Global Public Health have shown that communal group activities increase community resilience with increased interpersonal trust. The facts suggest that including water sports in wellness programs could enhance mental health and community well-being.

The blue economy primarily focuses on efficiently utilizing marine resources and has received massive contributions from the world of water sports. The United Nations Environment Programme (UNEP) estimates the annual value of this blue economy to be approximately US$ 2.5 trillion, making it the 7th largest global economy. Surfing, snorkeling, and many other ocean-based eco-friendly water sports activities create a significant source of high revenue-generating activities that indirectly support marine conservation. Some other nations that have integrated these practices into their blue economy initiatives include Costa Rica and the Maldives, which have created employment and generated funds (Sustainable Blue Economy n.d.). FinTechs are transforming how the water sports sector operates and interacts with the environment. Blockchain provides clean supply chains for green products to ensure that the raw materials, like recycled plastics, comply with environmental regulations. The BlueLedger platforms use blockchain technology to provide real-time financial statements on the economic and environmental impacts of decisions. With digital payment systems, access increases to the model of microfinance that allows involvement from the commonly underserved communities through water sports activities. The initiative needs to be scaled by policymakers and businesses, as the

European Commission's 2023 Blue Growth Strategy encourages. Ocean industries are pillars of a green transition (Sustainable Blue Economy n.d.).

The blue economy is a combination of all those economic activities regarding the oceans, seas, and coastal regions that strive for the sustainable exploitation of marine resources. It integrates economic growth with better livelihoods, employment generation, and ocean ecosystem protection (What is the Blue Economy? 2017; Admin 2023). Water sports serve as both recreational activities and significant economic drivers. Eco-tourism generates money that fuels conservation efforts within the marine environment. For instance, Costa Rica and the Maldives have combined surfing and snorkeling with a blue economy that is a beneficial source of employment and the means to raise money to restore the coral reefs. Environmental awareness and stewardship are the other areas through which water sports contribute to ecological responsiveness. Sailing harnesses wind energy, while windsurfing and wing foiling burn fewer fossil fuels, thus releasing fewer greenhouse gases (Warren 2015). The Blue Economy combines recreation and conservation and forms a symbiotic relationship between the enjoyment of human requirements and environmental preservation through sustainable practices for future generations using marine resources.

Water sports contribute to the blue economy through several economic, social, and environmental channels:

- **Economic Contributors**: Water sports events stimulate tourism, which promotes employment and income earnings. Surfing, kayaking, and sailing activities tend to boost tourism, thus increasing expenditure in the accommodation sector, foodstuffs, and other services. The Northern Forest Canoe Trail in the United States is associated with increased environmental awareness and benefits to communities and residents who stay near it (Zhang 2020).
- **Social Contributions**: Participating in water sports contributes to socialization with the community and social contact. Events and activities attract different segments of people for teamwork and cultural exchange. Social benefits from events and activities assist in enhancing the local society's development through social capital and community resilience (guoSerena 2023).
- **Environmental Contributions**: Water sports contribute to growing environmental awareness and promote the sustainable exploitation of marine resources. Participants often start caring for the aquatic ecosystems more profoundly and will become conservation champions. Moreover, the sustainable features of water sports include eco-friendly equipment, among other factors, which reduces pollution in water environments (Admin 2024).

However, these contributions could harm this industry's sustainability and inclusiveness,

- **Environmental Impact**—Pollution and habitat interference brought about in the aquatic world due to participating in water sports activities is a significant concern. The other one is equipment constructed of nonbiodegradable products, causing pollution that affects the environment and greenhouse emissions caused by water-based sports vehicles (Campos et al. 2024).

- **Funding Gaps**: A good amount of funding is in place for sports social inclusion initiatives, which just so happens to be the limitation of the cause of aquatic sports becoming inclusive to diverse groups. The feature of a shared lack of understanding of social inclusion in sports raises even more complexities concerning resource allocations (Saranova 2023).
- **Lack of inclusivity**: Water activities must be open to all ages, genders, capabilities, and economic statuses by implementing proper education and training with an equal opportunity policy (Richter 2024).

2 Research Objectives

One of the ways to reduce the environmental impact to a considerable level is through digital transformation by embracing FinTech applications as one of the solutions wherein all the stakeholders could enhance financial access and encourage sustainable activities, thereby contributing towards the environmental goals of the blue economy. FinTech means applying technology to transform and automate the delivery and usage of financial services. Applications include electronic payments, online lending, blockchain, and investment platforms to make operations in finance as easy and streamlined as possible.. FinTech innovations like digital payment systems and blockchain can streamline business, reduce carbon footprints, and make it accessible to diverse communities when it comes to water sports. For instance, FinTech solutions can create microfinance models for participation in water sports activities among the underserved population. FinTech would fill a critical financing gap in sustainable ocean initiatives such as marine conservation projects or eco-friendly water sports infrastructure. It would be about putting financial practices into environmental goals; this would make a sustainable blue economy.

The study focuses on two objectives:

- To address these environmental impact issues and discuss appropriate ways to resolve them.
- To study the role of Fintech in access to water sports and environmental sustainability

The following discussion traces actionable strategies for stakeholders, policymakers, businesses, and community leaders on how FinTech can enrich sustainability and sustain inclusion in water sports through sustainable practices, innovative funding models, and promoting inclusivity for a higher reach.

- **Sustainable Practices**: Using green materials and technology would reduce the impact of water sports on the environment. For example, using biodegradable equipment, that is, cleanliness in boating practice, can reduce harm to marine life and preserve aquatic ecosystems

- **Innovative Funding Models**: Blockchain is a financial technology that ensures transparency in supply chains and crowdfunding platforms and helps mobilize more resources to fund conservation efforts and community engagement.
- **Promotion of Inclusivity**: Create programs that target diversified groups and ensure that the water sports facilities are accessible equally to attract various populations. Hence, education or training needs to be provided, especially to the underprivileged, and implementing pro-diversity policies needs to be facilitated.

3 Water Sports Initiatives Contributing to the Blue Economy

Water sports activities largely contribute to economies' growth and push sustainable development within coastal and marine ecosystems. This growth is mainly due to employment, the enhancement of income-generating potential for tourism, and the responsible utilization of aquatic resources. The modern-day application of innovations such as AI, IoT, and FinTech pushes the boundary of environmental protection and gives financial access to the residents living in coastal communities. To quote a few,

- **Mediterranean Blue Economy initiates**: The Mediterranean region has pioneered the integration of water sports into blue economies. It visualizes the potential to achieve sustainable economic growth while preserving marine conservation. Countries in the area, including Spain, Italy, and Greece, have already initiated some initiatives. Such programs focus on activity promotion such as sailing, diving, and kayaking but do so in compliance with rigorous environmental regulations. In this respect, eco-friendly diving schools would teach visiting tourists about marine biodiversity. Sustainable marinas are environmentally friendly boating facilities that use clean energy, reduce waste, and protect the environment. They contribute to local communities by creating jobs and attracting international tourists (Richter 2024; Plan-bleu: Environnement et développement en Méditerranée 2021).
- **Floating Treatment Wetlands in South Asia**: Floating treatment wetlands (FTWs) have emerged as a revolutionary solution in Nepal, India, and Bangladesh to mitigate water pollution and create a sustainable water sports project. These wetlands, made of locally available waste materials like bamboo and recycled plastic, serve as cheap options for reactivating polluted lakes and rivers. FTWs work as a natural biofilter by removing any extra nutrients, heavy metals, and toxins, which ensures higher quality and more aquatic biodiversity in the water bodies. This success strengthens the statement that water-based recreation activities like rowing, kayaking, open-water swimming, and ecological restoration will likely create long-term environmental and economic benefits. This affordable and scalable platform (FTWs) has successfully been implemented in Nagdaha, Nepal, and urban waterways in the US, including Baltimore's Inner Harbor and Boston's

Charles River (Polluted lakes are being cleansed using floating wetlands made of trash 2024; Sailing Towards a Sustainable Future: The Rise of Green Maritime Facilities 2023).

- **National Aquatics Centre in Brisbane**: A proposed national aquatics center in Brisbane, almost at a staggering $650 million, is soon to become the ultimate destination in Australia for aquatic sports. This exciting venue will prepare Brisbane to host the games as part of the pre-Olympic games venture earmarked for 2032, accommodating one Olympic-sized swimming pool, world-class training facilities, and spectator-friendly infrastructure. The Institute will be the area for the development of high-quality Australian swimming and diving talent supported by high-technology equipment and coaching, sport science, programs of rehabilitation, and the training of world-class coaches on the job in preparation for athletes. It engages the broader community facilities for swimming courses, water safety education, and recreational activities in the water. It has a gigantic economic impact by boosting construction employment, facilities management, and sports tourism (Bold $650m Plan for National Aquatics Centre at Victoria Park n.d.).

4 Challenges and Opportunities

The water sports industry is exposing the water to many environmental-related challenges. Pollution and habitat destruction are significant among them. This could be through boating, jet skiing, fuel spills, waste dumping, and noise pollution, negatively impacting aquatic ecosystems. Most machine-intensive water sports cause damage to sensitive marine habitats through physical alteration and increased sedimentation, primarily to coral reefs and seagrass beds. Also to environmental problems, the industry faces fiscal and access-related issues. Besides environmental issues, the industry faces problems concerning money and access, including securing sufficient money for infrastructure investments, environmental activities, and community involvement (Fisher 2025). Likewise, the new capital investment for equipment, training, and access fees makes aquatic sports unattainable to disadvantaged communities and thus requires methods to make them accessible. Yet, with all these issues, the industry leaves enormous room for growth if it takes an eco-friendly approach: renewable energy for water transport or biodegradable materials. The other area is a technological advancement that includes artificial intelligence and data analytics to improve safety measures and optimize resource use in the industry. In wellness tourism, like adventure sports, there is the scope of upgrading the market by creating new products and services targeted towards other kinds of customer needs to make a market with growth (Austin 2025). Innovations, for instance, can stimulate sustainable development and address key environmental concerns and access issues in water sports industries.

5 FinTech and Sustainable Practices

FinTech has significantly impacted the water sports industry in the following ways: It simplifies the payment systems and streamlines transactions related to activities, equipment, and membership, thereby simplifying the processes as well as customer experience along with operational efficiency. The FinTech tools improve the water sports business budget; they will plan better financially, and cash flow will be managed appropriately, thus enhancing the venture's financial health with sustainable long-term outcomes (Patil 2024). Other FinTech applications, such as digital means of payments and microfinance programs, would provide better access to water sports as they make products more affordable and available to a wider variety of users. With FinTech penetrating every function within the sector, companies will become more efficient and inclusive and grow much faster than they did during the previous period. They would mirror more significant economic and technological changes.

Just like in FinTech, the uptake of the technology in water sports has seen the assimilation of many innovative tools for improving operational efficiency, sustainability, and user experience. Blockchain is a decentralized and immutable ledger system, ensuring that all transactions are transparent and accountable. Blockchain traces the environmental impact by proving it and ensures that the event's carbon footprint aspects respect the set sustainability goals. It makes it safe to register for events and buy merchandise. Research has indicated that blockchain-based decision support systems for water management can enhance efficiency and transparency in any water project (Alsharabi et al., Using blockchain and AI technologies for sustainable, biodiverse, and transparent fisheries of the future). Predictive analytics involves data, statistical algorithms, and machine learning techniques. This can be used to predict equipment and facility demands with optimized resource distribution in water sports activities. It helps to enhance safety as it foretells weather and water quality problems that lead to proactive safety measures. Predictive analytics is used to plan flawless events. The history of past events is examined to predict the upcoming trends and possible changes in the preparation of meeting the trend's demand.

5.1 Fintech and Financial Funding

Crowdfunding is one of the fastest-growing sources of funding for grassroots initiatives in water sports. Crowdfunding encourages small participatory contributions from huge populations, which support community-based projects funded by the contributions from local water sports events, facility upgradation, and environmental conservation programs. In this context, crowdsourcing engages enthusiasts by involving them in the success of a water sports development program, thereby facilitating financial accessibility for projects that would otherwise be beyond the reach of conventional financing. This study examines blockchain-based decentralized crowdfunding platforms, clarifying how the technologies may make funding ventures more transparent and efficient (Winning With Digital Payments in Sports Industry n.d.). The increasing adoption and introduction of digital payments in the water sports industry enhances convenience and accessibility by accelerating financial transactions, making them faster and safer than traditional channels. For instance, they facilitate paying for the rental of water sports equipment, entry payments to events, and merchandise purchases. The uptake of e-payments makes access possible since subscribers from different geographic locations can now join without necessarily using physical means of money. The systems further enhance financial handling since they offer detailed transaction reports, thus enabling businesses in financial planning and analysis. In this regard, digital payment systems can streamline and improve engagement in the water sports sector. Therefore, through these FinTech solutions, the water sports industry can further increase sustainability, strengthen its in-house performance efficiency, and provide an advanced experience for those engaging and interacting with its stakeholders.

5.2 *Promoting Environmental Sustainability*

FinTech plays a massive role in making the environmental footprint of water sports minimal through the efficiency of sustainable practice, providing transparency, and creating green initiatives. The Green FinTech companies are offering their services through green digital banking and payments, providing the service of digital banking focusing on sustainability; therefore, these services enable water sports companies to adopt a more environmentally conscious approach to their finances, ensuring that environmental accountability is integrated into their financial operations. Also, sustainable investment and asset management services provide data and analysis of the companies' environmental and social responsibility against Environment, Social and Governance (ESG) criteria. This makes investments in water sports' infrastructure and operations sustainable for long-term ecological balance (Syedhasnaatabbas 2024). The second FinTech-driven initiative is green crowdfunding and lending, which now finances directly ecologically focused projects. Other FinTech applications are nearly the same regarding the protection of transparency by blockchain technology in water sports event applications on environmentally friendly sources because the stakeholders can trace the carbon footprint for the respective event and whether sustainable ways are used. Predictive analytics optimises resources by providing data-informed insights to minimize waste and maximize the efficiency of water sports activities. Crowdfunding inspires grassroots water sports because it has channels and platforms through which a community can raise funds for conservation projects, eco-friendly infrastructure, and events. Electronic payment systems are available since they transform microfinance models by which the deprived can involve themselves in water sports. This means the monetary obstacles that would have barred people from involving themselves in eco-friendly water sports are not problems. These FinTech solutions help water sports reduce their operations to be significantly more sustainable while being environmentally friendly and inclusive globally.

5.3 *Democratizing Access to Water Sports*

Water sports should be accessible equitably for the betterment of the society. FinTech's involvement in funding and access matters is crucial as it fosters social equity and inclusion. One of the key ways that FinTech has dismantled financial barriers is through microfinance and peer-to-peer lending at Kiva (Home - Kiva Sports 2024), which provides individuals with loans intended for personal development, such as those for water sports. Financial empowerment meant community members acquired the capacity to invest in their interests and skills. For instance, digital payment systems such as M-Pesa (What Is M-PESA? n.d.) make it easier to transact because they do not need the traditional banking infrastructure. Crowdfunding sites like GoFundMe (GoFundMe | the #1 Crowdfunding and Fundraising Platform n.d.) have allowed individuals and organizations to crowdfund money for water sports projects. However, in this aspect, social integration enhances the relationship between individuals and creates social avenues for interactions among people. With FinTech solutions, stakeholders can make a water sports environment more inclusive. Financial restrictions cannot limit anyone from engaging in and enjoying these activities for their sense of belonging and well-being in the community.

Beautiful Ocean Coin (BOC), launched in October 2024, uses blockchain for transparency and funding in ocean conservation. Decentralizing this platform ensures that every donated amount is traceable and constructively used, thus increasing the trust of the donors and other stakeholders. This innovation strategy creates more financial responsibility and catalyzes further participation in the marine conservation business (Beautiful Ocean Coin 2024).

The other significant activity in which WWF collaborates with BCG Digital Ventures is the OpenSC Project. OpenSC uses blockchain to track fish locations to ensure that these routes are not used for water sports activity. Beyond that, blockchain has a critical role in ocean cleanup through traceable and credible recording and reporting of the collection and recycling of plastics from the sea. It will ensure that systematic documentation of cleanup efforts occurs, and the organizations can be held accountable for encouraging further participation in environmental restoration activities. Crowdfunding platforms revolutionize the modalities of arrangement for funding into community water sports programs as they allow the donor to engage with their project.

The Sportfunder was the first to raise million-dollar funds to develop water sports after 2012. It has thus opened up this opportunity to fund developing athletes and grassroots community organizations that might help augment participation in and sustain water sports (Loukoianova et al. 2024). MAKEACHAMP, a sports crowdfunding platform, has successfully assisted athletes and teams involved in water sports in raising funds for their respective activities. MAKEACHAMP is crucial because having passionate individuals and a dedicated group of backers connected ensures that no financial barriers halt the development of this talent in water sports (https://sportfunder.com/). Crowdfunding in water sports allows investors to support maritime ventures, boat designs, and marine conservation projects. Beyond economic aid, it provides the feeling of belonging among the fans, who then feel they have a more substantial backing for activities on sustainable nautical innovation.

6 Water Sports and Sustainable Development Goals (SDGs)

Water sports activities will give FinTech a unique contribution to various Sustainable Development Goals by encouraging inclusive, sustainable, and environmental conservation through innovative financial solutions.

- **Sustainable Cities and Communities (SDG 11)**: FinTech then pays back cities for their investment in water sports infrastructure into sustainable water by investing in 'green' facilities and responsible consumption and production patterns. Initiatives in FinTech's opportunities, like the green bonds and impact investing platforms, serve as conduits for investors to secure funding for sustainable water sports centers. These practices make cities safe and sustainable. Crowdfunding allows local communities to raise funds for sporting activities related to green water sports involving full participation with no environmental losses and ensures accessibility to all (Graham et al. 2017)
- **Life Below Water (SDG 14)**: FinTech projects include ocean resource conservation and harvesting through responsible means. FinTech has facilitated crowdfunding websites and electronic payment systems to finance ocean cleanup initiatives, coral reef restoration, and plastic waste management. Blockchain technology is used for transparency of funding for marine conservation activities. The websites encourage water sports users to contribute to protecting marine ecosystems while ensuring the long-term sustainability of aquatic recreational activities (Goal 14 - Life Below Water | the World's Largest Lesson 2022).

FinTech solutions have made it possible to align the SDGs for the water sports industry with the goals of economic inclusion, sustainable urban development, and marine conservation. FinTech is transforming the water sports space into a friendly industry that promotes this venture as inclusive, environmentally conscious, and socially responsible (Luers and Whieldon, BigFintechs and their impacts on sustainable development).

- **Use of Data Science**: Data science is the new revolution for sports since it enables teams to analyze massive data sets for performance improvement, strategy development, and injury prevention. Football analytics firms help a club find an undervalued player, which leads to successful acquisitions. The need for interdisciplinary education is growing due to changes in the landscape of ocean industries such as offshore energy and aquaculture, which require knowledge from various disciplines to address complex problems in ways that sustain the marine environment (DiscoverDataScience.org, 2023).

- **Artificial intelligent management in water sports training**: Artificial intelligence can help optimize the water conditions and performance of athletes engaged in water sports. AI-based models could predict weather patterns so sailors can plan their competition accordingly. According to the article by ACS ES&T Water, AI, and IoT enhance the activities involved in the monitoring and management aspects of the assessment of water quality, which helps maintain optimal water conditions for sports (Das et al. 2025).
- **Digitalization of water sports facility management**: The water sports facilities will implement all digital technologies, including Smart Water Metering and Supervisory Control and Data Acquisition (SCADA) systems. Real-time monitoring and control of the quality and distribution of the water will be in place for the athletes and events. The article in Frontiers in Water published a review of implementing intelligent systems, such as IoT and blockchain, in urban water management that can apply to water sports facilities (Asgari and Nemati 2022).
- **Green digital banking and payments for water sports events**: FinTech innovations such as digital wallets and blockchain would make the payment system of the water sports event seem seamless, thereby enhancing the experiences of its users and the management of finances. According to Global Brands Magazine, FinTech is revolutionizing sports and entertainment, thus bringing about more fluidity in payments and engaging more fans. This is relevant to water sports events (De Lorenzo 2024).
- **Sustainable investment in water sports infrastructure**: FinTech platforms could serve as channels for investing in green water sports infrastructure. For instance, crowdfunding can finance aquatic centers that utilize eco-friendly renewable sources or energy-based facilities. As indicated by the IMF, climate fintech is the utilization of financial technologies to solve the challenges that face climate change, which may align with sustainable investments in water sports (Loukoianova et al. 2024).
- **IoT for water sports equipment and safety monitoring**: IoT in water sports equipment will enable safety and performance monitoring. For instance, the smart sensors will feed live data on the state of the equipment and the athlete's biometrics. A study in Expert Systems explains how IoT can be integrated with blockchain technology in agricultural field monitoring and applied similarly to tracking in the water sports environment (Zeng et al. 2021).
- **Blockchain for water sports transparency and sponsorships**: The sports industries implement technology to mitigate sustainability challenges, mixing new solutions to reduce environmental impacts (Pratiwi et al. 2024). Blockchain will ensure transparent sponsorship deals, equipment sourcing, and athlete contracts in the water sports industry. This shows that blockchain can manage complicated, multi-stakeholder situations like those in water sports. Such solutions in FinTech will help increase the sustainability of the water sports industry while simultaneously increasing efficiency by providing more excellent safety to athletes, organizers, and spectators.

The interconnectivity of FinTech with sports, sustainability, and the environmental sector opened up innovative synergies that produced positive outcomes in all sectors involved. Value additions are one-of-a-kind, but they work together to solve complex problems. FinTech can track or balance carbon footprints using financial technology from sports organizations to track and lessen environmental impacts. The United States Tennis Association has developed high-tech weather forecasting systems to prevent climate-related setbacks that make mega-events such as the US Open relatively sustainably conducted business activity. Further, technologies have reshaped event management with sustainability in mind; the Seattle Climate Pledge Arena has become a success in being trash-free by embracing renewable energy and reducing waste, among other things. FinTech fosters fan participation in sustainability issues since it enables carbon offsetting and energy usage tracking while equipping the fan bases with transport data and the entire environmental footprint. Collaboration between the FinTech, sports, and environment sectors is crucial as they integrate their expertise to develop products and services that benefit all parties involved (Stupicic 2023). Sharing resources and expertise enhances sustainability, as FinTech solutions improve resource usage efficiency with minimal environmental impacts on sports sustainability projects. Collaborations bring market expansion through market access and potential customer base access to the brand, improving brand reachability and offering alternative means for access to numerous audiences.

7 Role of Stakeholders in Driving Change

The success of technology-based sustainability will need active engagement from various stakeholders.

- **Academic scholars and researchers**: Integrating environmental science, technology, and event management can develop more proficient professionals who can effectively handle current problems. This can entail integrating knowledge about marine biology and sustainable event management to organize celebratory events for water sports and the marine environment. Interdisciplinary learning contributes to critical thinking, adaptability, and a holistic approach that is pivotal in this dynamic industry.
- **Industry leaders**: Industry leaders, such as corporations and their business executives, are the pivot for integrating this sustainability into practical applications. By embracing green technologies, investing in renewable energy sources, and developing industry standards for sustainability, they promote innovation and competition for green products and services. Through active engagement, they can assist in altering supply chains, rendering production processes more responsible, and setting an example for other organizations to become sustainable business practices rather than ethical decisions.

- **Policymakers and regulators**: Governments offer structures that encourage the use of green technology through rules, subsidies, tax advantages, and financing of research and development. As a result of strict environmental regulations and help turn industries into sustainability models, the policy is likely to speed up the spread of green innovations to most areas (Silva et al. 2019). Policymakers should ensure evasion-free regulated trading carbon credits and grant tax credits and subsidies to companies adopting blockchain-based sustainable sourcing, AI-backed marine conservation tools, and cashless digital platforms for eco-tourism.
- **Consumers and civil society**: Consumers and civil society significantly affect sustainability efforts through what they buy and support. Increasing public consciousness about the environment has sparked increased demand for greener products and ethical business activities. Businesses revise their strategies when consumers choose greener alternatives. Civil society helps by scrutinizing corporate performances and holding such corporations accountable by promulgating a more vigorous environmental policy. Civil societies are the foundation for active citizenship and collective actions that shape market trends, encouraging industries to adopt more environmentally friendly practices (Abnett and Blenkinsop 2025).

8 Future Directions

AI and IoT have revolutionized water sports as an impetus toward helping within a much broader scope of a blue economy that helps with these new technologies about improving data-driven decision-making; it provides new ways toward a sustainability effort coupled with efficient resource management of oceans. The Internet of Things has enabled real-time monitoring that reports temperature fluctuations and water-body pollution with IoT devices, underwater sensors, intelligent buoys, and networked drones. For ocean conservation, this integration of IoT is highly essential for promoting ocean health because improved strategies for more accurate and proactive management of environments will be adapted.

FinTech fosters environmental responsibility, financial inclusion, and the economic resilience of a blue economy, all of which contribute to environmental protection. For instance, the RFI Foundation organized an innovation contest, providing an opportunity for the exemplary FinTech innovator to assist banks in enhancing the financial inclusion of economic growth in coastal regions. FinTech plays a significant role in providing underrepresented populations in coastal communities access to financial resources. Mobile banking, digital payment platforms, and blockchain-based financial solutions fill the gap in accessing secure and accessible financial services. The integration of FinTech into the blue Economy Economy will enhance economic resilience, assist local communities with business-related issues, and bolster sustainable living structures along the coasts. This will combine AI, IoT, and FinTech, transforming the future of water sports and blue economies. AI and IoT will continue to improve environmental monitoring and conservation practices, while

FinTech will develop tools to provide financial services for sustainable development and economic inclusion.

Call to Action

Investments in environmentally friendly FinTech solutions safeguard marine ecosystems and help coastal communities achieve economic resilience.

- **Develop Smart Financial Platforms**: The time has come for financial institutions to invest in AI-powered FinTech solutions engineered to comprehend and counter climate change effects on marine industries. These include analyzing weather patterns, rising sea levels, and economic risks with possible policy and investment decisions.
- **Encourage green bonds and sustainable financing**: Green bonds should be encouraged among the funding areas of ocean protection, marine biological diversity, and sustainable aquaculture projects with the government and private funders.
- **Transparency into Carbon Credit Trading**: Blockchain and AI can even make carbon credit trading more transparent and more accountable—to the point where investments spur carbon emission reductions in maritime industries.
- **DeFi for the Coastal Communities**. DeFi can deliver low-cost, decentralized financial services to small-scale fishers and microenterprises along the coast, opening up their credit, insurance, and investment opportunities.
- **Cross-industry partnerships**: Investment collaborations between oceanography research institutions to produce real-time tools that indicate ocean health and biodiversity conditions. Tie-ups for digitalizing the water sports industry include green ticketing, loyalty programs, and donations to Marine conservation. The government could play an active role through a public–private partnership (PPP) model and promote a customized Fintech to preserve marine resources.
- **Support Startups**: Industries investing in startups can support startups that work on advanced digital and sustainable solutions for tracking and optimizing water sports event logistics, financial logistics, fundraising, and monitoring digital platforms. The industry should promote FinTech Insurance, which covers weather-risk insurance products that protect fishing, water sports, and water sporting marine industries from weather-induced disasters and economic damage.

9 Conclusion

Integrating FinTech in water sports and the broader blue economy offers transformational opportunities for growth and environmental stewardship. Because FinTech solutions like blockchain make things more efficient and more apparent, they help these areas get access to finance in the long run. They support the move to a "blue economy" by encouraging green investments and ways of doing things. With FinTech integrated into the emerging technologies of AI and IoT, much more significant optimization in resource management is expected, along with environmental monitoring and driving innovation in water sports and ocean conservation. The blue economy is a significant one, promising to be highly favorable for FinTech growth, further avenues of development, and protection. Sustainable, inclusive, and innovative solutions between academics, industry leaders, and policymakers require stakeholder cooperation to unlock the full potential of this. Wherein the oceans and waterways are healthy and whole for generations to come, perfect harmony between the growth of economic activities and sustainability in the future will be fulfilled by developing it from within the powers of FinTech.

10 Declaration

The authors developed the contents of this paper entirely on their own, without any support from generative AI or a large language model. We created all the information, analysis, and insight individually to avoid plagiarism and uphold integrity and originality. We have provided sources and their appropriate credit, backed up by citations. Grammarly was used to ensure precise English language use and lucid communication.

References

Abnett K, Blenkinsop (2025) Europe's plan to regain a competitive edge for industries. https://www.reuters.com/world/europe/europes-plan-regain-competitive-edge-industries-2025-01-29/

Admin (2023, May 29) Water sports and sustainability - Wind&Friends. Wind&Friends. https://windfriends.com/en/water-sports-and-sustainability

Admin (2024, July 5) The environmental impact of water sports: how to minimize your footprint. A&M Water Sports. https://amwatersports.ca/blog/water-sports/the-environmental-impact-of-water-sports-how-to-minimize-your-footprint/

Alsharabi N, Ktari J, Frikha T, Alayba A, Alzahrani AJ, Jadi A, Hamam H, Using blockchain and AI technologies for sustainable, biodiverse, and transparent fisheries of the future. J Cloud Comput Adv Syst Appl

Asgari M, Nemati M (2022) Application of distributed ledger platforms in smart water systems—a literature review. Front Water 4. https://doi.org/10.3389/frwa.2022.848686

Austin (2025) Key fintech trends to watch in 2025 - global brands magazine. https://www.globalbrandsmagazine.com/key-fintech-trends-to-watch-in-2025/

Bold $650m plan for National Aquatics Centre at Victoria Park (n.d.) https://www.couriermail.com.au/news/queensland/incredible-plan-to-turn-brisbane-into-a-global-aquatic-powerhouse/news-story/b3990c1107a7603300e38388ebf349d5

Campos MJ, Oblak VP, Massart A, Ljubotina P, Perényi S, Farkas J, Sarmento H, Doupona M (2024) Listening to stakeholders' voices on funding social inclusion in sport for people with disabilities—proposal for criteria. Sports 12:147

Coin BO (2024) Beautiful ocean coin: revolutionizing marine conservation through blockchain Technology. https://www.globenewswire.com/news-release/2024/10/18/2965668/0/en/Beautiful-Ocean-Coin-Revolutionizing-Marine-Conservation-Through-Blockchain-Technology.html

Das S, Khondakar KR, Mazumdar H, Kaushik A, Mishra YK (2025) AI and IoT: supported sixth generation sensing for water quality assessment to empower sustainable ecosystems. ACS ES&T Water.https://doi.org/10.1021/acsestwater.4c00360

De Lorenzo M (2024, October 3) Revolutionizing sports: the power of financial technology sponsorships in global sports. https://www.linkedin.com/pulse/revolutionizing-sports-power-financial-technology-marco-de-lorenzo-429rf/

DiscoverDataScience.org. (2023, May 4). Data Science and Sports | Impact of Big Data & Data Analytics. https://www.discoverdatascience.org/industries/sports/

Fisher, G. (2025, April 1). Obstacles & Barriers to be aware of when running a watersports business - TripShock Partners. TripShock Partners. https://partners.tripshock.com/obstacles-barriers-to-be-aware-of-when-running-a-watersports-business/

Fundraising for sports - crowdfunding athletes & teams. https://sportfunder.com/

Global Market Insights Inc.: Water sports products market to hit USD 58.5 billion by 2032, says Global Market Insights Inc. https://www.globenewswire.com/news-release/2024/10/15/2963055/0/en/Water-Sports-Products-Market-to-hit-USD-58-5-billion-by-2032-s

Goal 14 - Life Below Water | The World's Largest Lesson. (2022, March 22). The World's Largest Lesson. https://worldslargestlesson.globalgoals.org/goal/life-below-water/

GoFundMe | the #1 crowdfunding and fundraising platform. (n.d.). https://www.gofundme.com/

Graham M, Hjorth I, Lehdonvirta V (2017) Digital labour and development: impacts of global digital labour platforms and the gig economy on worker livelihoods. Transf Eur Rev Lab Res 23:135–162. https://doi.org/10.1177/1024

guoSerena (2023, October 18) The environmental impact of water sports and how to minimize it: navigating the waves responsibly. SUMARPO_EU. https://eu.sumarpo.com/blogs/news/the-environmental-impact-of-water-sports-and-how-to-minimize-it-navigating-the-waves-responsibly?srsltid=AfmBOoo

Home - Kiva sports (2024, December 4) Kiva Sports. https://kivasports.net/

Loukoianova E, Natalucci FM, Wang YD, Kanada S (2024) Fintech applications for boosting climate finance. IMF eLibrary.https://doi.org/10.5089/9798400293740.066.A001

Luers A, Whieldon E, Technical paper 1.1: BigFintechs and their impacts on sustainable development – annexes, pp 1–6

Nik, Nik, Nik, Obstacles & Barriers to be aware of when running a watersports business - TripShock partners. https://partners.tripshock.com/obstacles-barriers-to-be-aware-of-when-running-a-watersports-business/

Patil NSB (2024) Using blockchain for water rights in irrigation management. Water Irrig Rev 44:01–06. https://doi.org/10.52783/iwi.v44i1.82

Plan-bleu: Environnement et développement en Méditerranée (2021, October 12) Blue economy in the Mediterranean: case studies, lessons and perspectives. Plan-bleu: Environnement Et Développement En Méditerranée. https://planbleu.org/en/publications/blue-economy-in-the-mediterranean-case-studies-lessons-and-perspectives/

Polluted lakes are being cleansed using floating wetlands made of trash (2024) https://www.globalplasticaction.org/resources/publications/polluted-lakes-are-being-cleansed-using-floating-wetlands-made-of-trash/ffbbf641614005df504fef1d138c2bdbaa119627

Pratiwi RAI, Fani LA, Kusasi F (2024) Blockchain technology in fisheries industry: a systematic literature review. BIO Web Conf 134:05004. https://doi.org/10.1051/bioconf/202413405004

Richter H (2024, July 25) Polluted lakes are being cleansed using floating wetlands made of trash. WIRED. https://www.wired.com/story/planet-pioneers-nagdaha-small-earth-nepal-soni-pradhanang-ftws-floating-treatment-wetland-systems-water-cleaning-pollution/

Sailing towards a sustainable future: the rise of green maritime facilities (2023, September 28) WSP. https://www.wsp.com/en-kw/insights/sailing-towards-sustainable-future-rise-of-green-maritime-facilities

Saranova I (2023, September 13) Navegando hacia un futuro más verde. Forbes España. https://forbes.es/opinion/335012/navegando-hacia-un-futuro-mas-verde/

Da Silva LM, Bitencourt CC, Faccin K, Iakovleva T (2019) The role of stakeholders in responsible innovation: a meta-synthesis. Sustainability 11:1766. https://doi.org/10.3390/su11061766

Stupicic M (2023, October 11) Cross-Industry collaboration in sports. SportsHack. https://www.sportshack.io/cross-industry-collaboration/

Sustainable Blue Economy (n.d.) https://www.unep.org/topics/ocean-seas-and-coasts/ecosystem-based-approaches/sustainable-blue-economy

Syedhasnaatabbas (2024) Blockchain in ocean cleanup: community-driven solutions for marine conservation. https://medium.com/%40syedhasnaatabbas/blockchain-in-ocean-cleanup-community-driven-solutions-for-marine-conservation-cda9a77af30b

Tang Z, Wang Y, Liu J, Liu Y (2022) Effects of aquatic exercise on mood and anxiety symptoms: a systematic review and meta-analysis. Front Psychiatry 13. https://doi.org/10.3389/fpsyt.2022.1051551

Warren N, Shimoda R, Waters C, Mellen L, Wrede E (2015) An economic argument for water trails

What is M-PESA? (n.d.) [Video]. Vodafone.com. https://www.vodafone.com/about-vodafone/what-we-do/m-pesa

What is the Blue Economy? https://www.worldbank.org/en/news/infographic/2017/06/06/blue-economy

Winning with Digital Payments in Sports Industry (n.d.) https://www.kashier.io/blog-posts/winning-with-digital-payments-in-sports-industry

WWF-Australia and OpenSC - WWF-Australia|WWF-Australia and OpenSC|WWF Australia. https://wwf.org.au/get-involved/panda-labs/wwf-australia-and-opensc/

Zeng H, Dhiman G, Sharma A, Sharma A, Tselykh A (2021) An IoT and Blockchain-based approach for the smart water management system in agriculture. Expert Syst 40. https://doi.org/10.1111/exsy.12892

Zhang B (2020) The social benefits of water sports events and their impact on environmental pollution. J Coast Res 104. https://doi.org/10.2112/jcr-si104-020.1

Green Finance: A Bibliometric Review

Marc François Richter, Andreia de Bem Machado, João Rodrigues dos Santos, and Antònio Miguel Pesqueira

Abstract The global effort to transform society into a more sustainable world will require trillion-dollar investments over the years, a challenge that is directly related to the role of green finance. Green finance, also known as sustainable finance, refers to investment and lending practices that seek to support projects and activities that have positive environmental and social benefits. This type of financing is aligned with the goal of promoting sustainability and mitigating negative impacts on the environment. More efficient and transparent governance needs to be built, in line with ESG (Environmental, Social and Governance) principles. In this context, the objective of this chapter is to explore and outline the term sustainable finance, also known as green finance, in the light of a bibliometric review. To do this, a bibliometric review was carried out using two databases: Web of Science and Scopus, with the purpose of addressing the following questions: What is green finance? What are the benefits for companies? The results show that the benefits of financing are linked to the adoption of sustainable practices with optimized use of resources like energy and water, reducing expenses over time. These results highlight the theoretical and practical implications for the field of sustainable finance, underlining its relevance in promoting a stronger and more sustainable global economy that prioritizes environmental responsibility, as well as its strategic role in the formulation of innovative business policies and practices.

Keywords Green financing · Corporate sustainability · ESG governance · Bibliometric review · Resource efficiency · Sustainable investments

---

M. F. Richter (✉)
State University of Rio Grande Do Sul (Uergs), Porto Alegre, RS, Brasil
e-mail: marc-richter@uergs.edu.br

A. de Bem Machado
Federal University of Santa Catarina (UFSC), Florianópolis, SC, Brasil

J. R. dos Santos
Instituto Superior de Educação E Ciências (ISEC Lisboa), Lisboa, Portugal

A. M. Pesqueira
Instituto Universitário de Lisboa (Iscte), Lisboa, Portugal

V. Sharma et al. (eds.), *FinTech for Sustainable Finance and a Well-Blue Economy*,
World Sustainability Series, https://doi.org/10.1007/978-3-031-92390-6_19

1 Introduction

Shifting towards a sustainable development model is among the major challenges confronting modern society. Faced with growing environmental pressures, social inequalities and the need for resilient economic growth, sustainable finance has emerged as an essential tool for making this transformation possible. The role of finance is structural in the global economy, providing the necessary resources to drive social and economic progress. However, traditional financing models do not always take into account long-term environmental and social impacts, and can exacerbate problems such as environmental degradation and social inequality (Ng et al. 2021).

In this context, sustainability becomes a fundamental principle, incorporating economic, environmental, and social dimensions into the decision-making process. The central idea of sustainability is to fulfill present needs while ensuring that future generations can meet their own, requiring a profound restructuring of economic and financial systems. Sustainable financing responds to this need by directing investments towards projects with positive environmental and social benefits, as well as promoting more efficient and transparent governance in line with ESG (Environmental, Social and Governance) principles (Hariram et al. 2023).

The transition to a sustainable economy requires significant investments in renewable energy, green infrastructure, water management, sustainable agriculture and technological innovations. According to estimates by international organisations, trillions of dollars will be needed annually to achieve the Sustainable Development Goals (SDGs) by 2030. This scenario reinforces the need to mobilise and allocate resources strategically, relying on collaboration between governments, the private sector and financial institutions (Mohieldin et al. 2023).

Incorporating sustainability principles into the financial sector is not merely a matter of social responsibility but also a valuable economic opportunity. Companies that adopt sustainable practices demonstrate greater resilience, adaptability and attractiveness to investors. In addition, consumers and investors are increasingly aware of the social and environmental impacts of their choices, putting pressure on companies and financial institutions to adopt more sustainable business models (Ingaldi and Ulewicz 2024).

However, significant challenges still limit the effective implementation of sustainable finance. The lack of standardisation in sustainability criteria, the difficulty in measuring and monitoring environmental and social impacts, and the need for robust regulatory mechanisms are all major obstacles. In addition, the asymmetry of information between investors and companies can make it difficult to allocate resources efficiently to truly sustainable projects (Panagopoulos and Tzionas 2023).

The global environmental crisis calls for urgent changes in economic models, and Green Financing has emerged as an essential mechanism for making sustainable investments viable. This financing approach directs resources toward projects that align with the SDGs, such as the Blue Economy, which advocates for the sustainable use of oceans, balancing economic growth with environmental conservation.

However, without solid financial mechanisms, sectors such as sustainable fishing, offshore renewable energy and marine biodiversity conservation may face structural limitations and a lack of long-term investment.

This chapter explores the significance of Green Financing in the transition to sustainable economic models, highlighting how it drives responsible practices and financial innovations. The bibliometric analysis presented makes it possible to identify trends and challenges, reinforcing the need for effective regulations to guarantee the responsible allocation of resources. Furthermore, the intersection between Green Finance and Blue Economy highlights the strategic role of sustainable investments in building a resilient economy.

This study advances to the discussion of sustainable finance, demonstrating its influence on the Blue Economy and highlighting ways to integrate sustainability and economic growth efficiently. The analysis will contribute to the academic and practical debate, offering insights that can guide the formulation of public policies, business strategies and the mobilisation of financial resources for a more sustainable future.

2 Green Financing

Green finance has been consolidating itself as an essential tool in the transition to sustainable economies, playing a leading role in the global agenda to mitigate climate change and promote sustainable economic practices. Green finance "*acts by increasing the flow of funds to environmentally beneficial projects, essentially reducing their costs, as well as by heightening awareness of the financial risks related to environmental change*" (Ehlers and Packer 2017; Kaur and Sharma 2024). This concept covers a broad range of financial products that are intended to promote initiatives and activities that have a beneficial environmental impact, including green bonds, environmentally conscious investment funds, microfinance, and green loans. Based on the conceptual essentials of the concept of "green", Su and collaborators highlight that "*green innovation has emerged as a critical pathway to address the current resource and environmental constraints*" (Su et al. 2025).

In recent years, the climate emergency and pressure to meet the UN's SDG have significantly boosted green finance's prominence and growth in both established and emerging nations. In this regard, Sachs and collaborators (2019) emphasize that "*the developing world, in particular, also needs to invest substantially in water and sanitation projects in fast-growing urban areas*". In fact, green finance is essential for bridging the financing gap for environmental projects, and is also an opportunity to align the economic and ecological interests of financial institutions and governments on a global scale.

The mobilization of financial resources for projects that support environmental sustainability is more often referred to as "green finance". This type of financing aims to minimize the environmental impact of economic activities by channelling capital into projects such as renewable energy, sustainable infrastructure, waste management, ecosystem protection and adaptation to climate change. However, in this regard, there are also aspects that continue to deserve care and attention. Liang and Renneboog's (2020) concern remains valid: "*ESG characteristics need to be measured more accurately by focusing on real ESG investments related to the UN's Sustainable Development Goals*". By taking social and environmental factors into account when making investment decisions, the idea of "green finance" actually goes beyond the limitations of conventional financial markets.

Green financial instruments are diversified and adaptable to the different needs of the market and the agents involved:

- Green Bonds, for example, are debt instruments that exclusively provide funding for environmentally beneficial initiatives like wind and solar farms and sustainable public transportation systems. Ehlers and Packer (2017) state that green bonds aim to "*internalize environmental externalities and adjust risk perceptions for the sake of increasing environmentally friendly investments*".
- Green Loans are credit lines aimed at business or community initiatives that contribute to environmental sustainability. Kim and collaborators (2021) state that green loans "*are loans where the proceeds are earmarked to exclusively finance environmental and climate-friendly projects*".
- Investment Funds are funds that channel capital to companies or projects that meet ESG criteria. Hartzmark and Shue (2022) attest that the "*sustainable investment funds do reward firms that improve their impact*".
- To give another example of green financing instruments, Green Microfinance is a particularly relevant instrument for communities in emerging economies, financing small projects such as efficient irrigation systems or domestic solar energy.

In fact, green financing is a key element in achieving greater efficiency in resource management. For this reason, it is a determining process in achieving a number of SDGs, mainly: SDG 7 (Clean and Affordable Energy), SDG 11 (Sustainable Cities and Communities), SDG 13 (Climate Action), and SDG 15 (Life on Land). This process makes it possible to accelerate the energy transition and therefore reduce economies' persistent dependence on fossil fuels. Precisely in this context, Sachs and collaborators (2019), reinforcing the crucial importance of the SDGs, highlight that ongoing climate global warming and climate change "*are mainly caused by greenhouse gas (GHG) emissions is now indisputable and the increases in GHG levels merit government intervention, which is to be guided by the Sustainable Development Goals*".

Within the scope of the SDGs, the International Capital Market Association's report (from 2023) stands out, which, according to Chen and coworkers (2025), "*provides a framework for evaluating how green bond financing objectives correspond with the SDGs*".

With regard to the difficulties faced by green finance, although it has virtually infinite transformative potential, it also faces considerable challenges. Wainwright, for example, highlights that the "*gap between climate commitments and actual funding*" is a difficult problem, referring especially to the process of encouraging more advanced economies to contribute to climate action. In less developed financial markets, such as in some emerging economies, the issuance of green instruments is limited by a lack of liquidity and, of particular concern, an adequate regulatory infrastructure. The lack of a consensus definition and standardized criteria also makes it difficult to identify and classify green projects. In this context, Nerlich and collaborators (2025) highlight as very important that "*well-designed green public investment can be expected to act as a catalyst through the crowding-in of green private investment*".

Another major obstacle to traditional investors' participation in green finance is the evaluation and management of climate change risks. Sachs et al. (2019), referring to a study by Krueger and coworkers (2019), report that 40% of institutions "*believe that climate risks have financial implications for their portfolio companies*".

To overcome these challenges, it is essential to develop robust public policies, create tax incentives and improve the mechanisms for evaluating and monitoring green projects. In addition, international collaboration, as demonstrated by the initiatives of the UN and the World Bank, is key to aligning global efforts around financial sustainability. As early as 2014, Nath and collaborators highlighted that the World Bank was forming Environmental & Social (E&S) "*norms for financial institutions to define their role in reducing pollution and managing their adverse social and environmental impact on the economy*". These organizations have long been on the ground implementing measures and programs to strengthen sustainability (Nath et al. 2014).

Emerging technologies are central to the success of these programs and measures. Artificial intelligence and blockchain technologies have the potential to transform green finance, but they require investment. Precisely in the sense of "*green achievement transformation*", Su and coworkers (2025) highlight the inherent expense, emphasizing that "*the transformation stage inputs include new product development funds and technology introduction and adoption expenditures*". But the return is rewarding. For instance, blockchain technology can be used to increase transparency in the issuing and tracking of green bonds, mitigating the risks of greenwashing. In this regard, Chen and collaborators (2025) point out that "*greenwashing concerns underscore the importance of a comprehensive framework that will ensure effectiveness of green initiatives*". AI can also optimize the assessment of climate and environmental risks, improving the efficiency of investment decisions.

However, the transition to green production requires, in many cases, investment in new technologies and equipment. Hartzmark and Shue (2022) emphasize that "*these new green investments are unlikely to pay off in the very short run and should be less attractive to firms in financial distress*".

Although the return does not appear to be in the short term, green finance is emerging as an indispensable mechanism for tackling global climate challenges and promoting more sustainable economies. Of great interest Phan and coworkers (2020) "*found a positive relationship between Sustainable Development Practices (SDPs) and financial performance*". In addition to helping reduce the effects of climate change, the mobilization of financial resources for ecologically conscious projects fosters innovation and sustainable economic growth. Dharwal and Agarwal (2013), cited by Nath and coworkers (2014), even highlight that green financial products are essential to mitigate the reputational risk of institutions.

Challenges remain, but the transformative potential of green finance is clear. Nerlich and coworkers warn that "*despite recent progress, global clean energy investment falls considerably short of achieving the net-zero goal by 2050*" (Nerlich et al. 2025).

Nevertheless, with appropriate public policies, harmonized international standards and the incorporation of advanced technologies, green finance can become one of the central pillars of the global transition to a low-carbon and climate-resilient economy. Very pertinent to this topic, Kim and collaborators (2021) concluded "*that sustainability-linked and green loans both flourish under stricter environmental regulations, consistent with the idea that these loans arise as lenders and borrowers respond to heightened stakeholder pressure*".

3 Benefits of Green Financing with Relation to the ESG Criteria and Social Responsibility

The distribution of funds to initiatives and undertakings that support environmental sustainability is known as "green finance" have gradually positioned itself as a central element in debates on corporate social responsibility (CSR), ESG criteria and the transition to more resilient economic models. This concept, whose relevance is evident in contemporary discussions on climate change and the sustainable management of natural resources, underscores the pressing need to balance economic growth, environmental preservation, and social equity. Despite the current debates that mark its development, green finance has consolidated itself as a heterogeneous field of research, crossed by theories, methodologies and critical visions that shape the way we understand the benefits and limits of this form of investment.

Financial institutions are crucial in advancing sustainable projects, positioning green finance within the broader context of rising concerns about corporate social responsibility. In this sense, initial studies have emphasised the emergence of instruments such as green bonds, environmentally oriented investment funds or micro-credit for renewable energies, framing them as pragmatic responses to environmental challenges (Baker et al. 2018). Consequently, as ecological concerns have gained greater public and political visibility, the pressure for capital markets to respond through responsible financial products has also grown, opening up space for a clearer interconnection with CSR principles.

The intersection with the ESG criteria has reinforced the importance of green financing, given that these criteria incorporate the environmental dimension (E), but also the social (S) and governance (G) aspects. If, on the one hand, the ecological focus makes it possible to channel resources towards clean energy projects, less polluting transport and energy efficiency, on the other hand, the ESG approach proposes a broader vision, integrating aspects such as gender equality, social inclusion and business ethics (Schoenmaker and Schramade 2019). In this way, green finance is not limited to an isolated environmental strategy, but is part of an integrated framework of corporate responsibility, in which it is recognised that sustainability cannot be dissociated from fair labour practices and transparency in governance. This holistic perspective also allows for greater convergence between economic, social and environmental priorities, boosting the legitimacy of companies that demonstrate serious and measurable commitments in the area of sustainability.

In theoretical terms, the literature often uses Stakeholder Theory to explain the growing acceptance and valorisation of green finance. This theory emphasises that companies must consider not only the interests of shareholders, but also those of all parties affected by their activities, including employees, local communities and the environment itself (Freeman and Phillips 2002). In the light of this approach, green finance emerges as a mechanism for generating long-term value and strengthening market confidence by mitigating reputational risks and ensuring corporate legitimacy. Also, within this conceptual framework, Institutional Theory complements explanations of the adoption of green instruments, arguing that companies and financial institutions incorporate them to respond to regulatory, cultural or normative pressures, seeking to align themselves with globally accepted standards and expectations. These two strands of theory converge in that they highlight the relational and socio-economic nature underlying green finance, as well as the need for organisations to adjust to the contextual transformations imposed on them.

In the methodological field, studies on green financing reflect a plurality of approaches, ranging from quantitative analyses that investigate the relationship between green bond issuance and carbon emission reductions to qualitative case study-based research that assesses the institutional and cultural impact of ecological practices in companies. For instance, empirical works employing regression models and time series analysis (Haas and Popov 2019) explore the correlation between green investments and macroeconomic indicators of energy efficiency, seeking to isolate variables such as government regulation or the availability of clean technologies. In parallel, other research adopts mixed methodologies, combining sustainability reports, climate data, and financial indicators to provide an integrated analysis of the effects of green financing (Wu et al. 2021). Despite the expansion of these approaches, the challenge of harmonising criteria and performance indicators persists, as the absence of unified standards for reporting and impact assessment complicates comparisons between studies and regions.

Although green financing has seen remarkable growth in recent years, reflected in the rising volume of green bond issuances and the expanding range of financial instruments focused on sustainability, the literature highlights certain obstacles and dilemmas that urgently need to be addressed. One of the most debated aspects concerns the rigorous assessment of the environmental benefits generated. While various studies indicate that these instruments contribute to climate change mitigation projects, there remains a lack of consensus on methodologies that can reliably determine the extent to which effective emission reductions or measurable environmental improvements are achieved (Tolliver et al. 2020). Adding to this challenge is the risk of "*greenwashing*", that is, the practice of labelling projects as green without providing solid evidence of their sustainable contribution (Zhang 2022). Such practices threaten the credibility of green financing itself and underscore the need for oversight mechanisms, certification processes, and transparency measures capable of preventing the improper use of the "*green*" label.

Another important gap involves the alignment between the environmental dimension and the social and governance aspects promoted by ESG criteria. Some argue that an exclusive focus on the green aspect might overshadow more pressing social issues, such as income inequality or the inclusion of vulnerable (Schoenmaker and Schramade 2019). Similarly, companies with poor records on labour practices or those accused of human rights violations communities can, paradoxically, issue green bonds, raising ethical and consistency concerns. Thus, the integration of environmental, social, and sound governance dimensions becomes crucial, as it lends greater legitimacy to this type of financing and helps ensure a more systemic and less fragmented approach.

Considering these tensions and challenges, the future of green financing necessarily involves the consolidation of robust regulatory frameworks that establish reporting standards and unified metrics. The European Taxonomy for Sustainable Activities represents a concrete step towards clarifying which types of investments can be classified as green, helping to reduce uncertainties and standardise practices (OECD 2019). Furthermore, the adoption of emerging technologies, such as blockchain, presents an opportunity to enhance the traceability and transparency of capital flows, ensuring that funded projects are effectively conducted in accordance with sustainable criteria (Zhang et al. 2021). Simultaneously, incorporating artificial intelligence can facilitate can enable more predictive and detailed analysis of environmental risks, helping to direct capital towards sectors and regions where the positive impact can be most significant.

By bringing together principles of social responsibility, ESG criteria, and cutting-edge technologies, green financing is likely to establish itself as a catalyst for profound changes in the way we produce and consume. However, it requires increasingly rigorous and comprehensive approaches. Companies and financial institutions that genuinely embrace this model can benefit not only from enhanced reputations but also from opportunities for innovation and access to growing markets, such as renewable energy and sustainable mobility. Moreover, the academic community plays an indispensable role, producing critical and methodologically robust analyses that can improve knowledge of the benefits, risks, and restrictions of green funding. This convergence of efforts will ultimately contribute to legitimising these financial instruments and ensuring that the commitment to sustainability becomes not just a label but a structural transformation capable of addressing the challenges of our time.

4 Methodology

In order to map scientific publications related to green finance, a bibliometric analysis was carried out based on data obtained from the global database platforms widely used for academic research: Clarivate Analytics' Web of Science (WoS) and Scopus (Bem Machado et al. 2023). The study employed a structured methodology consisting of three key stages: planning, data collection, and bibliometric analysis. To address the central research question, a systematic search was conducted in the selected online databases, followed by an in-depth analysis of the obtained results. The R package Bibliometrix, called Biblioshiny, has the most extensive and appropriate set of techniques among the tools researched for bibliometric analysis (Moral-Muñoz et al. 2020).

Bibliometrics, which comes from the information sciences, uses mathematical and statistical methods to explore bibliographic records in databases (Linnenluecke 2017). This methodology makes it possible to identify aspects such as: the volume of productions by region; the periodicity of publications; the distribution of research between different areas of knowledge; the number of citations related to a specific study; and the impact of a scientific article, among other factors that help to organise the results and reduce biases when investigating a given topic. The bibliometric analysis was structured into three essential phases—planning, data collction, and result analysis—conducted in an integrated manner to respond to the guiding research question: "What is green finance?".

Planning took place during December 2024, the period in which the study was conducted. Important rules were established during this phase, such as the sole use of electronic databases and the exclusion of physical library catalogues because the selected source included a sufficient number of documents. Still in the planning stage, the search terms were delimited: 'green financing' OR 'green financing' OR 'green financing', considering that different variations of expressions fit into the same concept, depending on the context in which they are applied. It was decided that the search terms would be entered in the 'title, abstract and keywords' fields, with no time, language or additional restrictions.

5 Results and Discussion

In order to analyse the bibliometric data of the 726 papers in the Scopus database and the 504 in the WOS database, the Bibliometrix software was used, using the R Bibliometrix package, called Biblioshiny, which has the most extensive and appropriate set of techniques among the tools researched for bibliometric analysis (Moral-Muñoz et al. 2020). This information is summarised in Table 1.

Table 1 Summary of information found in the databases

Item	Scopus	Web of science
Timespan	2008:2025	2010:2025
Sources (journals, books, etc.)	309	182
Documents	719	504
Document contents		
Keywords plus (ID)	2193	645
Author's keywords (DE)	1774	1462
Authors		
Authors	1708	1197
Document types		
Article	519	437
Article book chapter	1	27
Article conference paper	1	4
Book	9	ND
Book chapter	72	ND
Conference paper	49	ND
Conference review	1	ND
Editorial	1	4
Erratum	12	ND
Letter	1	2
Note	2	ND
Retracted	25	2
Review	20	12
Short survey	1	ND
Proceedings paper	0	15

Source Created by authors
ND: no information obtained

The study started by examining the general rate of increase in the quantity of scientific publications. With 280 publications in Scopus and 196 in the WOS database, the topic saw a notable uptick in 2023 (see Fig. 1). This was due to the fact that this year's COP28 brought attention to how important green finance is to reaching the objectives outlined in the Paris Agreement, like keeping global warming to 1.5 °C over pre-industrial levels.

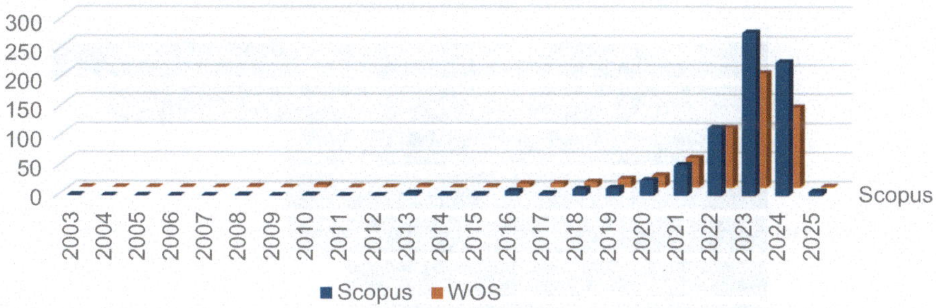

Fig. 1 Distribution of scientific publications from 2003 to 2025—Scopus and WOS. *Source* Created by authors

First, the ten most pertinent authors on the topic of this study were found in the Scopus and WOS databases; they are shown in Figs. 2 and 3, respectively.

The most cited author in the Scopus database is Farhad Taghizadeh-Hesary from Tokai University, Japan, with 16 citations. Li, Ying from Tianjin University of Commerce with 12 citations in the area, is the author most frequently cited in the WOS database.

There is a wide variety of authors, organizations, and nations that stand out in green finance research among the 504 publications in the WOS database and the 726 papers in the Scopus database. China is the most cited nation in both databases when looking at the top ten countries in the region by number of citations, with 6030 citations in the Scopus database and 6172 in the WOS database, as Fig. 4 illustrates.

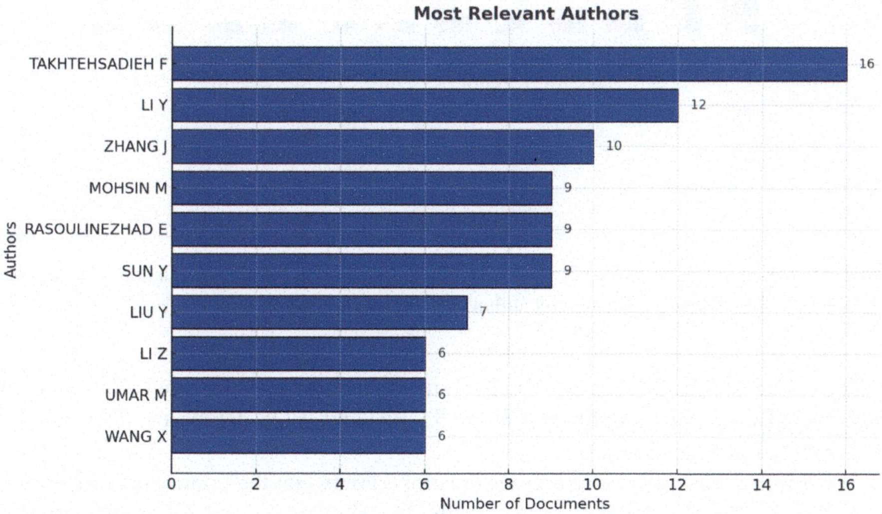

Fig. 2 Most relevant authors via the Scopus database. *Source* Created by authors

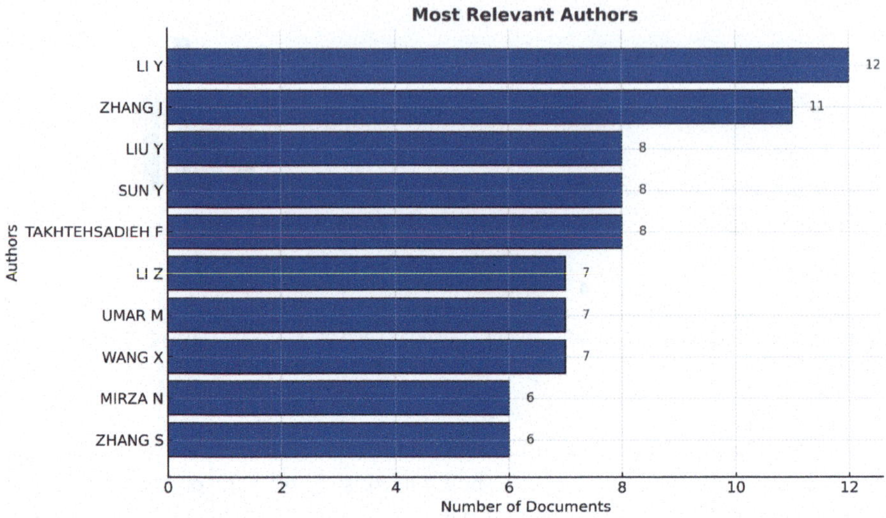

Fig. 3 Most relevant authors via the web of science database. *Source* Created by authors

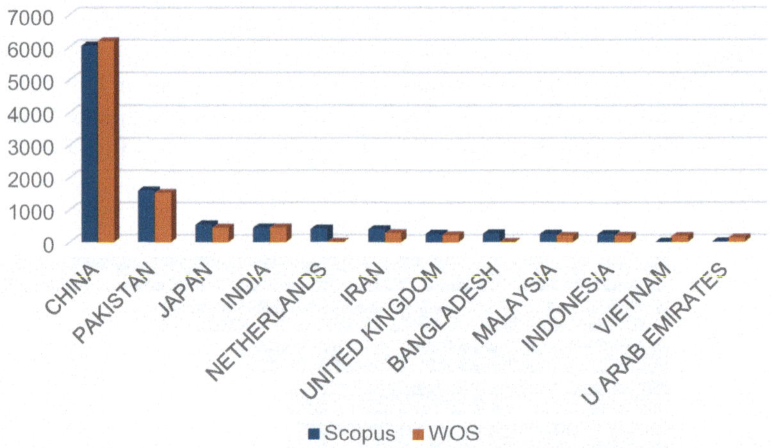

Fig. 4 Distribution by country of the papers published based on the results obtained via Scopus and WOS databases. *Source* Created by authors

The intensity of publications by nation and the correlation between them based on citations between papers published in the Scopus and WOS databases, respectively, are presented in Figs. 5 and 6.

According to scientific interactions developed between the continents of Europe, Asia, North Central America, and South America, based on citations between published articles in the database, the intensity of publication is localized in these regions.

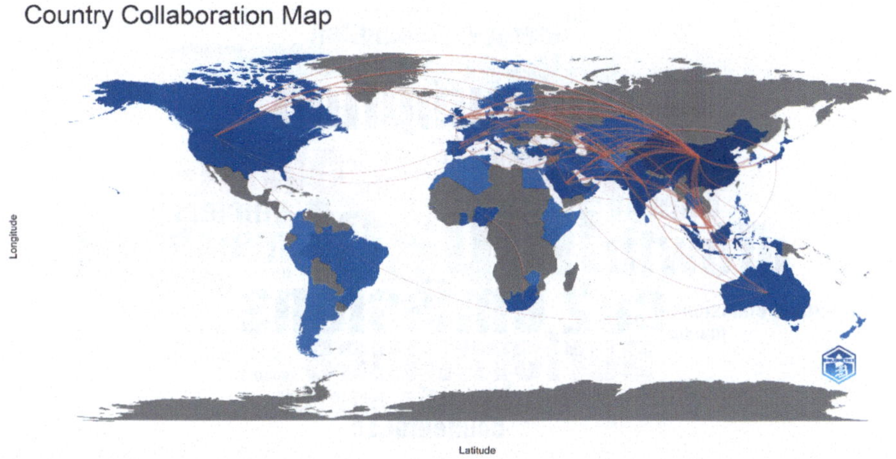

Fig. 5 Country collaboration map via Scopus database. *Source* Created by authors

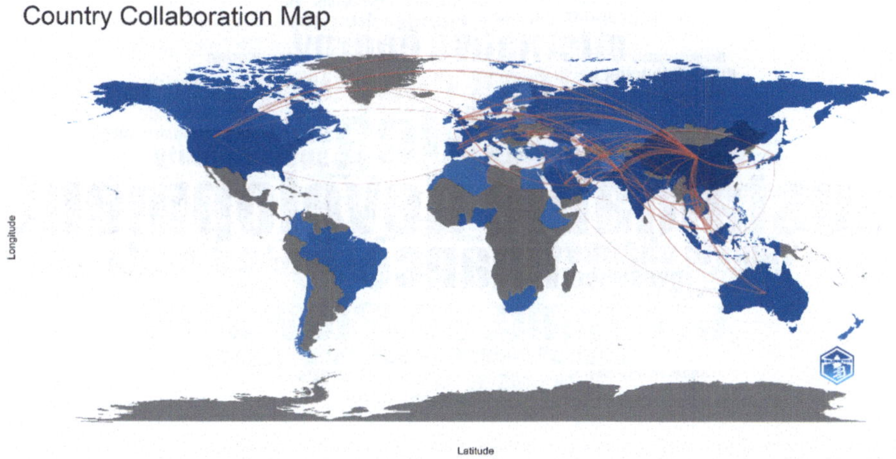

Fig. 6 Country collaboration map via web of science database. *Source* Created by authors

Based on the collection of articles that were obtained from both databases, the bibliometric analysis revealed that the Scopus database had 1774 keywords while the Web of Science database contained 1462 keywords. The keywords that stood out in the first database were: "sustainable development" with 200 occurrences in the Scopus database, as detailed in Fig. 7, and "impact" with 81 occurrences in the WOS database, as shown in Fig. 8.

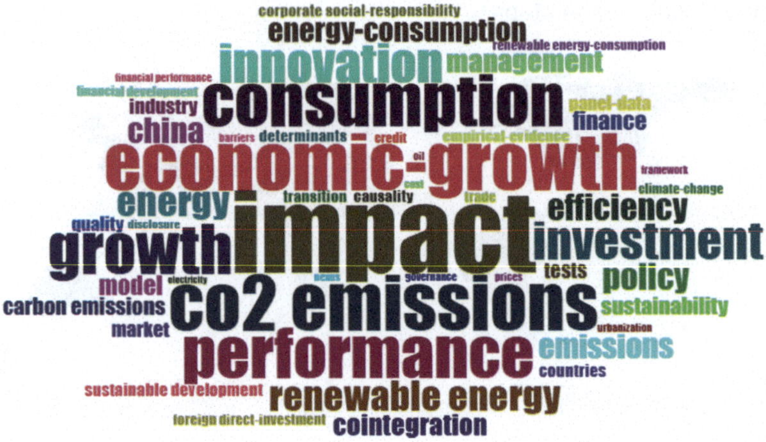

Fig. 7 Tag cloud via Scopus database. *Source* Created by authors

Fig. 8 Tag cloud via web of science database. *Source* Created by authors

The bibliometric analysis revealed a growth in publications on green finance, driven by events such as COP28, which reinforced its role in meeting the goals of the Paris Agreement (Schleussner et al. 2016). The search for sustainable investments has been consolidating globally, with China leading the way regarding the quantity of articles and references in the databases analysed. The main terms identified—such as 'sustainable development' and 'impact'—demonstrate how sustainability, green finance, and environmental impact mitigation are related.

In the context of corporate sustainability, it was observed that companies are progressively adding sustainable practices to their plans, not only out of social responsibility, but also for competitive advantage. Companies that adopt sustainable business models tend to reduce operating costs, especially with resources such as energy and water, while at the same time increasing their attractiveness to investors aligned with ESG criteria. ESG governance has proven to be an essential pillar for making responsible and transparent business decisions (Kandpal et al. 2024).

Another important element that has been highlighted is resource efficiency. Using new technology and cleaner processes enables economic and environmental gains, making it a promising field for research and innovation. However, there is still a shortage of empirical studies that assess in detail how industrial sectors can maximise resource efficiency (Ahmad et al. 2023). Sustainable investments, although growing, still face challenges such as the lack of standardisation in impact measurement and transparency in sustainability criteria. The lack of clear regulations can make it difficult to compare initiatives and increase the risk of greenwashing (Popescu et al. 2021).

The analysis also highlighted the importance of public policies to boost sustainable financing. Effective regulations and government incentives can stimulate responsible financial practices and amplify the sector's positive impact. Finally, the convergence between academia and the market demonstrates an opportunity for the development of innovative solutions, ensuring that sustainable finance is strengthened as an essential instrument for the shift to a more robust and balanced economy.

The results of the bibliometric review on Green Finance indicate an essential connection with the Blue Economy concept, since both share the objective of aligning financial resources with sustainable practices, minimising environmental impacts and promoting regenerative economic development. The Blue Economy emphasises the conservation of marine and coastal ecosystems through sustainable investments in sectors such as responsible fishing, ecological tourism and marine renewable energies. In this sense, the green financing strategies analysed in this study show significant potential for enabling such practices, especially through the issuance of green bonds, sustainable investment funds and tax incentives for businesses committed to preserving ocean biodiversity. The need for more robust regulatory guidelines and standardised metrics for evaluating the effectiveness of sustainable investments, highlighted in the literature on green finance, also applies to the advancement of the Blue Economy, ensuring that resources earmarked for sustainable marine projects are transparent and effective.

6 Final Considerations

The bibliometric analysis carried out in this chapter highlights the growing importance of green finance as an essential tool for promoting sustainability in the global economy. According to the reviewed literature, this method strengthens the incorporation of ESG considerations into financial choices by playing a crucial role in allocating resources to sustainable projects. A change in the markets toward more accountable and robust processes is reflected in the growth of sustainable financial products like green bonds and ESG investment funds.

Despite the progress made, challenges remain, especially in standardising metrics to assess environmental and social impacts. The risk of greenwashing jeopardises the sector's credibility, requiring greater regulation and transparency. In addition, the unequal distribution of green finance between developed and emerging economies highlights structural and regulatory barriers that need to be overcome to ensure an equitable impact.

Technological innovation has emerged as an ally in this process, with blockchain and artificial intelligence boosting the traceability and efficiency of sustainable financing. However, the adoption of these technologies requires investment and public policies to incentivise their implementation. The intersection of green finance with corporate sustainability reinforces its transformative potential. Companies that adopt sustainable practices show greater resilience and attractiveness in the market, but there are still challenges in balancing environmental, social and governance objectives.

A key component of the shift to a low-carbon economy is clearly green finance, but its success depends on joint efforts between governments, the financial sector and academia. Harmonised regulatory standards, interdisciplinary research and international cooperation are essential to maximise its impact and consolidate it as a pillar of global sustainability.

In light of the discussions presented, it is clear that the strengthening of the Blue Economy depends directly on the advancement of Green Finance mechanisms. Sustainable financing not only enables the transition to more responsible economic models, but also provides solutions to mitigate the impacts of human activity on marine ecosystems. The connection between the blue economy and green finance reinforces the importance of financial mechanisms that incentivise practices such as sustainable fishing, offshore renewable energy generation and the protection of marine biodiversity. Future research could therefore explore the interdependence between these concepts in greater depth, assessing how innovative financial instruments can not only boost the green economy, but also guarantee the health of the oceans and their resilience to climate change.

References

Ahmad MI, Naseem MA, Battisti E, Rehman RU, Giovando G (2023) Resource efficiency actions and financial performance: Exploring the moderating role of production cost. BEER 34(1):69–80. https://doi.org/10.1111/beer.12587

Baker HS, Millar RJ, Karoly DJ, Beyerle U, Guillod BP, Mitchell D, Shiogama H, Sparrow S, Woollings T, Allen MR (2018) Higher CO2 concentrations increase extreme event risk in a 1.5 C world. Nat Cli Change 8(7):604–608. https://doi.org/10.1038/s41558-018-0190-1

de Bem Machado A, Domingos L, Sousa MJ, Rocha Á (2023) Developing an innovative digital learning environment with metaverse. In: Digital technologies and transformation in business, industry and organizations, vol 2. Springer Nature Switzerland, Cham, pp 147–164. https://doi.org/10.1007/978-3-031-40710-9_8

Chen Y, Erlwein-Sayer C, Mamon R, Spagnolo F, Spagnolo N (2025) Impact of green bonds on issuer's environmental improvements. Available at SSRN 5092465. https://doi.org/10.2139/ssrn.5092465

Dharwal M, Agarwal A (2013) Green banking: an innovative initiative for sustainable development. ACCMAN Inst Manag Articl 2(3):1–7

Ehlers T, Packer F (2017) Green bond finance and certification. BIS Q Rev 1–14. https://ssrn.com/abstract=3042378

Freeman RE, Phillips RA (2002) Stakeholder theory: a libertarian defense. BEQ 12(3):331–349. https://doi.org/10.2139/ssrn.263514

De Haas R, Popov AA (2019) Finance and carbon emissions. Available at SSRN 3459987. https://doi.org/10.2139/ssrn.3459987

Hariram NP, Mekha KB, Suganthan V, Sudhakar K (2023) Sustainalism an integrated socio-economic-environmental model to address sustainable development and sustainability. Sustainability 15(13):10682. https://doi.org/10.3390/su151310682

Hartzmark SM, Shue K (2022). Counterproductive sustainable investing: the impact elasticity of brown and green firms. https://doi.org/10.2139/ssrn.4359282

Ingaldi M, Ulewicz R (2024) The business model of a circular economy in the innovation and improvement of metal processing. Sustainability 16(13):5513. https://doi.org/10.3390/su16135513

Kandpal V, Jaswal A, Santibanez Gonzalez ED, Agarwal N (2024) Corporate social responsibility (CSR) and ESG reporting: redefining business in the twenty-first century. In: Sustainable energy transition: circular economy and sustainable financing for environmental, social and governance (ESG) Practices. Springer Nature Switzerland, Cham, pp 239–272. https://doi.org/10.1007/978-3-031-52943-6_8

Kaur A, Sharma V (2024) Nexus between climate change and mitigation approaches for sustainable development: a bibliometric review. Glob Knowl Mem Commun

Kim S, Kumar N, Lee J, Oh J (2021) ESG lending. In: Proceedings of Paris December 2021 finance meeting EUROFIDAI - ESSEC, European Corporate Governance Institute – finance working paper no. 817/2022. https://doi.org/10.2139/ssrn.3865147

Krueger P, Sautner Z, Starks LT (2019) The importance of climate risks for institutional investors. Swiss Finance Institute research paper no. 18–58, European Corporate Governance Institute – finance working paper no. 610/2019. https://doi.org/10.2139/ssrn.3235190

Liang H, Renneboog L (2020) Corporate social responsibility and sustainable finance: a review of the literature. European Corporate Governance Institute – finance working paper no. 701/2020. https://doi.org/10.2139/ssrn.3698631

Linnenluecke MK (2017) Resilience in business and management research: a review of influential publications and a research agenda: resilience in business and management research. Int J Manag Rev 19(1):4–30. https://doi.org/10.1111/ijmr.12076

Mohieldin M, Wahba S, Gonzalez-Perez MA, Shehata M (2023) Business, government and the SDGs. Springer Books. https://doi.org/10.1007/978-3-031-11196-9

Moral-Muñoz JA, Herrera-Viedma E, Santisteban-Espejo A, Cobo MJ (2020) Software tools for conducting bibliometric analysis in science: an up-to-date review. El Profesional De La Informaci6n 29(1):e290103. https://doi.org/10.3145/epi.2020.ene.03

Nath V, Nayak N, Goel A (2014) Green banking practices – a review. IMPACT: Int J Res Bus Stud Manag 2(4):45–62. https://ssrn.com/abstract=2425108

Nerlich C, Köhler-Ulbrich P, Andersson M, Pasqua C, Abraham L, Bańkowski K, Emambakhsh T, Ferrando A, Grynberg C, Groß J, Hoendervangers L, Kostakis V, Momferatou D, Rau-Goehring M, Rariga E-J, Rusinova D, Setzer R, Spaggiari M, Tamburrini F, Simon JMV, Vinci F (2025) Investing in Europe's green future: Green investment needs, outlook and obstacles to funding the gap. ECB occasional paper no. 2025/367. https://doi.org/10.2139/ssrn.5088767

Ng AW, Nathwani J, Fu J, Zhou H (2021) Green financing for global energy sustainability: prospecting transformational adaptation beyond Industry 4.0. SSPP 17(1):377–390. https://doi.org/10.1080/15487733.2021.1999079

OECD (2019) An OECD learning framework 2030. The future of education and labor, pp 23–35
Panagopoulos A, Tzionas I (2023) The Use of Sustainable Financial Instruments in Relation to the social impact investment: ESG policies, capital markets' approach and investors' protection: an innovative perspective for a global surveillance authority. Int J Bus Adm. https://doi.org/10.5430/ijba.v14n1p87
Phan LT, Nguyen TV, Luong QC, Nguyen TV, Nguyen HT, Le HQ, Nguyen TT, Cao TM, Pham QD (2020) Importation and human-to-human transmission of a novel coronavirus in Vietnam. New Engl J Med 382(9):872–874. https://doi.org/10.1056/NEJMc2001272
Popescu IS, Hitaj C, Benetto E (2021) Measuring the sustainability of investment funds: a critical review of methods and frameworks in sustainable finance. J Clean Prod 314:128016. https://doi.org/10.1016/j.jclepro.2021.128016
Sachs JD, Woo WT, Yoshino N, Taghizadeh-Hesary F (2019) Why is green finance important? In: ADBI working paper, vol 917. https://doi.org/10.2139/ssrn.3327149
Schleussner CF, Rogelj J, Schaeffer M, Lissner T, Licker R, Fischer EM, Knutti R, Levermann A, Frieler K, Hare W (2016) Science and policy characteristics of the Paris Agreement temperature goal. Nat Clim Change 6(9):827–835. https://doi.org/10.1038/nclimate3096
Schoenmaker D, Schramade W (2019) Investing for long-term value creation. J Sustain Financ Inv 9(4):356–377. https://doi.org/10.1080/20430795.2019.1625012
Su X, Wang Z, Yang Y, Wu B, Jiang Y (2025) Does common ownership affect green innovation efficiency? Evidence from China. https://doi.org/10.2139/ssrn.5092853
Tolliver C, Keeley AR, Managi S (2020) Drivers of green bond market growth: the importance of Nationally Determined Contributions to the Paris Agreement and implications for sustainability. J Clean Prod 244:118643. https://doi.org/10.1016/j.jclepro.2019.118643
Wu X, Sadiq M, Chien F, Ngo QT, Nguyen AT, Trinh TT (2021) Testing role of green financing on climate change mitigation: evidences from G7 and E7 countries. Environ Sci Pollut Res 28:66736–66750. https://doi.org/10.1007/s11356-021-15023-w
Zhang D (2022) Green financial system regulation shock and greenwashing behaviors: evidence from Chinese firms. Energy Econ 111:106064. https://doi.org/10.1016/j.eneco.2022.106064
Zhang C, Bengio S, Hardt M, Recht B, Vinyals O (2021) Understanding deep learning (still) requires rethinking generalization. Commun ACM 64(3):107–115. https://doi.org/10.1145/3446776

The manufacturer's authorised representative in the EU is Springer Nature Customer Service Centre GmbH, Europaplatz 3, 69115 Heidelberg, Germany. If you have any concerns regarding our products, please contact ProductSafety@springernature.com

Printed and bound by CPI Group (UK) Ltd, Croydon, CR0 4YY
30/09/2026
02225584-0014